Erratum

The cover of this book should read:

The Evolutionary History and a
Systematic Revision of Woodrats of the
Neotoma lepida Group

James L. Patton, David G. Huckaby, and
Sergio Ticul Álvarez-Castañeda

The Desert Woodrat, *Neotoma lepida lepida*. Photograph taken by D. G. Huckaby, on 24 July 1991, near Pioneertown, San Bernardino Co., CA (locality CA-342, see Appendix); image made available by the Mammalian Image Library of the American Society of Mammalogists (image number 1252).

The Evolutionary History and a Systematic Revision of Woodrats of the *Neotoma lepida* Group

James L. Patton[1], David G. Huckaby[2], and Sergio Ticul Álvarez-Castañeda[3]

[1]Museum of Vertebrate Zoology, University of California, Berkeley, CA 94720

[2]Department of Biological Sciences, California State University, Long Beach, CA 90804-3702

[3]Centro de Investigaciones Biológicas del Noroeste, La Paz, Baja California Sur, México

UNIVERSITY OF CALIFORNIA PRESS
Berkeley • Los Angeles • London

University of California Publications in Zoology, Volume 135

Editorial Board: Carla Cicero, Douglas A. Kelt, Eileen A. Lacey, David S. Woodruff

University of California Press

Berkeley and Los Angeles, California

University of California Press, Ltd.

London, England

Library of Congress Cataloging-in-Publication Data

The evolutionary history and a systematic revision of woodrats of the Neotoma lepida group / James L. Patton ... [et al.].
   p. cm. -- (University of California publications in zoology ; v. 135)
  Includes bibliographical references and index.
  ISBN 978-0-520-09866-4 (pbk. : alk. paper)
  1. Desert wood rat--North America. I. Patton, James L.
  QL737.R666E96 2008
  599.35'73--dc22
               2007051010

Contents

Figures

Tables

Acknowledgments

This work could not have been accomplished without the combined support of numerous colleagues and organizations, both government and private, who have allowed us access to specimens in museum collections, who have expended time in the field helping to collect specimens, who have helped to generate the data we summarize, who have read and commented on all, or part, of this manuscript, and who have given us permission to collect recent materials.

At the risk of forgetting the contributions of some, we are especially grateful to Michael Carleton, Alfred Gardner, and Robert Fisher (National Museum of Natural History, Smithsonian Institution), Jim Dines (Los Angeles County Museum of Natural History), Philip Unit (San Diego Natural History Museum), Kathy Molina (Dickey Collection, University of California, Los Angeles), Judy Chupasko (Museum of Comparative Zoology, Harvard University), Donald Hoffmeister (University of Illinois), and Ned Gilmore and Leo Joseph (Academy of Natural Sciences, Philadelphia) for permitting us to examine specimens in their respective collections. We are particularly appreciative of Mike Carleton and Kathy Molina for allowing us to retrieve residual tissue adhering to the skulls of some specimens in the USNM and UCLA collections for DNA extraction and sequencing, and Bob Fisher for providing a printout of the USNM holdings of desert woodrats.

The fish and game agencies of the states of Arizona, California, Nevada, and Utah graciously provided collecting permits to JLP, those of California and Utah to DGH, and the country of Mexico to STA-C. JLP acknowledges Melissa Kreitbaum, Scientific Collecting Permit Administrator-Nongame Branch, Arizona Game and Fish Department, for annually both facilitated the reporting process and expedited permit renewal. The system she administered during the course of our studies is a model that all other state and federal institutions should follow. For permission to conduct fieldwork on federal or other lands within Arizona and California, we also thank Jeff Cole (Navajo Fish and Wildlife Branch, Navajo Nation), Vergial Hart and Curt McCasland (Cabeza Prieta National Wildlife Refuge and Barry M. Goldwater Gunnery Range, Arizona), Tim Tibbets (Research & Endangered Species Coordinator, Organ Pipe National Monument, Arizona), Eric Oswald and Peter Holm (Luke Air Force Base, Arizona), David Clendenin (Director, Wind Wolves Preserve, The Wildlands Conservancy, Kern Co.,

California), Robert Parker and Hector Villalobos (Ridgecrest Resource Area, Bureau of Land Management, Ridgecrest, California), and Dennis Mullins and Shirlene Barrington (Tejon Ranch Corporation, Kern Co., California). Felisa Smith and Jake Goheen (University of New Mexico) kindly provided tissue samples from Death Valley National Park from their on-going population studies.

Carol Patton, Patrick Kelly, Shana Dodd, and Hanna Shohfi aided JLP in his fieldwork, and Nichole Patrick, Bridget Sousa, and Maria Svensson competently generated a substantial portion of the molecular data we include in this report. Maria Svensson, in particular, obtained the sequence data from museum specimens of the extinct insular populations of woodrats that proved so crucial to our analyses. Michelle Koo patiently trained JLP in the use of Arc-Map© to produce many of the figures used herein. STA-C thanks Patricia Cortés-Calva, Mayra de la Paz, Evelyn Rios, Ana Trujano, and Anahid Gutiérrez for help in his fieldwork. Funding for his activities was provided by the Consejo Nacional de Ciencia y Tecnología de México (CONACYT grants I25251N, 39467Q; SEMARNAT-2002-CO1-019), and the University of California MEXUS-CONACYT project and grant for a faculty visit to UC-Berkeley. Susanna Guizar drew the camera lucida illustrations of the glans penis. We thank Craig Moritz and Jim McGuire, in particular, for discussions of both methods of analysis and interpretations of data, and Adam Leaché for providing help with the Bayesian analyses. Finally, we owe Doug Kelt, Phil Myers, Enrique Lessa, and Alfred Gardner more than just a debt of gratitude for their careful and insightful reviews of parts or the entire manuscript, efforts that without doubt enhanced the final version. These individuals, however, should not be held responsible for its contents.

Abstract

We review the evolutionary history and systematic status of species and subspecies of the desert woodrat complex of the *Neotoma lepida* group. Currently, this complex comprises six taxa currently recognized as species from western North America, two "continental" (*Neotoma lepida* Thomas and *Neotoma devia* Goldman) and four from islands on both the Pacific and gulf sides of Baja California (*Neotoma anthonyi* [Todos Santos], *Neotoma martinensis* [San Martín], *Neotoma bryanti* [Cedros], and *Neotoma bunkeri* [Coronados]). In this review, we examined more than 4600 museum specimens for morphological characters, both qualitative and quantitative craniodental, male phallic, and colorimetric variables, analyzed mitochondrial DNA (mtDNA) sequence data for the cytochrome-b gene and allelic variation for 18 nuclear microsatellite loci from more than 1000 individuals, and nuclear DNA sequences (nucDNA) from intron 7 of the β-fibrinogen gene (*Fgb-I7*) from 166 specimens. We analyzed morphological data by a combination of univariate and multivariate methods to define discrete groups in nature and to document patterns of variation across geography. We applied phylogenetic analyses to delineate geographic clusters that are evolutionarily independent and examined the concordance between these lineages and morphological groupings. We used population genetic methods to determine the degree to which there is genetic exchange between phylogenetic and morphological groups where they co-occur in nature. We then used coalescent approaches to develop hypotheses about the timing and processes that underlie diversification of the molecular and morphological groups that we identified. Finally, we examined a set of testable, objective criteria that can be used to bound species groups in nature, and we rearranged the taxonomy of this group of woodrats according to those criteria.

Our analyses, applications, and results confirm the inadequacy of the current systematics of the *Neotoma lepida* group. We define four species: (1) *Neotoma bryanti* Merriam, which is distributed along coastal California and throughout Baja California, including all islands on both sides of that peninsula occupied by woodrats except one; (2) *Neotoma insularis* Townsend, from Isla Ángel de la Guarda in the northern Gulf of California; (3) *Neotoma lepida* Thomas, which occurs throughout the Colorado, Mojave, and Great Basin deserts west and north of the Colorado River; and (4) *Neotoma devia* Goldman, distributed south and east of the Colorado River in Arizona and northwestern Sonora, Mexico. Each of these species is defined as a unique and independent phylogenetic lineage established by molecular sequences and diagnosed by a number of discrete qualitative morphological craniodental and male phallic characters as well as by multivariate analyses of craniodental and colorimetric variables. Each of these species, with the exception of the insular *N. insularis*, is also composed of two or three well-defined molecular subclades. While subclade structure indicates deep and complex histories, nuclear genetic markers suggest that individuals of separate mtDNA subclades within each of these species are both completely interfertile and continue to interbreed freely at points of contact.

Both a molecular clock based approach and the use of coalescent parameters provide estimates of the timing of species and clade diversification. All splits occurred within the Pleistocene, with timing ranging from about 1.6 Ma for the basal split within the group to approximately 50-100 Ka for the most terminal splits among molecular subclades within *N. lepida*. These dates typically fall well after the major vicariant geological processes that have been suggested to underlie the diversification of other co-distributed species of vertebrates and invertebrates. We also employ coalescent methods and Nested Clade analysis to develop hypotheses of the past population history of each molecular clade and subclade defined. The subclades of *N. bryanti*, for example, have undergone combinations of geographic expansion on one margin of their current ranges while experiencing fragmentation on another. Each of these subclades is older than those of *N. lepida* or *N. devia*. In contrast, the two subclades of *N. lepida*, and particularly the geographically widespread subclade 2A, have experienced recent and rapid spatial expansion throughout the central deserts of the United States, a process that is perhaps still in progress.

Limited hybridization with backcrossing does occur at two areas of contact of the coastal *N. bryanti* and desert *N. lepida* (Morongo Valley, San Bernardino Co., California, and Kelso Valley, Kern Co., California), but evidence for introgression from 18 microsatellite loci is limited to the contact populations and does not extend into the parental ranges of either species. Thus, although the two species are not reproductively isolated, the lack of introgression beyond the point of contact suggests lowered fitness of hybrid individuals and thus the genetic isolation of both species.

INTRODUCTION

This study examines the distribution, biogeographic history, and systematics of woodrats of the *Neotoma lepida* group from the western United States and northwestern Mexico. Collectively referred to as desert woodrats, these are ubiquitous occupants of dryland habitats from western Arizona to coastal California and from southern Idaho and Oregon to the cape region of Baja California Sur in Mexico. Their historical record is widespread and temporally deep. As we detail below, the taxonomic history of this group has been complicated, but most authors of the past half-century have viewed the complex to include one or two mainland species and four insular ones off the Pacific and gulf coasts of Baja California (e.g., Hall, 1981; Musser and Carleton, 2005).

As Verts and Carraway (2002) detail in their synopsis of the population ecology and behavior of *Neotoma lepida*, these rats are important components of the small mammal fauna throughout their range. They construct stick nests that serve as refuges for a variety of other taxa, both vertebrate and invertebrate; they serve as important prey for avian and non-avian reptiles as well as other mammals; they play a critical role in nutrient cycling; and they provide, with other woodrat species, perhaps the best historical record of vegetation community change of the late and post Pleistocene (Betancourt et al., 1990; Grayson, 1993). Members of the *lepida* group range in habitat from desert scrub communities below sea level in Death Valley to the Mediterranean scrub or oak woodland of coastal California to piñon-juniper woodlands at elevations above 7,000 feet in the Great Basin. They are dominant members of the Baja California mammalian fauna, occurring in all major vegetation communities including the pine-oak woodland of the Sierra La Laguna in the Cape Region; they also occur on five islands along the Pacific Coast and eight within the Sea of Cortez.

Our interest in this complex of woodrats began with Jim Mascarello's 1978 analysis of chromosomal, allozymic, and morphological differentiation among population samples across the lower Colorado River. This study established the set of species boundaries currently recognized (e.g., Musser and Carleton, 2005), although others (e.g., Hoffmeister, 1986) have challenged his taxonomic conclusions. One of us (DGH) then began a more thorough geographic review of one of the character suites employed by Mascarello – bacular morphology and the soft anatomy of the glans penis. This culminated in an unpublished manuscript that

1

detailed the taxonomic history of the complex and drew attention to additional questions about species boundaries as well as the correct applicability of available names based on standard nomenclatural rules. Planz (1992), in an unpublished PhD dissertation, also addressed the issue of species boundaries within the *lepida* group through a geographically limited use of restriction fragment length analysis of mitochondrial DNA sequences. He summarized some of his views in a generalized treatment on North American mammals (Planz, 1999). Finally, Patton and Álvarez-Castañeda (2005) undertook a more thorough analysis of variation in mitochondrial DNA sequences that, in conjunction with these previous studies, supported revisions in the current systematics of the *lepida* group. Because each of the studies had been limited both geographically and in character dataset, we decided to combine efforts and provide the thorough review necessary to resolve these lingering systematic issues.

The primary, or fundamental component of biodiversity is the definition of species boundaries and the delimitation of subspecies, which are the unambiguously diagnosable geographic units within species. To understand species and subspecies boundaries within this complex of woodrats, we use a combination of traditional univariate and multivariate morphological analyses of museum specimens coupled with molecular markers from both the mitochondrial and nuclear genomes. We then build hypotheses of a second fundamental component of biodiversity, which is the history of diversification and range occupation of evolutionary lineages over the past millennia. We end with a synopsis of the nomenclatural history of the taxa we recognize, with a rationalization of why we make the choices we do with regard to species and subspecies definitions.

We recognize that this type of intensive systematic study has lost favor in the past decade, particularly with the burgeoning and now, nearly sole use of molecular genetic applications to investigate biodiversity and systematic questions. We believe, however, that such limited analyses, although exceedingly powerful and unparalleled for their insight into evolutionary history, nevertheless run the risk of losing sight of the organism in nature. We hope that our combined character and analytical approaches provide the reader with a useful understanding not only of the biological diversity of this complex of woodrats but also a view of these taxa as the naturally occurring organisms that they are.

TAXONOMIC HISTORY OF THE *Neotoma lepida* GROUP

Our concept of the *Neotoma lepida* group, and thus the taxa included in this monograph, follows Goldman (1932), but excludes *N. goldmani* and *N. stephensi*.

This narrowed view also excludes *N. fuscipes* and *N. macrotis*, sister species to those of "*N. lepida*" based on molecular phylogenetic analyses and placed, as such, in a broadened *lepida* species group by Edwards and Bradley (2002; see also Matocq et al, 2007).

Merriam (1887) named the first taxon of the *Neotoma lepida* group, *N. bryanti*, based on a single specimen from Isla Cedros (= Cerros), Baja California, Mexico that was singed in a fire deliberately set to drive it from its nest. Six years later, Thomas (1893) named *N. lepida*, based on a specimen that was, according to Goldman (1910, p. 79), obtained by the British Museum from the U. S. National Museum and originally identified as *N. cinerea*; he gave the type locality as "Utah." Rhoads (1894) named *N. intermedia* from Dulzura, San Diego Co., California and *N. intermedia gilva* from Banning, San Bernardino Co., California. In May of that year, Price (1894) named *N. californica* from Bear Valley, San Benito Co., California. The following July, Merriam (1894a) named *N. desertorum* from Furnace Creek, Inyo Co., California and *N. desertorum sola* from San Emigdio, Kern Co., California. He suggested that both *californica* Price and *gilva* Rhoads were the same as typical *N. intermedia*, placed *N. desertorum* and *N. intermedia* together as the only two members of the *desertorum* group, and placed *N. arizonae*, which he had described in 1893, and *N. lepida* together in his *arizonae* group. In September 1894, Merriam (1894b) listed *N. lepida* as a synonym of *N. arizonae*, albeit with a question mark. Allen (1898) named *N. arenacea* from San Jose del Cabo, Baja California Sur, Mexico and *N. anthonyi* from Isla Todos Santos, Baja California, Mexico. He considered *N. arenacea* related to *N. fuscipes macrotis* but considered *N. anthonyi* to have no close relatives within the genus. Bangs (1899) named *N. bella* from Palm Springs, Riverside Co., California, synonymized *N. desertorum* with *N. lepida*, and put *N. bella* into an "*intermedia-lepida*" group. Elliot (1903) named *N. bella felipensis* from San Felipe, Baja California, Mexico, and referred specimens from numerous localities in the northern part of Baja California to *N. intermedia*. Elliot (1904), without describing any new forms, assigned specimens to *N. desertorum* and *N. d. sola* and remarked that he did not agree with Bangs (1899) that *N. lepida* and *N. desertorum* were the same animal. Elliot (1904) also reduced *N. bella* to a subspecies of *N. intermedia* and listed both that form and *N. intermedia gilva* from a single locality, Whitewater, Riverside Co., California. Goldman (1905) named *N. martinensis* from Isla San Martin, Baja California, Mexico and *N. nudicauda* from Isla Carmen, Baja California Sur, Mexico. He wrote that *N. martinensis* resembled *N. anthonyi* in color but not in skull morphology and that *N. nudicauda* resembled *N. arenacea* and *N. albigula*. Goldman (1909) named *N. intermedia pretiosa* from Matancita, *N. i. perpallida* from Isla San José, *N. i. vicina* from Isla Espíritu Santo, and *N. abbreviata* (which he placed in the *intermedia* group) from Isla San Francisco, all four localities in Baja California Sur, Mexico. In February 1910, Taylor named *N. nevadensis* from Virgin Valley, Humboldt Co., Nevada and considered it related to, but specifically distinct from, both *N. desertorum* and *N. lepida*.

In October 1910, Goldman published his revision of the genus in which he recognized seven species groups in the nominate subgenus. Species in his *intermedia* group inhabited coastal southern California and virtually all of Baja California and included the species *N. abbreviata*, *N. anthonyi*, *N. bryanti*, *N. intermedia* (with the 5 subspecies: *arenacea*, *gilva*, *perpallida*, *pretiosa*, and *vicina*), *N. martinensis*, and *N. nudicauda*. He arranged *N. californica* as a synonym of *N. intermedia* and both *N. desertorum sola* and *N. bella felipensis* as synonyms of *N. intermedia gilva*. His *desertorum* group included as full species *N. desertorum*, *N. goldmani*, and *N. lepida*, with *N. stephensi* arranged as a subspecies of *N. lepida* and both *N. bella* and *N. nevadensis* listed as synonyms of *N. desertorum*. Goldman did not consider his *intermedia* group to be particularly closely related to his *desertorum* group. Furthermore, he believed the type locality of *N. lepida* to be unknown, not "Utah" as identified by Thomas (1893), probably because he believed the name represented a taxon that occurred only south of the Utah-Arizona border. He also treated *N. arizonae* as a subspecies of *N. cinerea*, where it has remained.

Townsend (1912) named *N. insularis* from Isla Ángel de la Guarda, Baja California, Mexico and considered it most closely related to *N. intermedia gilva*. Townsend based his description on a female deposited in the AMNH; a few years later the holotype was transferred to the USNM (catalog number 198405). To our knowledge, all subsequent references to the holotype still list it as in the AMNH.

Grinnell and Swarth (1913), based on specimens collected in the vicinity of the San Jacinto Mountains in southern California, suggested that *N. intermedia* intergraded with *N. desertorum* and arranged the latter as a subspecies of the former. At least in part, their conclusion rested on the earlier assumption by Goldman (1910) that *N. bella* from Palm Springs represented the same species as *N. desertorum* from Furnace Creek. Goldman (1927) accepted the conclusion of Grinnell and Swarth and named the sample from Tanner Tank, Coconino Co., Arizona as *N. intermedia devia*. He gave the range of this new taxon as western Arizona east of the Colorado River. Nelson and Goldman (1931) named *N. intermedia ravida* from Comondú, Baja California Sur, Mexico and gave its distribution as the volcanic region of southern Baja California from the Sierra de la Giganta north to latitude 28°.

Goldman (1932) reviewed the entire complex and concluded that all specimens that he (1910) previously referred to *N. lepida*, except the type, belonged to a different species, the oldest name for which is *N. stephensi* Goldman (1905). He considered the specimens he had listed in 1910 as nominate *N. lepida* (following the custom of the time, he had not used trinomials for nominate subspecies) to be subspecifically distinct from true *N. stephensi* and named them as *N. s. relicta*, with the type locality as Keams Canyon, Navajo Co., Arizona. Part of the confusion resulted from difficulties with determining the type locality of *N. lepida*. Goldman concluded that the type specimen had been collected on the Simpson expedition that started at Camp Floyd (= Fairfield), Utah and ended in Carson City, Nevada but could not determine the locality more exactly. He considered *N. lepida* as the oldest name for all forms previously listed as subspecies of either *N. intermedia* or *N. desertorum* and arranged *arenacea*, *devia*, *felipensis*, *gilva*, *intermedia*, *notia*, *perpallida*, *pretiosa*, *ravida*, and *vicina* as subspecies of *N. lepida*, while retaining *N. anthonyi*, *N. abbreviata*, *N. bryanti*, *N. insularis*, *N. martinensis*, and *N. nudicauda* as full species. He treated the names *bella*, *desertorum*, and *nevadensis* as synonyms of *N. l. lepida*; treated *sola* as a synonym of *N. l. gilva*; and listed *californica* as a synonym of *N. l. intermedia*. He named *N. l. monstrabilis* from Ryan, Coconino Co., Arizona as a new subspecies of *N. lepida*, giving its range as southern Utah and Arizona north of the Colorado River. Finally, he retained *N. goldmani* as a full species in his *lepida* group.

Burt (1932) named *N. lepida marcosensis* from Isla San Marcos, *N. l. latirostra* from Isla Danzante, and *N. bunkeri* from Isla Coronados, all three localities in Baja California Sur, Mexico. He placed *N. bunkeri* in the subgenus *Homodontomys* and considered the skull similar to that of *N. fuscipes macrotis*. Blossom (1933) named *N. auripila* from near Papago Well, Agua Dulce Mts., Pima Co., Arizona and considered it related to *N. lepida devia*. Orr (1934) named *N. lepida egressa* from El Rosario, Baja California, Mexico giving its range as the Pacific coast between 30° 03' and 31°N. Blossom (1935) named *N. lepida bensoni* from Papago Tanks in the Pinacate Mts., Sonora, Mexico with its range restricted to that region. Later that same year Benson (1935) reviewed geographic variation in *N. lepida* in Arizona; he named *N. l. flava* from Tinajas Altas, Yuma Co. with a range restricted to the Tinajas Altas Mts., reduced *auripila* to a subspecies of *N. lepida*, and referred all other specimens from the state to *N. l. devia* or *N. l. monstrabilis*. Huey (1937) named *N. l. aureotunicata* from Punta Peñascosa, Sonora, Mexico and *N. l. harteri* from south of Gila Bend, Maricopa Co., Arizona, both then known only from their respective type localities. Von Bloeker (1938) resurrected *californica* as a subspecies of *N. lepida* with a range along the inner coast ranges of California from Santa Clara Co. south to Monterey Co. and named *N. l. petricola* from Abbott's Ranch, Arroyo Seco, Monterey Co., California and gave its distribution as the Santa Lucia and Sierra de Salinas mountains. Goldman (1939) named *N. lepida marshalli* from Carrington Island, Tooele Co., Utah, known only from its type locality. Hall (1942) named *N. lepida grinnelli* from north of Picacho, Imperial Co., California, and defined its range as the western side of the Colorado River in Nevada and California. Huey (1945) named *N. lepida molagrandis* from Santo Domingo Landing, Baja California with a range along the northern and western coastal section of the Vizcaino Desert region of the peninsula. Kelson (1949) named *N. lepida sanrafaeli* from Rock Canyon Corral, near Valley City, Grand Co., Utah with a range in eastern Utah north of the Colorado River. The most recently described taxon in this complex is *N. lepida aridicola*, which Huey (1957) named from El Barril, Baja California, Mexico and occurs on the Gulf side of the peninsula from San Francisquito Bay to El Barril.

Relying solely on characters of the baculum, Burt and Barkalow (1942) suggested that *Neotoma bunkeri* was related to *N. lepida* and not to *N. fuscipes* and that the bacula of *N. bunkeri* and *N. lepida* differed sufficiently from those of other members of the species group to suggest separate subgeneric status. The taxonomic arrangement of *Neotoma* in Hall and Kelson (1959) summarized the numerous changes to that time. Hoffmeister and de la Torre (1959) concluded that the baculum of *N. stephensi* was more similar to either *N. mexicana* or *N. phenax* than to that of *N. lepida*. Burt (1960), in his monograph on the bacula of North American mammals, reiterated the conclusions of Burt and Barkalow (1942), considered *N. stephensi* more similar to *N. mexicana* than to *N. phenax*, and, based on the examination of one specimen, suggested that *N. lepida insularis* had an abnormal baculum. Hooper (1960), in his account of the soft anatomy of the glans penis of *Neotoma* and related genera, also stated that *N. lepida* was unique in the genus to the point of possibly requiring its own subgenus and that the glans penis of *N. stephensi* resembled those of *N. mexicana* and *N. phenax* more than the glans of *N. lepida*.

Baker and Mascarello (1969) documented differences among different populations of *N. lepida* based on standard karyotypes of non-differentially stained chromosomes. Mascarello and Hsu (1976) subsequently showed that the karyotypic variation, based on C- and G-banded chromosomes, was between populations on opposite sides of the Colorado River. Differences between karyotypes consisted of heterochromatic short-arm additions on two autosomes and a pericentric inversion in chromosome 2. They also decided that the karyotypes did not support putting *N. stephensi* in the same species group with *N. lepida*, a conclusion subsequently supported by a cladistic analysis of *Neotoma* banded karyotypes (Koop et al., 1985). Mascarello (1978) utilized characters of the glans, chromosomes, and isozymes to determine that *N. lepida* comprised three forms: one from Baja California and coastal California corresponding to the *N. intermedia* of Goldman (1910); one from the deserts of California ranging north into Nevada, Utah, Colorado, and Arizona north of the Colorado River corresponding roughly to *N. desertorum* (= *N. lepida*) of Goldman; and one occurring east of the Colorado River in Arizona and Sonora (= *N. devia*). Mascarello did not consider the first and second forms sufficiently different to warrant treating them as separate species but did suggest recognizing *N. devia* as a species.

Carleton (1980), in his study of phylogenetic relationships among taxa of the neotomine-peromyscine complex, once again concluded that *N. lepida* showed few close affinities with other members of the genus but declined to propose a new subgenus for it. Hall (1981) summarized all previous work except that of Mascarello (1978) by recognizing a single species, *N. lepida*, with 31 subspecies. Hall also retained as full species the four island forms not previously treated as subspecies of *N. lepida*, and did not recognize any species groups in the subgenus. Hoffmeister (1986), in his monograph on Arizona mammals, concluded that the glans penis characters used by Mascarello to separate *N. lepida* from *N. devia* did not hold and, without evaluating the isozyme or chromosomal data, considered the two forms as conspecific. Musser and Carleton (1993), in their review of the taxonomy of muroid rodents, followed Mascarello in recognizing *N. devia* as a species separate from *N. lepida*, but listed *monstrabilis* Goldman and *sanrafaeli* Kelson, both from north of the Colorado River in Arizona, Utah, and Colorado, as synonyms of *N. devia*. In their more recent synopsis, however, Musser and Carleton (2005) limited their concept of *N. devia* to only those samples south and east of the Colorado River in Arizona and Sonora, and included *aureotunicata*, *auripila*, *bensoni*, *flava*, and *harteri* as synonyms. They allocated both *monstrabilis* and *sanrafaeli* to *N. lepida*. Finally, these authors in both their 1993 and 2005 reviews continued to recognize the insular taxa *N. anthonyi*, *N. bryanti*, *N. bunkeri*, and *N. martinensis* as distinct species and treated *insularis* as a synonym of *N. lepida*.

Riddle et al. (2000a) supported Mascarello's (1978) suggestion that *N. lepida* itself might be a composite of two species based on mtDNA sequence data. Edwards and Bradley (2002) examined phylogenetic relationships among species of woodrats based on mtDNA cytochrome b gene sequences, and limited Goldman's *lepida* group to *N. lepida* and *N. devia*. Matocq et al. (2007) documented a phylogenetic sister relationship between *N. lepida* (including *devia*) and the *N. fuscipes-N. macrotis* complex based on evidence from both mtDNA and nuclear DNA sequences. None of these authors, however, examined the insular "species" of the *lepida* group, namely *N. bryanti*, *N. anthonyi*, *N. martinensis*, and *N. bunkeri*. Finally, Patton and Álvarez-Castañeda (2005) delineated *cyt-b* sequence variation throughout the range of the *lepida* group, as redefined by Edwards and Bradley, including the insular *N. bryanti* from Isla Cedros. They documented strong molecular clade structure, with populations from the coastal region of California and Baja California more differentiated from *N. lepida* proper then *N. lepida* is from populations of *N. devia* from east of the Colorado River in Arizona. Their analysis suggested the possible nomenclatural priority of

Merriam's (1887) name *N. bryanti* for the coastal California-Baja California complex, an action that awaited "...integrated morphological and molecular confirmation" (Musser and Carleton, 2005, p. 1056). Our objective here is to provide such confirmation based on a thorough analysis of craniodental and colorimetric data combined with qualitative morphological characters of the skull and glans along with molecular genetic data from both the nuclear and mitochondrial genomes.

MATERIALS AND METHODS

SPECIMENS AND ABBREVIATIONS

We examined more than 4,600 specimens of desert woodrats from 1,095 individual localities. Most of these are housed in the collections of the Museum of Vertebrate Zoology, which contains 3,419 specimens of woodrats of the *Neotoma lepida* group (see http://mvz.berkeley.edu/), including 3,004 from the United States and 415 from Mexico. More than 1,100 specimens were collected specifically for this study. The additional specimens we surveyed are housed in the mammal collections at the National Museum of Natural History, Smithsonian Institution (USNM, n = 683), Academy of Natural Sciences, Philadelphia (ANSP, n = 2), Museum of Comparative Zoology, Harvard University (MCZ, n = 3), Dickey Collection, University of California Los Angeles (UCLA, n = 69), Los Angeles County Museum of Natural History (LACM, n = 288), San Diego Society for Natural History (SDNH, n = 277), California State University, Long Beach (CSULB, n = 62), University of Washington Burke Museum (UWBM, n = 1), University of Illinois Museum of Natural History (UINHM, n = 17), University of North Texas (UNT, n = 11), and Centro de Investigaciónes Biológicas del Noroeste, La Paz, Baja California, Mexico (CIB, n = 131). We examined holotypes of 30 of the 38 named forms of the *lepida* group (considered as valid taxa or as junior synonyms; see Hall, 1981; Álvarez-Castañeda and Cortés-Calva, 1999; and Musser and Carleton, 2005): *abbreviata* Goldman (MCZ 12260), *aridicola* Huey (SDNHM 15595), *aureotunicata* Huey (SDNHM 10907), *bella* Bangs (MCZ 5308), *bryanti* Merriam (USNM 186481), *bunkeri* Burt (UCLA 19725), *desertorum* Merriam (USNM 33139/25739), *devia* Goldman (USNM 226376), *egressa* Orr (MVZ 50142), *flava* Benson (MVZ 62657), *gilva* Rhoads (ANSP 1665), *grinnelli* Hall (MVZ 10438), *harteri* Huey (SDMNH 11462), *insularis* Townsend (USNM 198405), *intermedia* Rhoads (ANSP 8343), *latirostra* Burt (UCLA 19718), *marcosensis* Burt (UCLA 20010), *marshalli* Goldman (USNM 263984), *martinensis* Goldman (USNM 81074), *molagrandis* Huey (SDNHM 14065), *monstrabilis* Goldman (USNM 243123), *nevadensis* Taylor (MVZ 8282), *notia* Nelson and Goldman (USNM 146794), *nudicauda* Goldman (USNM 79073), *perpallida* Goldman (USNM 79061), *petricola* von Bloeker

9

(MVZ 30203), *pretiosa* Goldman (USNM 140123), *ravida* Nelson and Goldman (USNM 140692), *sola* Merriam (USNM 43381/31516), and *vicina* Goldman (USNM 146803).

A complete list of localities for which we have examined specimens is provided in the Appendix. Each locality is assigned a unique number for reference in the lists of specimens used in the separate geographically based analyses. We obtained the georeferenced coordinates used to map localities from each museum collection database via the Mammal Networked Information Systems (MaNIS; http://manisnet.org). Because these data are dynamic and thus subject to change as locality coordinates are refined, all data from non-MVZ specimens are from 1 January 2005; those from MVZ are from 1 January 2006; and those from CIB, which is not yet a participant of the MaNIS network, are from 10 December 2005.

MORPHOMETRICS

We took external measurements from specimen labels, as follows:

TOL   Total length, from tip of nose to tip of terminal tail vertebra

TAL   Tail length, from dorsal flexure at base of the tail to tip of the last vertebra

HF   Hind foot length, from proximal margin of calcaneus to tip of longest claw

E   Ear height, from notch to top of pinna (only crown height is available for many specimens collected in the early part of the 1900s; these measurements were excluded from analyses)

We took twenty-one cranial dimensions with digital calipers to the nearest 0.01 mm (Fig. 1), as follows:

CIL   Condyloincisive length, from the anterior margins of the upper incisors to the posterior margins of the occipital condyles

ZB   Zygomatic breadth, greatest breadth across the zygomatic arches

IOC   Interorbital constriction, least distance across the roof of the skull between the orbits

RL   Rostral length, diagonal measurement taken from anterior margin of orbit to anterior margin of nasal bones

NL   Nasal length, maximum midline length of nasal bones

RW   Rostral width, taken across outside margins of the nasolacrimal capsule

OL   Orbital length, taken diagonally from the anterior to posterior margins of the orbit

D   Diastema length, from the posterior face of the upper incisors to the anterior edge of M1

MTRL   Molar toothrow length, alveolar length of maxillary toothrow

IFL   Incisive foramen length, length of maximal opening of incisive foramen

PBL   Palatal bridge length, from the posterior margins of upper incisors to anterior margin of mesopterygoid fossa

AW   Alveolar width, outside distance across the alveolae of the second upper molars

OCB   Occipital condyle breadth, outside distance between occipital condyles

MB   Mastoid breadth, greatest width of cranium across the mastoid bones

BOL   Basioccipital length, distance from ventral margin of foramen magnum to basioccipital-basisphenoid suture

MFL   Mesopterygoid fossa length, midline distance from anterior margin of posterior tip of hamular processes

MFW   Mesopterygoid fossa width, maximal width taken at suture of palatine and pterygoids bones

ZPL   Zygomatic plate length, taken at mid-height from anterior to posterior margins of zygomatic plate

CD   Cranial depth, vertical distance from plane determined by incisor tips and bullae and top of cranial vault

BUL   Bullar length, greatest length of tympanic bulla

BUW   Bullar width, greatest width of tympanic bulla

We estimated age by measuring the height of the hypoflexus on the first upper molar using an optical micrometer (M1H; Fig. 1). The molars of woodrats are coronally hypsodont, with elevated crowns that erupt and begin to wear before the tooth roots and growth ceases. Most individuals with the base of the hypoflexus still hidden by the bony alveolus were still in juvenile or subadult pelage and were considered very young and not measured. For all remaining individuals for which the height of the hypoflexus could be measured, we placed individuals into one of five equal groups: Age 5 (youngest): height 2.00-2.50 mm; Age 4: height 1.50-1.99 mm; Age 3: height 1.00-1.49 mm; Age 2: height 0.50-0.99; and Age 1 (oldest): height 0.00-0.49. Individuals in all five age-categories had adult pelage, although they may not have been post-reproductive. We included all age 1-5 individuals in the analyses we undertook. Exceptions to this "age" scheme are the very large-bodied individuals from Baja California, where animals in adult pelage and with evidence of reproductive activity may also have incompletely erupted molars wherein the base of the hypoflexus is below the bony alveolus. These specimens were measured and tested to determine if they could be included in an "adult" category for statistical comparisons.

We performed statistical analyses with one of three commercially available programs for the personal computer. Univariate summaries of morphometrics variables were performed primarily with StatView® (version 5.0; SAS Institute Inc.). Principal components and discriminant function analyses designed to compare samples pooled by locality were performed with JMP® (version 5.0, SAS Institute Inc.) or Statistica® (StatSoft, Inc.). We examined nongeographic variation (due to sex and age as estimated from the tooth height categories, and their interaction) by two-way analysis of variance (ANOVA, random-effects model to accommodate unequal sample sizes), and all univariate comparisons between populations or geographic areas used one-way ANOVAs, again with a random-effects model. Both sets of analyses used StatView. All multivariate analyses used $\log_{10}$ transformations of the original cranial variables. Finally, to explore the effects of age, as measured by hypoflexus height, on multivariate analyses, we performed separate analyses with the raw transformed variables and the residuals taken from the regression of each original transformed variable on hypoflexus height. This "correction" for age had no effect on the patterns of variation or on the conclusions that stem from the interpretation of those patterns.

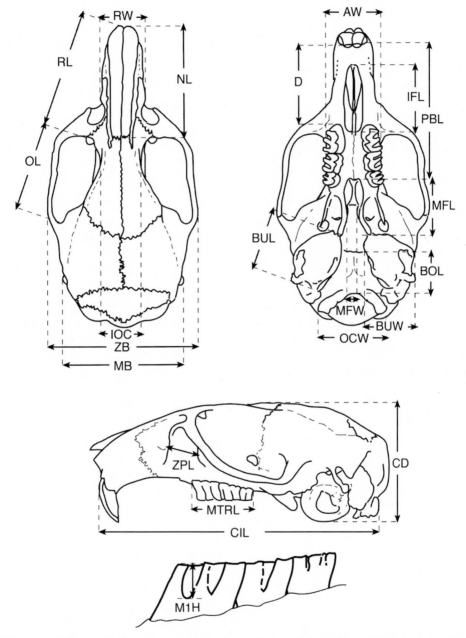

Figure 1. Views of the cranium (dorsal, ventral, and lateral) and labial view of the measurements taken from woodrat specimens examined. Abbreviations are defined in the text.

We summarize variation in both external and craniodental variables by standard descriptive statistics (mean ± standard error, sample size, and range) for all samples and use multivariate principle components (PCA) and canonical variates analyses (CVA) to document trends in character variation across geography and at particular regions of sharp transition. Because of the pronounced size variation exhibited among both peninsular and insular samples from Baja California, for these samples we employed a size-free canonical discriminant analysis (CDA) following the methodology outlined by Patton and Smith (1990) and dos Reis et al. (1990). The first step in this procedure is to perform a PCA on the within-group variance-covariance matrix on the log-transformed craniodental variables. The resulting first PC axis can be considered a multivariate size vector if all variables load positively and are significantly correlated with the values of their respective cranial characters (Strauss, 1985). Residuals are then obtained from the regression of each original craniodental variable on PC-1 scores and entered into a CDA with sample groups (taxa or geographic groups) identified a priori. The resulting distribution of these groups in multivariate space is then based on size-free cranial proportions or a measure of overall cranial shape in our comparisons among groups. We then assessed how individual characters might influence the separation of these sample groups by transforming canonical coefficients into correlation vectors calculated from the correlation between individual scores for the canonical variables and the actual values of the characters for each individual (Strauss, 1985).

Finally, we determined empirically the sample groups used in analyses of geographic trends in an iterative process, using both each individual univariate character as well as scores on PC-1 and PC-2 axes. Individual sample localities were initially grouped arbitrarily by close geographic proximity and commonality of taxon assignment. If fewer than 5 of the 21 craniodental univariate characters and neither PC-1 and PC-2 scores were found to be significantly different among this set of geographic samples (based on one-way ANOVA, using Fisher's PLSD posterior multiple pairwise test and applying the Bonferroni correction; Rice, 1989), this set of samples was then joined as a single, pooled sample and compared in a similar fashion to other pooled samples geographically adjacent and currently assigned to the same taxon, based on present taxonomy (e.g., Grinnell, 1933; Hall, 1981; Álvarez-Castañeda. and Cortés-Calva, 1999). The final groups thus included clusters of geographically adjacent localities that were statistically uniform in the characters examined, at both the univariate and multivariate levels. Because we wished to examine the veracity of current infraspecific taxonomy, we also made sure that pooled geographic samples included only localities currently allocated to a single subspecies (as mapped by Hall, 1981).

GLANS PENIS, INCLUDING BACULUM

One of us (DGH) prepared glandes originally dried on museum skins by clearing them with 2% potassium hydroxide for no longer than 48 hrs and staining them with alizarin red S followed by storage in glycerin prior to examination. We also had available a large number of specimens originally preserved in formalin and maintained in 70% ethanol; most of these were not cleared and stained. Of the total examined, 606 specimens from 216 localities proved complete enough for detailed analysis. Many of the glandes dried on the skins no longer contained the all-important tip (see below). We noted that many of the specimens that had been cleared and stained by other workers were stretched and excessively cleared, with the result that the spines were lost and the glans greatly distorted. Some workers processed the specimens only to use the baculum; in these cases almost all soft tissues were lost. We recommend that anyone wishing to preserve glandes of these species place them in formalin in the field, since drying them on the skin risks damaging the tip. We also recommend that dried glandes not be left in KOH longer than 48 hrs, with an additional day for staining; any longer risks major damage to the specimen.

We used characters 52 through 67 of Carleton (1980) to describe the glandes. Characters 63 through 67 refer to measurements. The varying methods of preservation of these elongate penial structures have produced much variation in the length and width of the glans, and additionally, the curved nature of the baculum makes its length very difficult to determine accurately. Therefore we based decisions on character states on direct comparisons rather than on actual measurements. Further, the tip-type was classified according to Mascarello (1978, Fig. 5).

We examined glandes from specimens referable to all of the 31 subspecies of this complex recognized in Hall (1981) except *N. l. aureotunicata* and *marshalli*. In addition, we examined topotypes or near topotypes of all other named forms, even those not currently recognized at the subspecies level, except *lepida*, *desertorum*, *grinnelli*, *intermedia*, *sola*, *bella*, and *egressa*. Our list includes specimens from all but three of the islands known to have populations of this complex; all three of these island forms for which no glandes are available were originally named as full species (*N. anthonyi*, *N. martinensis*, and *N. bunkeri*) and retained as such by all subsequent workers. Unfortunately, we know of no specimens of *N. bunkeri* collected since its description or of either *N. anthonyi* or *N. martinensis* collected since the mid-1920s (specimens in the MVZ); each of these populations may be extinct (see Mellink, 1992a, b).

COLORIMETRY

We employed an X-Rite Digital Swatchbook® spectrophotometer (X-Rite, Inc., Grandville, MI, USA) to measure color on a total of 3,406 study skins of woodrats from 835 specific localities. Specimens examined span the entire range of the *Neotoma lepida* group and include all insular and mainland taxa. We set the spectrophotometer to compare measured colors to the CIE (Commission Internationale de l''Eclairage, or International Commission on Illumination) Standard Illuminant F7 for fluorescent illumination, which represents a broadband daylight fluorescent lamp (6500 K). We chose this standard because all measurements were taken indoors under fluorescent ambient lighting. The instrument provides a reflectance spectrum (390-700 nm) of the object being measured as well as tristimulus color scores (CIE X, Y, and Z) that can be directly compared to scores from the Munsell or other color references (Hill, 1998).

We measured color with a 3 mm diameter port placed at four topographic positions on each individual specimen: (1) on the dorsal surface at mid-rump, (2) on the dorsal stripe of the tail about 1/3 its length from the base, (3) just above the lateral point of contact between the dorsal and ventral color on the flank, and (4) at mid-chest on the ventral surface. Dorsal and tail color is generally uniform for any given individual specimen, and the exact placement of the colorimeter resulted in little variation in the measurements obtained. The color of the sides of a specimen, where there is always an abrupt shift from the dorsal to ventral color, may simply involve a gradual lightening of the dorsal color laterally or may be more complex with a distinct lateral line of a color different from that of either the dorsal or ventral surfaces. Consequently, care was taken in all lateral measurements to place the colorimeter port on the flank just above the contact point with the ventral color, thereby ensuring measurement of any lateral line color if present. The color and pattern of color across the venter varies greatly, both with regard to the degree of exposure of the gray base of individual hairs and of the different colors of the hair tips. Because it was not possible to record all of this variation in a single measurement, we always measured the mid-chest between the axillary regions of the forearms. We examined only adult, non-molting specimens with non-oily fur. Finally, to determine the repeatability of the instrument, we took 10 separate measurements from each of the four sites on the skin for an initial set of 10 specimens from each of three different populations whose dorsal color was easily distinguished by eye and examined the mean and variance of each. Because the variance was less than 3% in all cases, we subsequently took three separate measures from each specimen and used the means of these as input data for analysis.

We also took color measurements from Munsell glossy finish colors (Munsell, 1976) so that direct comparisons could be made between our measurements of color for any sample of woodrats to this standard color system. We chose the series of Munsell colors that we determined by visual comparison to be closest to the range of dorsal colors exhibited in museum skins of desert woodrats.

We examined variation in X, Y, and Z variables both separately as well as combined for each of the four topographic areas of the skin by standard univariate and multivariate statistics using the StatView, JMP, or Statistica software programs. We then compared in a single analysis the range of measured color for each topographic region to same variable taken from the Munsell color chips. This allows one to associate more formally the Munsell color system with woodrat colors than by simple visual notation. Finally, we examined color among samples organized for each of the geographic transects concomitantly with variation in craniodental and molecular data. Because melanism is common among many desert woodrat populations, particularly those inhabiting basalt flows of even limited geographic extent (e.g., Leiberman and Lieberman, 1970), we undertook separate analyses to compare the color characteristics of melanic populations occurring on different lava fields and encompassing separate molecular clades. This analysis examined the degree of phenotypic similarity among melanic individuals, regardless of geographic area or hypothesized phyletic origin of their respective populations.

ENVIRONMENTAL VARIATION

We examined the relationship between morphological attributes, both each separate craniodental or colorimetric variable as well as their multivariate PCA summaries, and environmental variables in our geographic analyses, below. To do this, we used the 19 bioclimatic (Bioclim) variables derived from the monthly temperature and rainfall values from weather stations in the western United States and northern Mexico. These data are archived in the WorldClim database that is accessible at http://www.worldclim.org/bioclim.htm. We then generated data layers for each environmental variable in ArcView 3.2 (http://www.esri.com/), with point data interpolated and extracted for each geographic sampled locality (see Appendix) using DIVA-GIS, version 5.2 (http://www.diva-gis.org/). These data were then subjected to a principal components analysis to reduce the large number of correlated individual bioclimatic variables to a reduced set of orthogonal axes. We then used correlation analyses to relate craniometric and colorimetric variables to bioclimatic scores on the first two PC axes, which combine to explain 73% of the total pool of variation among the 19 individual Bioclim variables.

MOLECULAR SEQUENCE METHODOLOGY

We extracted genomic DNA from liver or ear biopsies either preserved originally in 95% ethanol or frozen in liquid nitrogen in the field and maintained at -80°C in the lab, using either Chelex® (Walsh et al., 1991) or DNAeasy kits (Qiagen Inc.). We specify the methods employed for each molecular marker system we used in the three following sections.

mtDNA cytochrome b gene sequence

The mitochondrial cytochrome b gene (*cyt-b*) in *Neotoma* is 1143 base pairs (bp) in length (Edwards and Bradley 2001, 2002; Edwards et al. 2001). We amplified the entire gene in two fragments of approximately equal length that overlapped by about 350 bp. Primer pair MVZ05-MVZ16 amplified the initial 800+ bases of the gene, and primer pair MVZ127-MVZ108 amplified the terminal 700+ bases (primer sequences in Smith and Patton, 1999; Leite, 2003). We purified double stranded DNA using the QIAquick PCR Purification kit (Qiagen, Valencia, CA), and then cycle-sequenced this template with MVZ05 and MVZ127 for the light strand and MVZ108 and MVZ16 for the heavy strand using the Taq FS kit. We generated all sequences on either an ABI 377 slab gel or ABI 3730 capillary automated sequencer following manufacturer protocols. We aligned and edited all sequences using the Sequence Navigator software (Applied Biosystems, Inc.). Both strands of the entire gene were sequenced for an initial set of 203 specimens to ensure the constancy of sequence for each individual, but only the light strand was then obtained for subsequent specimens, which comprised the majority of the sequences examined.

We also extracted DNA from museum specimens of four taxa of the *Neotoma lepida* group now believed to be extinct (*anthonyi* [USNM 137173, 137201], *bunkeri* [UCLA 19720], *insularis* [UCLA 19911], and *martinensis* [USNM 139030]), following established guidelines for "ancient" DNA (e.g., Gilbert et al., 2005). In each case, a small piece of skin was removed from the edge of the ventral incision with sterilized instruments, hair was carefully removed by a sterile scalpel blade, the skin fragment was subsequently soaked in sterile ddH$_2$0 overnight, with extraction then performed with the DNAeasy kits in the same fashion as tissue samples for freshly collected samples. All procedures took place in a "DNA clean room" under a pressurized hood to eliminate opportunities for contamination. We amplified extracted DNA from these samples in smaller fragments, averaging about 400 bp in length, using a combination of published primer pairs (MVZ03, MVZ04, MVZ103, and MVZ14; Smith and Patton, 1991, 1993) and others designed from multiply-aligned sequences that we had previously generated from fresh specimens of the *N. lepida* group. The latter included primer pair Neo66F (5'—CYA CCC CAC CCA ACA TCT CAT CAT G—3') and Neo66R (5'—TTG TRA TAA CNG TGG CYC CTC AGA ARG—3'), which amplified a 376 bp fragment beginning at position 66 in the *Neotoma cyt-b* genome, and primer pair Neo365F (5'—CCG TAA TAG CAA CAG CAT TTA TAG G—3') and Neo365R (5'—GCT GGG GTG TAG TTG TCT GG—3'), which produced a 411 bp fragment beginning at position 365. All extraction and PCR procedures were done in a clean room physically separated from laboratory areas where modern samples are routinely processed. We sequenced each sample on multiple occasions using separate extractions and amplification reactions, as well as in both directions. Each sequence we report here and list in GenBank was thus confirmed independently by at least three separate amplification and sequencing reactions.

We obtained the entire 1143 bp cytochrome b sequence for 500 specimens and the initial 801 bp fragment of this same gene from an additional 648 individuals, for a total of 1148 specimens from 198 localities of the *Neotoma lepida* group. These data include topotypes or near topotypes (defined as specimens collected from within approximately 1 km of the type locality) from 25 of the 35 named taxa currently recognized within this complex (following Hall, 1981; Table 1). The only taxon not sampled by us is *flava* Benson, from southwestern Arizona. Singleton specimens represent seven taxa, six of which are insular races (*bryanti, bunkeri, insularis, marshalli, martinensis,* and *vicina*). Excluding these singletons, the average number of specimens per taxon sampled is 39.4 (range = 2 [*anthonyi*] to 410 [*lepida*]). Not surprisingly, the most completely sampled taxa are those with the broadest distributions (*lepida* [n = 410, 43 localities] and *gilva* [n = 243, 48 localities]), which are also more heavily sampled because of our desire to examine contact points between them. The total number of localities sampled per taxon ranged from one to 48 (mean = 8.06), with an average of 5.97 individuals per locality (range 1 – 66, Fig. 2). The fewest specimens and localities in proportion to the number of described taxa are those from Baja California, although samples are available from the length of the peninsula and from all of the islands on both sides that historically, if not presently, contain woodrat populations (Álvarez-Castañeda and Cortés-Calva, 1999).

Forty-two of the 1148 individuals examined lack museum vouchers; these came from one of three localities in California (Deep Canyon, Riverside Co. [n = 1]; Freeman Canyon, Kern Co. [n = 21]; and Furnace Creek, Death Valley, Inyo

Co. [n = 20]) where on-going mark-recapture population studies precluded sacrificing specimens. For these, ear biopsies were taken and preserved in 95% ethanol in the field from living individuals that were subsequently released. We used sequences for most other species in the genus, obtained from GenBank based on data published in Edwards et al. (2001) and Edwards and Bradley (2001, 2002) or obtained by us, as outgroups in phylogenetic analyses (Table 1). Unique complete cytochrome b sequences from 188 individuals representing each taxon and each mtDNA clade and subclade (see below) are deposited in GenBank as accession numbers DQ781064-DQ781305.

Table 1. List of taxa of the *Neotoma lepida* group (from Hall, 1981) and outgroups used in the phylogenetic analyses, including sample sizes for mtDNA sequences.

Taxon	$N_{individuals}$	$N_{populations}$	Topotypes*
ingroup			
lepida abbreviata	6	1	yes
lepida arenacea	16	6	-
lepida aridicola	3	1	yes
lepida aureotunicata	8	1	yes
lepida auripila	17	42	yes
lepida bensoni	4	1	yes
lepida californica	37	6	yes
lepida devia	20	3	yes
lepida egressa	10	5	-
lepida felipensis	1	1	yes
lepida flava	-	-	-
lepida gilva	245	48	yes
lepida grinnelli	49	4	-
lepida harteri	4	2	yes
lepida insularis[†]	1	1	yes
lepida intermedia	15	5	-
lepida latirostra	6	1	yes
lepida lepida	410	43	-
lepida marcosensis	10	1	yes
lepida marshalli	1	1	-
lepida molagrandis	23	14	-

Table 1 (continued)

lepida monstrabilis	31	7	yes
lepida nevadensis	10	3	yes
lepida notia	4	1	yes
lepida nudicauda	4	3	yes
lepida perpallida	10	1	yes
lepida petricola	10	1	yes
lepida pretiosa	16	11	-
lepida ravida	18	6	-
lepida sanrafaeli	20	6	yes
lepida vicina	1	1	yes
lepida ssp.**	108	3	-
anthonyi[†]	2	1	yes
bryanti	1	1	yes
bunkeri[†]	1	1	yes
martinensis[†]	1	1	yes

Outgroups	$N_{individuals}$	GenBank number	MVZ number
Neotoma albigula	1	AF186828	
Neotoma cinerea	1	AF186799	
Neotoma floridana	1	AF186818	
Neotoma fuscipes	1	DQ781303	MVZ 195212
Neotoma goldmani	1	AF186830	
Neotoma macrotis	1	DQ781304	MVZ 198597
Neotoma mexicanus	1	AF305569	
Neotoma micropus	1	AF186827	
Neotoma stephensi	1	DQ781305	MVZ 197170

*   Topotypes are considered specimens collected within 1 km of the type locality.
[†]   Presumed extinct (see Álvarez-Castañeda and Ortega-Rubio, 2003).
**   Specimens from contact points between recognized taxa.

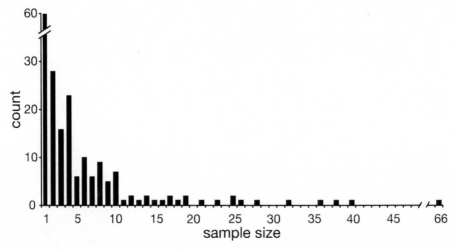

Figure 2. Histogram of sample sizes for the 195 separate localities examined for variation in the mtDNA cytochrome b gene.

For the cytochrome b dataset, we performed two levels of analyses, each based on a separate data set. First, we used the set of 500 complete cytochrome b sequences of 1143 bp to establish hierarchical relationships among haplotypes, geographic areas, and taxa using phylogenetic methods. Second, we used the complete set of 1148 reduced sequences 801 bp in length to examine phylogenetic structure within clades defined from the complete sequence analysis, to describe population genetic parameters of haplotype diversity, and to estimate measures of population connectedness, coalescent growth models, and the hierarchical apportionment of molecular diversity as a function of phylogenetically defined clades or currently recognized subspecies. We give details of each set of analyses in their respective results sections below. Prior to all analyses we identified redundant haplotypes using the program Collapse 1.2 (Posada, 2005; http://darwin.uvigo.es/).

We examined the hierarchical relationship of unique haplotypes of the complete *cyt-b* 1143 bp data set by the construction of minimum length trees, using the maximum parsimony (MP) criterion as implemented in PAUP* 4.0b10 (Swofford, 2002). We treated all sites as equal and unordered, and we employed a heuristic search option with stepwise addition of taxa and tree bisection-reconnection (TBR) branch-swapping. Due to the very large number of sequences, we performed only a single heuristic search. We represented the topological relationships among haplotypes as the strict consensus of all minimum length trees obtained. Finally, we used bootstrap re-sampling, with 1000 pseudoreplicates and the same settings as for the heuristic search, to assess the robustness of the resulting tree topology. We included only unique complete *cyt-b* sequences, representing each sampled taxon and 122 separate localities, in the MP analysis.

We employed Bayesian methods in a second analysis (reviewed in Huelsenbeck et al., 2001; Lewis, 2001) run with MrBayes 3.1.1 (Huelsenbeck and Ronquist, 2001; Ronquist and Huelsenbeck, 2003). Here, we used the best-fit model determined by the hierarchical likelihood ratio test employed by MrModelTest, version 2.2 (Nylander, 2004). This program selected the most parameter rich GTR+I+G model (log likelihood = -9196.0312, K = 10, AIC = 18412.0625), with proportion of invariable sites (I) of 0.5149, Gamma distribution shape parameter of 0.7719, and base frequencies of A = 0.3605, C = 0.3162, G = 0.0810, and T = 0.2423. The analysis was run with site-specific rate variation partitioned by codon position, with substitution rates estimated separately for first, second, and third codon positions, in keeping with a protein-coding gene. We initiated the analysis with a random starting tree and ran it for 2×10^7 generations. Four Markov chains were sampled every 1000 generations. We then computed the 50% majority-rule consensus tree after excluding those trees sampled prior to a stable equilibrium, with the posterior probability of nodal support given by the frequency of the recovered clade (Rannala and Yang, 1996; Huelsenbeck and Ronquist, 2001).

We used the software package Arlequin (version 3; Excoffier et al., 2005) to calculate gene and nucleotide diversities, the mean pairwise differences between all unique haplotypes, Tajima's D and Fu's Fs (to test for deviations from neutrality and/or historical demographic change), and histograms of the total number of pairwise differences among all 1148 individuals for the 801 bp dataset. We analyzed data separately for each clade identified with bootstrap values greater than 80% but excluded samples from contact zones where independence might be compromised. We then compared the histograms of pairwise differences, or "mismatch distributions," to the distribution expected in an expanding population (Slatkin and Hudson 1991; Rogers and Harpending 1992). Approximate 95% confidence intervals for this distribution were obtained by a parametric bootstrap approach (Schneider and Excoffier 1999). Finally, we obtained the "raggedness index" of Harpending (1994) for each distribution, a measure of the "stationarity" of population history. Large values for this index characterize multimodal distributions commonly found in populations that have been stable for long periods of time or that are mixtures of regionally differentiated groups; lower indices characterize unimodal and smoother distributions typical of expanding populations.

Nuclear gene sequences

We also sequenced a 609 bp fragment of intron 7 of the β-fibrinogen gene (*Fgb-I7*) for 166 specimens of the *Neotoma lepida* group, using the primers published by Wickliffe et al. (2003). To compare with published data, we also resequenced and included in our analyses each of the eight MVZ specimens for which Matocq et al. (2007; GenBank accession numbers DQ180031-DQ180038) reported *Fgb-I7* sequences. For geographic coverage we obtained sequences from at least 2 individuals from between two and 13 locality samples of each of the mtDNA clades we identified (Table 2; see below, and Patton and Álvarez-Castañeda, 2005). We sequenced most individuals in both directions and considered a position heterozygous if two bases exhibited overlapping peaks of equivalent height in the electropherogram. In such cases, we scored the heterozygous base position by the appropriate IUPAC nucleic acid code. We aligned each of our *Fgb-I7* sequences by comparison to those published for the *Neotoma lepida* group (Matocq et al., 2007).

Table 2. Sample sizes for sequences of the *Fbg-I7* gene, arranged by mtDNA clade and for each of three contact localities (identified separately, below).

mtDNA clade	N_{pop}	N_{ind}
1A	3	3
1B	4	22
1C	4	31
2A	13	85
2B	4	9
2C	4	7
2D	4	7
2E	2	2
Locality	Clades in contact	
Joaquin Flat	1C – 2A	32
Kelso Valley	*	27
Morongo Valley	1B – 2A	57

* All individuals in Kelso Valley are mtDNA clade 2A, but both "coastal" and "desert" morphological and microsatellite groups are present at this locality (see Tehachapi Transect).

MICROSATELLITE ANALYSES

We examined variation at 18 microsatellite loci for 1034 specimens from 140 separate localities. Sousa et al. (2007) described loci, primers, amplification, and other laboratory methods. Five loci are dinucleotide repeats; the remaining 13 are tetranucleotide repeats. The localities we have examined span nearly the complete range of the *Neotoma lepida* group, from southern Baja California (including four insular taxa from the Gulf of California) to central Nevada and from coastal California to western Arizona. Sample sizes varied from singletons (31 localities) to 66, with a mean of 7.9 individuals per locality. Seventeen localities have sample sizes of 15 or greater, 11 have sample sizes of 20 or more, and 21 have sample sizes of at least 10. These loci were constructed specifically to provide insights into mating patterns within populations where individuals belonging to separate mitochondrial DNA clades co-occur. Consequently, the majority of our analyses involved pooled samples in a series of transect analyses we describe in detail below. However, in a general summary section that follows, we provide global data on allelic variation (allele richness, observed and expected heterozygosities, deviations from Hardy-Weinberg expectations, and linkage disequilibrium) for those localities where n > 10. Because it was not possible to genotype every locus for each individual, the mean sample size is given for each sample for these summary statistics.

We analyzed these data with a variety of software programs now widely available, depending upon the specific set of questions asked. For general diversity measures within and among loci for individual populations or pooled geographic samples, including Hardy-Weinberg equilibrium and linkage disequilibrium estimates, we used GENEPOP on the Web (Raymond and Rousset, 1995), the Genetic Data Analysis (GDA; Lewis and Zaykin, 2002), FSTAT 2.9.3 (Goudet, 2001), Cervus (Marshall et al., 1998), and/or Arlequin3 (Excoffier et al., 2005). For specialized analyses involving assignment tests, we employed the model-based method described by Pritchard et al. (2000) and implemented in the program STRUCTURE, version 2 (Pritchard and Wen, 2003). The model probabilistically assigns individuals to source populations (or jointly to two or more in case of admixture) on the basis of their genotypes without using a priori information regarding population origin. Allele frequencies and the assignment of individuals to populations are inferred simultaneously using a Bayesian approach. We used a parameter set with a burn-in length of 2500 generations, Markov chain Monte Carlo (MCMC) repetitions after burn-in of 100,000, an admixture model with default settings, and correlated allele frequencies, again with the default settings. We varied k (the parameter for the number of populations) in separate analyses from the number of mtDNA clades present (k = 2 in contact zones, for example) to the number of actual geographic subsamples included in the particular anlysis. We then compared the posterior probabilities across these analyses for consistency. Data from k = 2 analyses are typically reported because of high consistency. Finally, in two analyses that include contact points between separate morphological groups and/or mtDNA clades and where hybridization might be present, we used the NewHybrid program, version 1.1 beta (Anderson and Thompson, 2002), to compute the posterior probability that individuals in the sample fall into parental or different hyrid categories (F1, F2, or first-generation backcross hybrids) based on their combined allelic states across all loci. This analysis also uses an MCMC framework.

MOLECULAR PHYLOGENETICS AND PHYLOGEOGRAPHY

mtDNA SEQUENCE VARIATION

In this section we examine the hypothesis of the monophyly of members of the *Neotoma lepida* group with respect to other species in the genus *Neotoma*. Once monophyly is established, we then address the degree of geographic structure at the molecular level, or the phylogeography of internal molecular clades delineated by phylogenetic analyses. Although we organize the presentation below separately for the mtDNA and nucDNA sequence and microsatellite datasets, all molecular data are fully concordant with respect to both the monophyly of the complex as well as in the major elements of internal clade structure. In subsequent sections we analyze both the expanded database of haplotype and diversity in the mtDNA *cyt-b* gene and 18 microsatellite loci to address the temporal depth and historical population history of the clades and subclades identified here, using coalescence methodologies.

To establish the phylogenetic structure within the *Neotoma lepida* group, we use the 309 unique haplotypes among the 500 complete cytochrome b sequences we obtained. These sequences fit the pattern typical of a mitochondrial, protein-coding gene, with a low frequency of G and nearly even frequencies of the three other bases (mean base frequencies: G = 12.19%, A = 32.82%, T = 27.06%, and C = 27.89%). Moreover, as expected, most changes occur at the third position, with the overall number of observed differences (transitions plus transversions) at first positions averaging 8.69 (15.42%), at second positions 0.98 (0.02%%), and at third positions 46.67 (82.28%). There are an average of 22.361 ± 2.725 (standard deviation) observed amino substitutions among the nine outgroup species but only 5.613 ± 1.184 among all 487 *lepida* group sequences. However, the number of 0-fold, 2-fold, and 4-fold degenerate sites is similar between both outgroup and ingroup sequences (705 vs. 715, 183 vs. 183, and 155 vs. 157, respectively).

Plots of p-distances versus K2-p distances are linear for both first and second positions and only slightly curvilinear for third positions, with deviations not surprisingly only present at the highest degrees of divergence (Fig. 3). Hence, saturation is not a major factor in any comparison among sequences, including those of the *Neotoma lepida* group as well as the nine outgroup species used in phylogenetic analyses. As a result, the maximum parsimony phylogenetic analysis includes equal weighting for base substitutions at each codon position. Bayesian analyses incorporated separate partitions for each codon position.

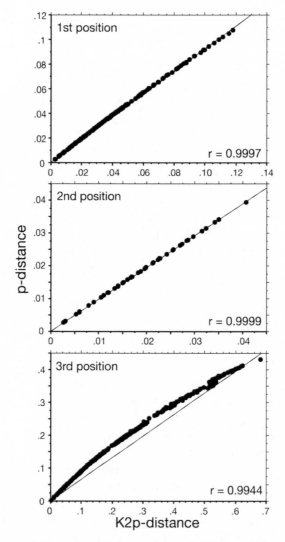

Figure 3. Bivariate plots of p-distances versus Kimura 2-parameter distances among all nine outgroup sequences and each of the 309 unique, complete *cyt-b* sequences from individuals of the *Neotoma lepida* group. Lines are x=y.

mtDNA phylogenetic clade structure

Our expanded phylogenetic analysis is completely consistent with previous studies that document the monophyly of a *Neotoma lepida* group within the genus *Neotoma* (Edwards and Bradley, 2001; Matocq et al., 2007). These prior studies, however, did not include all of the taxa of the *Neotoma lepida* group we analyze here (notably the insular *N. anthonyi*, *N. bryanti*, *N. bunkeri*, *N. insularis*, and *N. martinensis* as well as a number of subspecies of *N. lepida* itself). We present an unrooted strict consensus MP tree in Fig. 4 to illustrate the unity of all sequences of members of the *Neotoma lepida* group relative to those of other species in the genus. We use this as confirmation of the monophyly of taxa comprising the *lepida* group as we define this group.

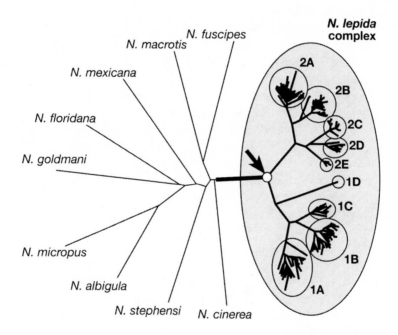

Figure 4. Unrooted 50% majority-rule consensus maximum parsimony tree of 309 unique and complete mtDNA cytochrome b sequences of the *Neotoma lepida* group and representative sequences for nine other species in the genus (Table 1). All *N. lepida* group sequences unite at a single node (arrow) with a bootstrap value of 100, an average p-distance of 0.0896, and collectively average a p-distance of 0.1326 from all other species of woodrats examined. Nine subclades within the *N. lepida* group are identified and are described in greater detail immediately below.

Both the MP and Bayesian analyses delineated the same series of subclades (Fig. 5), most of which had been previously defined by Patton and Álvarez-Castañeda (2005) and Matocq et al. (2007). These latter two studies differ from the results presented here only in our addition of subclade 1D (the single individual of *N. l. insularis* from Isla Ángel de la Guarda) and in the slightly different topologies of some subclades within both major clades (detailed below). In all of these studies, including the present analysis, the *N. lepida* group is divisible into two major clades, each of which in turn is subdivided into 4 or 5 subclades (Clade 1, subclades 1A, 1B, 1C, and 1D, and Clade 2, subclades 2A, 2B, 2C, 2D, and 2E), respectively. We use the terms "Clade 1" and "coastal clade" as well as "Clade 2" and "desert clade" interchangeably, reflecting the general geographic positions of each, following our initial study (Patton and Álvarez-Castañeda, 2005). Both clades and each subclade within them are strongly supported, with a bootstrap of 100 in the MP analysis and a posterior probability of 1.0 in the Bayesian analysis. The two methods of phylogenetic reconstruction also yield the same topologies of relationship among the subclades within each clade, with two exceptions. Within Clade 1, the MP analysis supports a sister relationship between subclades 1B and 1C relative to subclade 1A while the Bayesian analysis results in an unresolved polytomy of these three subclades. However, the bootstrap support for this sister relationship in the parsimony analysis is relatively low at 88. And, within Clade 2, the MP analysis supports a sister relationship between subclades 2C and 2D while the Bayesian analysis suggests that subclades 2D and 2E are sisters. In both cases, the support for the depicted relationship is again relatively low, with a bootstrap of 79 and a posterior probability of 0.78, respectively. Matocq et al.'s (2007) study, based on exemplar singleton sequences with a combined dataset of 4242 bp from 4 mitochondrial and 4 nuclear genes, generated the same topology within Clade 1 as our MP analysis, with subclade 1A basal to a sister pair comprised of subclades 1B and 1C, with a bootstrap value of 93 and a Bayesian posterior probability of 1.0. However, their analyses of relationships within Clade 2 provide yet a third possible topology to the subclade 2C-2D-2E triad, with subclades 2C and 2E apparent sisters relative to subclade 2D. Both the bootstrap (63) and Bayesian probabilities (ranging from 0.76 to 0.90, depending on data partition employed) for the linkage of subclades 2C and 2E are relatively low. Full resolution among the three subclades to the east and south of the Colorado River (Fig. 6), thus, remains for future analyses.

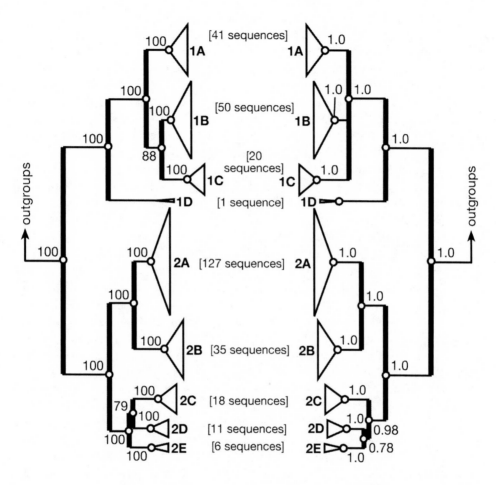

Figure 5. Left, strict consensus maximum parsimony topology and, right, 50% majority rule Bayesian topology based on the GTR+I+G model, depicting of phylogenetic relationships among 309 complete, unique mtDNA cytochrome b sequences of the *Neotoma lepida* group. Both trees used nine other species in the genus as a collective outgroup (see Fig. 4, above). Terminal triangles are proportional to the number of sequences in the cluster, as identified. Numbers above nodes in the parsimony tree are bootstrap resampling values; those in the Bayesian tree are posterior probabilities. The parsimony analysis resulted in 81,401 equally length trees, each of 1270 steps, CI = 0.377, RI = 0927, RC = 0.353, and HI = 0.623. The Bayesian analysis summarizes 13,883 final trees.

The two clades and their subclades are geographically structured, with that structure fully consistent with the description and mapped ranges given by Patton and Álvarez-Castañeda (2005) where that study overlapped with the expanded analyses here (Fig. 6). Clade 1 is distributed along coastal California and throughout Baja California, including all of the Pacific coast and Gulf islands that harbor woodrats. The four subclades in this phylogenetic unit are serially distributed from south to north, with subclade 1A occupying nearly all of Baja California, from approximately San Felipe (BCN-39; Appendix) on the Gulf coast and Punta Prieta (BCN-74) on the Pacific side south to the Cape region. Subclade 1B occurs along the Pacific coast of Baja from at least El Rosario (BCN-56 and 57) north to Ventura Co. in southern California and east as far as the western margin of the San Bernardino Mts. and eastern edge of the Peninsular ranges. Subclade 1C is present throughout central coastal California, from at least San Luis Obispo Co. north to Alameda Co. Finally, subclade 1D comprises the single sample from Isla Ángel de la Guarda in the northern Gulf of California. In contrast, samples of Clade 2 occur throughout the interior deserts of eastern California, Nevada, Utah, and Arizona. The largest proportion of this entire range is occupied by subclade 2A, which is known from the Colorado Desert of southeastern California, throughout the Mojave Desert, and a substantial portion of the Great Basin Desert, all west of the lower Colorado River and Virgin River. Subclade 2B occurs in northern Arizona east of the Virgin River and north of the Grand Canyon, and throughout the Colorado River basin in southern Utah and adjacent Colorado. Subclade 2C occurs south of the Grand Canyon, from Navajo Bridge and north of Flagstaff west to Hoover Dam and south along the eastern side of the lower Colorado River to the north side of the Bill Williams River (boundary of La Paz and Mohave Counties in Arizona). Subclade 2D is apparently limited to the narrow strip along the lower Colorado River between the Bill Williams and Gila rivers, and subclade 2E is present south of the Gila River in southwestern Arizona and in northwestern Sonora from the Pinacate lava flows and Puerto Peñasco to the west (Fig. 6).

The distributional ranges of some subclades overlap at several points where we have trapped individuals of more than one subclade at the same locality. For example, individuals with haplotypes from subclade 1B and subclade 2A co-occur throughout Morongo Valley, San Bernardino Co., California (localities CA-340, CA-341) and near Red Mountain in Los Angeles Co. (locality CA-102). Haplotypes of subclade 1C and 2A are also present at two localities, near Three Points in Los Angeles Co. (locality CA-99) and Joaquin Flat in the Tehachapi Mts., Kern Co. (locality CA-64). Overlap also occurs between haplotypes belonging to subclades within each major clade, namely between subclades 1B and 1C near Fort Tejon, Kern Co. (locality CA-60) and near Gorman, Los Angeles Co. (locality CA-

97) as well as between subclades 2C and 2D at Burro Creek, Mojave Co., Arizona (locality AZ-65). Subclades 1A and 1B are juxtaposed geographically between Cataviña (BCN-64; subclade 1A) and Bahia San Luis Gonzaga (BCN-65; subclade 1B) on the northeast coast of Baja California, although individuals of these subclades are not as yet known from the same locality. We detail each of these zones of overlap and document the degree to which there is gene flow between populations with phyletically different haplotypes in the separate transect sections described below.

Differentiation between the two major clades is substantial, as the average p-distance between them is 0.0896 (± 0.0082, standard error). We provide divergence levels between all pairs of subclades in Table 3. Note that within-subclade divergence is typically < 0.01 in all cases, reaching a maximal level of only 0.014 within subclade 1A from Baja California. Differences between subclades within Clade 1 range from 0.031 [subclades 1B and 1C] to 0.059 (subclade 1D, from Isla Ángel de la Guarda, to all others), with an average among all four subclades of 0.0427. Divergence levels among subclades in Clade 2 are slightly less, both on average (mean p-distance = 0.0376) and range (minimal p-distance = 0.263 [between subclades 2D and 2E] and maximal distance = 0.0494 [between subclades 2A and 2C]).

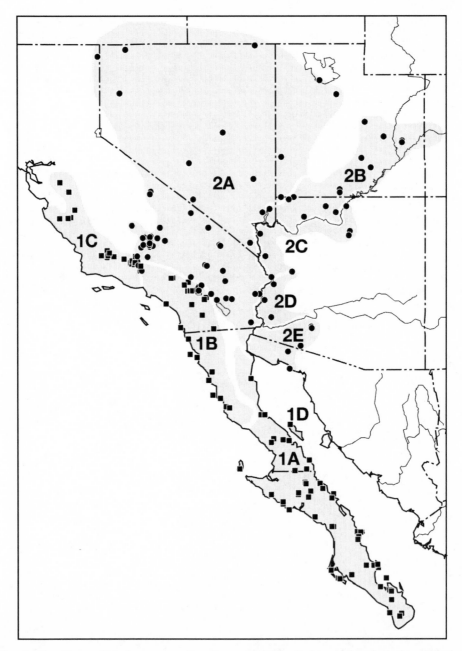

Figure 6. Generalized ranges of major mtDNA clades and subclades (Clade1, subclades 1A, 1B, 1C, and 1D [open circles] and Clade 2, subclades 2A, 2B, 2C, 2D, and 2E [solid triangles]).

Table 3: Average p-distance within and among nine mtDNA subclades (mean ± standard error [estimated by bootstrap method, with 500 replications beginning with a random seed number) of the *Neotoma lepida* group.

Subclade	1A	1B	1C	1D	2A	2B	2C	2D	2E
1A	0.0144± 0.0019	0.0361± 0.0044	0.0422± 0.0051	0.0613± 0.0067	0.0905± 0.0075	0.0875± 0.0077	0.0821± 0.0074	0.0812± 0.0074	0.0833± 0.0073
1B		0.0122± 0.0018	0.0312± 0.0045	0.0552± 0.0063	0.0882± 0.0077	0.0824± 0.0076	0.0776± 0.0076	0.0756± 0.0073	0.0766± 0.0073
1C			0.0057± 0.0011	0.0599± 0.0069	0.0813± 0.0075	0.0770± 0.0072	0.0727± 0.0071	0.0726± 0.0074	0.0758± 0.0075
1D				------	0.0841± 0.0083	0.0766± 0.0079	0.0744± 0.0080	0.0777± 0.0081	0.0758± 0.0085
2A					0.0100± 0.0012	0.0321± 0.0046	0.0494± 0.0061	0.0398± 0.0052	0.0442± 0.0062
2B						0.0074± 0.0012	0.0411± 0.0055	0.0422± 0.0054	0.0445± 0.0061
2C							0.0062± 0.0015	0.0277± 0.0044	0.0284± 0.0050
2D								0.0088± 0.0017	0.0263± 0.0047
2E									0.0039± 0.0012

Haplotype diversity within clades and subclades

We summarize standard estimates of molecular diversity in the mtDNA *cyt-b* haplotypes for the 801 bp dataset in Table 4. Gene diversity (the probability that two randomly chosen haplotypes are different) and nucleotide diversity (the probability that two randomly chosen homologous nucleotides are different) estimates are similar for both Clade 1 and Clade 2 sets of haplotypes (0.9883 and 0.0016 versus 0.9836 and 0.0017, respectively). For the most part, these measures are also consistent among the subclades within each clade, with the exception of subclades 2C, 2D, and 2E where gene diversity measures are substantially lower (0.3235 [subclade 2E] to 0.8718 [subclade 2D]). Nucleotide diversity in subclade 1C is less than half that of subclades 1A and 1B (0.0044 versus 0.0129 and 0.0102, respectively), and all subclades of Clade 2 exhibit lower nucleotide diversities, especially that of subclade 2E (0.0011). The number of pairwise differences for Clade 2 is less than that of Clade 1, only about 60% of the latter, and within Clade 1 pairwise differences are greatest for subclade 1A (10.32, on average, between all haplotypes) and least in subclade 1C (4.23). These measures for subclades in Clade 2 are rather uniform (ranging from 3.56 for subclade 2C to 6.42 in subclade 2A), except for subclade 2E, which is substantially lower with less than one difference per haplotype pair (Table 4). Overall, and not surprisingly, these measures mirror the p-distances in Table 3, above.

The pattern of apportionment of haplotype diversity within and among subclades is also similar for the two major clades. When we arranged localities into geographic groups within each subclade and performed an Analysis of Molecular Variance (AMOVA) using the Arlequin3 software (Excoffier et al., 2005), the total pool of variation in both clades was highest at the subclade level (68.33% versus 77.78% for Clades 1 and 2, respectively). These high and nearly equivalent numbers did not change appreciably under different geographic clustering of regional localities within subclades. In this analysis, variation among regions within subclades was small, ranging from 8.28% in Clade 1 to a low of 1.71% in Clade 2, while that within regions was moderate and nearly equivalent in both clades (23.39% for Clade 1 and 20.56% for Clade 2). This general pattern, particularly the high level of molecular variance among subclades, is expected, since subclade structure is strongly supported in the phylogenetic analyses of these same data (Fig. 5).

Table 4. Estimates of molecular diversity for each mtDNA clade and subclade of the *Neotoma lepida* group (mean ± 1 standard deviation). Summaries are for the 801 bp dataset of the *cyt-b* gene.

Clade/Subclade	Sample size	# Unique haplotypes	# Polymorphic sites	Haplotype diversity	Nucleotide diversity	$\Theta\pi$
Clade 1	406	199	199	0.9883±0.0016	0.0286±0.0131	21.31±10.41
subclade 1A	126	83	131	0.9874±0.0036	0.0129±0.0066	10.32±5.25
subclade 1B	179	80	102	0.9735±0.0043	0.0102±0.0052	8.14±4.20
subclade 1C	125	42	45	0.9365±0.0116	0.0053±0.0029	4.23±2.34
subclade 1D	1	1	----	----	----	----
Clade 2	717	264	194	0.9836±0.0017	0.0167±0.0083	13.37±6.66
subclade 2A	580	198	157	0.9767±0.0025	0.0080±0.0042	6.41±3.36
subclade 2B	59	40	57	0.9684±0.0135	0.0062±0.0034	4.94±2.70
subclade 2C	34	10	13	0.8039±0.0580	0.0044±0.0026	3.56±2.06
subclade 2D	27	13	23	0.8718±0.0465	0.0072±0.0040	5.79±3.18
subclade 2E	17	3	5	0.3235±0.1359	0.0011±0.0009	0.90±0.73

The difference in overall apportionment of haplotype diversity and measures of pairwise difference ($\Theta\pi$) between the two major clades extends as well to differences in geographic distribution of single haplotypes within each. For example, 22 haplotypes in Clade 1 occur at 2 or more localities, with an average distance among them of 28.8 linear miles (range = 1.5 – 151.8). This contrasts with 47 haplotypes in Clade 2 that are found at more than one locality, with an average between-locality distance of 104.4 miles (range = 0.7 – 366.5), 4 times greater than in Clade 1. This difference in geographic spread of single haplotypes is significant (ANOVA, $F_{(1,515)}$ = 44.888, p < 0.0001). Haplotypes within the geographically expansive desert subclade 2A are particularly widely distributed. Nineteen percent of all haplotypes within this subclade (37 of 198) are found at multiple localities, with an average distance among them of 107.2 miles and a maximum distance of 366.5 miles. While most of these haplotypes are distributed among only a few localities, several are very widely spread (up to an average of 135 miles) among a large number (up to 17) of sample sites.

Geographic structure within subclades

We determined the internal geographic structure within each subclade with separate Bayesian analyses using the unique 801 bp *cyt-b* sequences. We used the GTR+I+G model and ran each analysis for 1×10^7 generations with four Markov chains sampled every 1000 generations with randomly chosen sequences from the opposite clade as an outgroup. We computed the 50% majority-rule consensus tree after excluding a burn-in sample of 2500 trees. We accept as geographic clusters groups of haplotypes supported by a Bayesian posterior probability of 0.90 or greater.

The 83 haplotypes of subclade 1A form two clusters, each supported by posterior probabilities of 1.0 (Fig. 7). Each has geographic continuity, with the southern one ranging from near Santa Rosalia (locality BCS-16) to the Cape and a more northern group distributed from San Pedro de La Presa (locality BCS-73) to San Felipe (BCN-39). The southern cluster includes samples of the insular taxa *nudicauda* (Carmen), *latirostra* (Danzante), *perpallida* (San José), *abbreviata* (San Francisco), and *vicina* (Espíritu Santo). Each insular sample is linked phylogenetically to the closest mainland samples (localities BCS-16 and BCS-40, BCS-41 and BCS-84, and BCS-84, BCS-97, BCS-104, respectively), or in the case of *perpallida* and *abbreviata* to each other before connecting to the adjacent mainland (BCS-74). The northern cluster includes the insular taxa *bryanti* (Cedros) from the Pacific and *marcosensis* (San Marcos) and *bunkeri* (Coronados) in the gulf sides of the peninsula. Both *bryanti* and *marcosensis* are phylogenetically closest to mainland samples (locality BCN-72 and BCS-16, respectively). The phyletically closest haplotype to *bunkeri* is from San Juanico (BCS-39), some distance to the northwest along the Pacific coast (Fig. 7). These two clusters overlap broadly in the mid part of Baja California Sur, and haplotypes of each co-occur at one locality near Santa Rosalia (locality BCS-16).

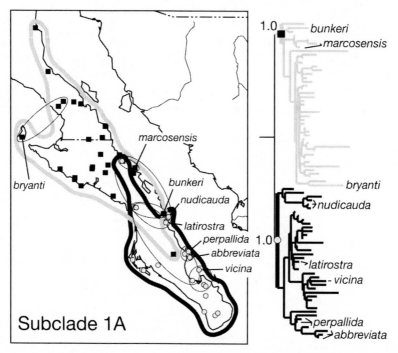

Figure 7. Locality map (left) and Bayesian tree (right) of 83 subclade 1A haplotypes in southern Baja California. Two phyletic clusters, each with posterior probability support of 1.0, are indicated on both the map and tree. The geographic and phyletic position of each of the eight insular taxa is indicated, and the linkage of each to mainland localities is indicated on the map by ellipses.

Three well-supported clusters (Bayesian posterior probabilities 0.94 or greater) group 71 haplotypes within subclade 1B, with all three forming a larger cluster supported by a probability of 0.92 (clusters "a", "b", and "c" in Fig. 8). These clusters, in turn, are nested within a basal set of nine other haplotypes, each from localities on the southern margins of the distribution of subclade 1B in Baja California, extending across the peninsula from near San Vicente (locality BCN-18) on the Pacific coast to Bahia San Luis Gonzaga (BCN-65) on the gulf coast.

The insular taxon *N. martinensis* (San Martín, BCN-49) is represented by one of these "unique" haplotypes. The three phyletic clusters are geographically nested. Cluster "a" includes all localities from near El Rosario in northwestern Baja California (locality BCN-57, and *N. anthonyi* from Isla Todos Santos, locality BCN-13) to Los Angeles and Kern counties in southern California (localities CA-60, 96, 97, 61, and 102). Clusters "b" and "c" are limited to the northern tier of localities in California, with the latter contained nearly completely within the range of the former. Haplotypes from each of these clusters are present at the same localities in several combinations: clusters "a" and "b" overlap through San Gorgonio Pass in Riverside Co. (localities CA-222, CA-230, CA-232, CA-247, and CA-261); clusters "a" and "c" are at Dana Point (locality CA-142) in Orange Co.; clusters "b" and "c" co-occur at Lone Pine Canyon (locality CA-324); and, finally, haplotypes of all three clusters are present in Morongo Valley (locality CA-338 and CA-341) and in the Santa Rosa Mts. (locality CA-281).

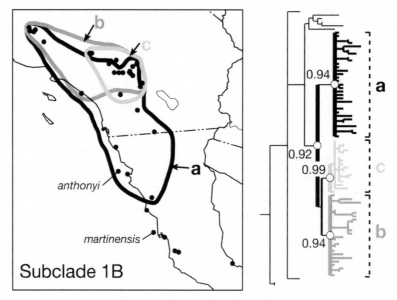

Figure 8. Locality map (left) and Bayesian tree (right) of 80 subclade 1B haplotypes in northern Baja California and southern California. Three phyletic clusters, each with posterior probability support of 0.94 or greater, are nested within a single cluster supported at 0.92. Nine more basal haplotypes not included in any cluster are apparent, including the insular *N. martinensis*.

In contrast to both subclades 1A and 1B, subclade 1C contains little internal geographic structure (Fig. 9), with two minor haplotype clusters supported by posterior probabilities of 0.90 or greater. These group haplotypes from localities from the northern part of the subclade's range in Alameda or Merced counties (localities CA-4 and CA-7, respectively) to the Temblor Range in San Luis Obispo Co. (locality CA-40). Haplotypes from these two clusters co-occur at Romero Creek in western Merced Co. (locality CA-7). The majority of the 42 haplotypes recovered within subclade 1C form a large basal polytomy.

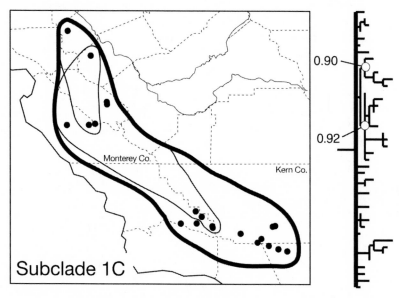

Figure 9. Locality map (left) and Bayesian tree (right) of 42 subclade 1C haplotypes in the central coast of California, from Alameda Co. in the north to Kern and Los Angeles counties in the south. Two phyletic clusters, each with limited numbers of included haplotypes but with posterior probability support of 0.90 or greater, are nested with the majority of haplotypes in a large basal polytomy.

There are 198 unique 801 bp haplotypes among the 580 specimens of subclade 2A, with 17 clusters supported by Bayesian posterior probabilities of 0.90 or greater. Only two clusters, however, include more than six haplotypes or are distributed across more than two geographically adjacent localities. Both of these clusters are supported by posterior probabilities of 1.0 (Fig. 10). Cluster "a" ranges widely, from southeastern California (Cargo Muchaco Mts., Imperial Co., locality CA-205) to northwestern Nevada (Virgin Valley, Humboldt Co., locality NV-30). Cluster "a" also includes 37 haplotypes that are each broadly distributed, with an average geographic span of 107 miles (maximum of 367 miles) among pairs of localities harboring them. One of these haplotypes, for example, is found at 16 separate localities with a mean inter-locality distance of 135 miles. We discuss the significance of these widely distributed haplotypes in a separate section, below. The second large cluster ("b" in Fig. 10) is more narrowly delineated geographically, limited to samples from the Tehachapi and Piute mountains and the western half of the Kern River Plateau in Kern Co. and eastern Tulare Co. (localities CA-64, CA-55, CA-65-66, CA-70-72, CA-77, and CA-80-81). Individuals of these clusters co-occur at one locality in Kelso Valley, on the east side of the Piute Mts. in Kern Co. (Whitney Well, locality CA-80).

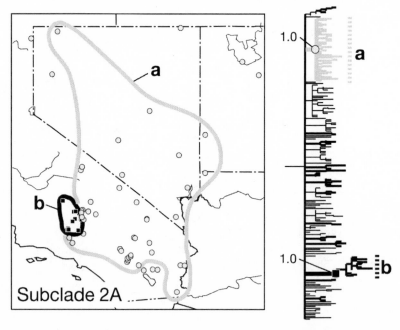

Figure 10. Locality map (left) and Bayesian tree (right) of 198 subclade 2A haplotypes in eastern California, Nevada, and western Utah. Seventeen clusters with posterior probabilities greater than 0.9 are indicated in thick lines and the two major clusters, "a" and "b", both with a probability of 1.0, are indicated by the very heavy lines in both the map and tree.

Of the remaining 15 "minor" clusters comprising 6 or fewer haplotypes, all but one are distributed among localities within the broad range of cluster "a." Each cluster is found in groups of geographically adjacent samples. The single exception to this pattern is a cluster of 4 haplotypes found in northeastern California and northwestern Nevada (Cedarville in Modoc Co., locality CA-424, and Gerloch in Pershing Co., locality NV-46), a distribution contiguous to that of cluster "a." While clearly delineated clusters of haplotypes are present within subclade 2A, the two major clusters and the 15 minor ones still assemble at a single and massive basal polytomy. Hence, no hierarchical pattern to their relationship is supported with the data currently available.

Subclade 2B has a limited distribution, with 40, 801 bp haplotypes found among 14 localities distributed north and west of the Colorado River from extreme southeastern Nevada and northern Arizona to east-central Utah (Fig. 11). The Bayesian analysis finds no well-supported clusters among this group of haplotypes, as the only apparent cluster has a probability of support of only 0.65. A single basal polytomy encompasses all haplotypes in the subclade in Fig. 11.

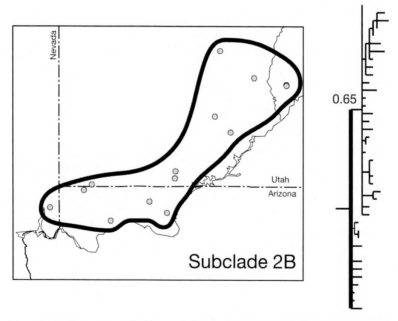

Figure 11. Locality map (left) and Bayesian tree (right) of 40 subclade 2B haplotypes north and west of the Colorado River in Nevada, Arizona, and Utah.

Finally, we depict haplotype clusters within each of the three subclades (2C, 2D, and 2E) in western Arizona in Fig. 12. Subclade 2E lacks internal geographic structure. Within subclade 2C, all haplotypes recorded at localities in Coconino Co. north of Flagstaff, Arizona (localities AZ-37, AZ-48, and AZ-49) cluster strongly, with a Bayesian posterior probability of 1.0. This eastern set of samples within Subclade 2C connect phyletically with one haplotype found at two localities in the western segment of the subclade range (near Hoover Dam, locality AZ-56, and Burro Creek, locality AZ-65) at a posterior probability of 0.9. Three small haplotype clusters are present within Subclade 2D; one couples two haplotypes found at the same locality and the other two connect localities in the northern and southern parts of the subclade's range, respectively. One of these haplotype clusters ("a" in Fig. 12) co-occurs with a haplotype of subclade 2C at Burro Creek, in Mohave Co. (locality AZ-65).

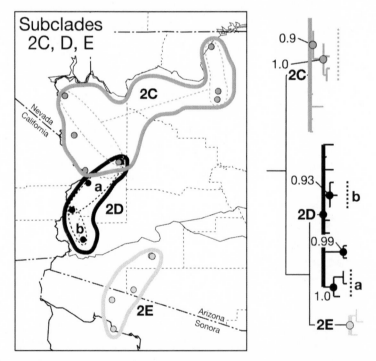

Figure 12. Locality map (left) and Bayesian tree (right) of 26 subclade 2C, 2D, and 2E haplotypes, distributed south and east of the Colorado River in Arizona. Haplotype clusters within each supported by Bayesian posterior probabilities of 0.90 or greater are indicated. Note that haplotypes from subclades 2C and 2B are found at one locality (Burro Creek, Mojave Co., locality AZ-65).

NUCLEAR DNA SEQUENCE VARIATION

We encountered only seven β-Fibrinogen-intron 7 (*Fbg-I7*) sequences among the 166 individuals of the *Neotoma lepida* group we examined, excluding individuals that were uniquely heterozygous at particular base positions or that were obviously heterozygous for two different haplotypes. Our data include representatives of each mtDNA clade and subclade defined above, with the exception of subclade 1D (*insularis*, from Isla Ángel de la Guarda). Each unique haplotype is identical to those published by Matocq et al. (2007; GenBank accession numbers DQ180031-180038) and, importantly, we obtained the same sequence independently for these same individuals. The seven haplotypes differ among each other at 24 sites; two (positions 118 and 248) require single deletion/insertion events to maintain alignment and the remainder are base substitutions. Haplotypes from mtDNA subclades 2C and 2D are identical to one another; all other subclades possess only one or two (subclade 1B) haplotypes among the individuals we sequenced.

We present a matrix of substitution differences for the *Fbg-I7* sequences in Table 5. *Fbg-I7* haplotypes 11 and 13 each differ from haplotypes 22 and 24, respectively, at 15 positions, while haplotypes 23 and 24 differ by a single change. Overall, haplotype 11 is the most divergent, averaging 12.1 differences in comparison to all other haplotypes, nearly as much as all haplotypes belonging to mtDNA Clade 1 differ from those of Clade (an average of 12.6 substitutions).

As Matocq et al. (2007) present a phylogenetic analysis of these seven *Fbg-I7* sequences, in conjunction with other nuclear and mtDNA genes, we make no attempt to do so here. Rather, we examine the degree of concordance between the geographic distribution of mtDNA *cyt-b* subclades and the *Fbg-I7* haplotypes (Fig. 13). In general, there is excellent correspondence between the distributions of haplotypes from both genes, with two exceptions. First, *Fbg-I7* haplotype 13, which is typically distributed throughout the range of mtDNA subclade 1C along the central coast of California extends east into the Kern River Plateau and southern foothills of the Sierra Nevada, where it is found in individuals of the "desert" mtDNA subclade 2A (localities CA-64 and CA-55). At one locality (Kelso Valley, CA-80l, top arrow, Fig. 13) heterozygotes between *Fbg-I7* haplotypes 13 and 21 were recovered along with homozygous 13/13 individuals. This is a complex area of genetic and morphological transition, which we describe in greater detail in the Tehachapi Transect section, below. *Fbg-I7* haplotype 13 also extends to the Morongo Valley region in San Bernardino Co. (localities CA-338 to CA-342), which is otherwise the area of contact between the coastal mtDNA subclade 1B and desert subclade 2A (see San Gorgonio Pass Transect analysis, below). Here, individuals with *Fbg-I7* haplotypes 12 and 13 co-occur with those that are heterozygotes between haplotypes 13 and 21 (bottom arrow, Fig. 13).

Second, *Fbg-I7* haplotypes 21 and 22, each otherwise concordant with mtDNA subclades 2A and 2B, respectively, are found in heterozygote combinations in the four mtDNA subclade 2B localities for which data are available (NV-138, UT-33, AZ-7, and AZ-15) where individuals were sequenced (composite white and gray triangles, Fig. 13). Otherwise, haplotype 21 is the sole allele found in mtDNA subclade 2A and haplotype 22 is similarly homozygous in mtDNA subclade 2C.

Table 5. Matrix of the number of base substitutions between all pairs of the seven unique β-Fibrinogen-intron 7 (*Fbg-I7*) sequences recovered from specimens of the *Neotoma lepida* group. The corresponding mtDNA clade in which each *Fbg-I7* haplotype was found is indicated.

Fbg-I7 haplotype	mtDNA haplotype	*Fbg-I7* haplotype						
		11	12	13	21	22	23	24
11	1A	---	10	13	13	15	11	12
12	1B		---	3	11	13	11	12
13	1C			---	12	14	14	15
21	2A/2B				---	2	4	5
22	2B/2C					---	6	6
23	2D						---	1
24	2E							---

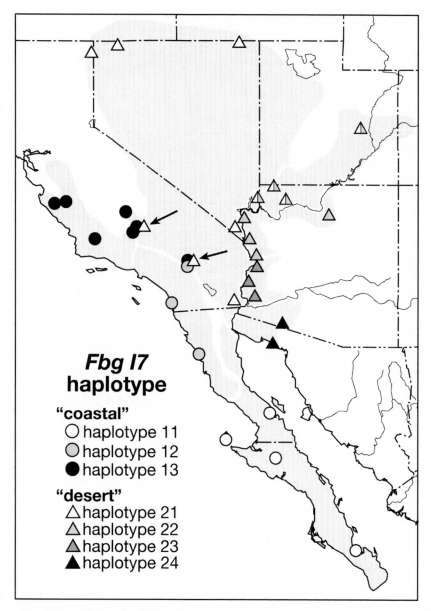

Figure 13. Map of the distribution of the 7 *Fbg-I7* sequences (11, 12, 13, 21, 22, 23, and 24) relative to the mtDNA clade structure ("coastal" and "desert" clades; gray tones, from Fig. 6, above). Localities where overlapping *Fbg-I7* haplotypes co-occur are indicated, as are those for which heterozygotes between two specific *Fbg-I7* haplotypes were found (arrows and half-toned circles or triangles).

VARIATION AT MICROSATELLITE LOCI

We summarize allelic diversity at 18 microsatellite loci for 21 population samples for which sample size is 10 or greater in Table 6. This set of samples includes at least one from six of the nine mtDNA clades and subclades, with only subclades 1A, 1D, and 2E lacking samples larger than 10 individuals (Table 7). Allelic richness varies extensively across loci, from a low of 7 alleles (*Nlep7*) to a high of 58 (*Nlep8*), both tetranucleotide repeat loci. Only for locus *Nlep17* does the mean observed heterozygosity (H_o) across all 21 samples differ from Hardy-Weinberg expectations (H_e).

Table 6. Repeat motif, allelic diversity, and average observed (H_o) and expected (H_e) heterozygosities for 18 microsatellite loci for 21 population samples of the *Neotoma lepida* group.

Locus	Repeat motif	Number of alleles (A)	Mean A (range)	H_o	H_e
Nlep1	CA	23	7.68 (1-12)	0.7548	0.7366
Nlep2	TG	17	7.95 (4.13)	0.8225	0.7895
Nlep3	CA	26	9.81 (5-14)	0.8324	0.8277
Nlep4	CA	52	11.23 (3-22)	0.8344	0.8339
Nlep5	CA	30	9.18 (4-19)	0.8244	0.8283
Nlep6	TAGC	24	7.68 (1-16)	0.7055	0.7182
Nlep7	TGTA	7	3.82 (2-6)	0.5871	0.5785
Nlep8	CATA	58	9.09 (1-14)	0.7363	0.6856
Nlep9	TGTA	5	2.64 (1-5)	0.3167	0.2985
Nlep10	TATG	20	7.82 (3-12)	0.8108	0.8085
Nlep11	CATA	10	4.73 (2-8)	0.6118	0.6285
Nlep12	TACA	10	3.55 (1-5)	0.4264	0.4321
Nlep13	AGAT	19	5.50 (2-8)	0.6863	0.6786
Nlep14	TGTA	28	9.55 (3-16)	0.8429	0.8417
Nlep15	TAGA	57	13.77 (6-20)	0.9062	0.9347
Nlep16	GATA	21	6.91 (2-11)	0.7398	0.6862
Nlep17	TCTA	12	4.64 (2.8)	0.6593	0.4427**
Nlep18	GATA	26	7.50 (3-14)	0.7640	0.7137

** $p < 0.01$

Our analyses detected few departures from Hardy-Weinberg expectations at individual loci for any of the 21 population samples for which data are summarized in Table 7. Furthermore, only three cases of significant departures remained following Bonferroni correction, all involving locus *Nlep17* for three desert localities (Freeman Canyon, CA-92; Little Lake, CA-381; and Halloran Spring, CA-367). These minor deviations could result from null alleles or population substructure due either to non-random mating or the mixing of locally differentiated subunits. Null alleles seem unlikely since only three of the 21 samples, particularly ones that are separated by 50 miles or more, exhibit departures from expectation. Similarly, since each of these samples was taken during a single trapping effort spanning one or two nights, with the sample taken over a very limited area of no more than a hectare, substructure on this scale seems unlikely. Finally, deviations due to biased mating should affect all loci, which is not the case here. Consequently, we assume the deviations are stochastic artifacts of sampling and not due to more directed processes.

The mean number of alleles per locus is weakly correlated with sample size ($r = 0.446$, Z-value $= 2.037$, $p = 0.0416$) but the total number of alleles, number of private alleles, and both observed and expected heterozygosities are not ($p > 0.05$ in all comparisons). At the population level, all measures of diversity vary widely across the 21 sample localities (Table 7). The three measures of the number of alleles are uniformly highest at the subclade 2A locality from the Orocopia Mts., Riverside Co., California (locality CA-300) and lowest at the subclade 2C locality near Tanner Tank, Coconino Co., Arizona (locality AZ-49). Expected heterozygosity is nearly highest and absolutely lowest at these two localities as well. Overall, however, desert samples (those of mtDNA Clade 2) harbor larger numbers of total alleles on average (ANOVA, $F_{(1,19)} = 4.548$, $p = 0.0462$), alleles per locus ($F = 4.568$, $p = 0.0458$), and mean expected heterozygosity ($F = 4.573$, $p = 0.0457$), considerably more so if the three samples of subclades 2C and 2D that are especially low in these measures are excluded from the comparison (p decreases to between 0.0012 and 0.0001). While the number of private alleles varies from 0 (King City, pooled sample including localities CA-20 and CA-34) to 11 (Mokaac Wash, locality AZ-7), there is no geographic trend apparent. We will examine patterns to the distribution of allelic diversity trends more explicitly in the geographic analyses presented below.

Table 7. Genetic diversity indices calculated for 18 microsatellite loci for 21 localities of the *Neotoma lepida* group for which sample size is 10 or greater. Shown are sample size (n), numbers of alleles (total number, per locus, and "private" [*i*]), polymorphic information content (PIC), and observed (H_o) and expected (H_e) heterozygosities averaged across all loci.

mtDNA Clade / Locality (locality number)	N	Number of alleles Total	Per locus	*i*	PIC	H_o	H_e
1B-Jacumba (CA-185)	13	133	7.4	8	0.672	0.723	0.702
1B-Lamb Canyon (CA-222)	15	118	6.9	2	0.633	0.690	0.609
1B-Banning (CA-225)	13	110	6.1	6	0.626	0.688	0.589
1C-Ft Tejon (CA-60)	32	93	5.4	2	0.556	0.611	0.585
1C-Joaquin Flat (CA-64)	36	108	6.3	6	0.605	0.657	0.661
1C-King City (CA-20, 34)	20	75	4.4	0	0.514	0.572	0.544
2A-Berdoo Canyon (CA-291)	20	167	9.3	7	0.735	0.780	0.788
2A-Orocopia Mts. (CA-300)	36	218	12.1	7	0.778	0.813	0.781
2A-Tumco Mine (CA-205)	26	177	9.8	4	0.766	0.808	0.790
2A-Hoffman Summit (CA-83)	25	188	10.4	5	0.769	0.811	0.793
2A-Freeman Canyon (CA-92)	23	155	8.6	5	0.711	0.762	0.760
2A-Halloran Spring (CA-367)	16	153	8.5	6	0.710	0.761	0.742
2A-Searchlight (NV-142)	12	154	8.6	2	0.729	0.782	0.768
2A-Little Lake (CA-381)	25	172	9.6	7	0.764	0.803	0.772
2A-Birch Creek (CA-388)	15	132	7.3	4	0.678	0.729	0.678
2A-Furnace Creek (CA-405)	18	133	7.4	3	0.696	0.750	0.709
2A-Delamar Mts. (NV-135)	10	117	6.5	3	0.675	0.741	0.771
2B-Mokaac Wash (AZ-7)	12	126	7.0	11	0.700	0.762	0.764
2C-Tanner Tank (AZ-49)	10	60	3.9	2	0.341	0.400	0.393
2C-Hoover Dam (AZ-56)	10	80	4.9	2	0.509	0.572	0.578
2D-Dome Rock Mts. (AZ-74)	17	122	6.8	10	0.650	0.699	0.650

The level of microsatellite divergence among populations is constrained by both the repeat size at which no additional slippage can occur and the maximum size of an allele. As a result, there is an expected high level of homoplasy of alleles and thus a limited ability to delineate phylogenetic structure except in cases of shallow evolutionary history (Takezaki and Nei, 1996; Angers and Bernatchez, 1998). There are, however, case studies where microsatellite markers have been successfully applied in deep phylogeographic contexts, including across species boundaries (Estoup et al., 1995; Queney et al., 2001). Given the marked divergence yet strong phylogenetic and phylogeographic structure among and within the members of the *Neotoma lepida* group in both mitochondrial and nuclear gene sequences, we asked whether there was identifiable structure in our dataset of 18 microsatellite loci as well. We addressed this question in two separate analyses. First, we apportioned allelic differentiation as a function of clade structure defined by mtDNA sequences, employing the analysis of molecular variation (AMOVA) approach in the Arlequin3 software. Second, we asked if there was visible phylogenetic structure concordant with that observed for both mtDNA and nucDNA sequence data presented in Figs. 5 and 13, with the tree constructed from a molecular distance (Fst) matrix among all population pairs. The results, on the surface, appear contradictory.

For the AMOVA analysis, we grouped 52 localities where sample size was five or greater into their respective mtDNA subclades and organized these into the two major clades. The results suggest that little phylogenetic signal is present, since the vast majority (78.5%; $F_{(7,1311)}$ = 8.10, p < 0.01) of the variation is contained within the individual population samples and only a limited amount (14.5%; $F_{(46,1311)}$ = 13.19, p < 0.001) is distributed among clades or subclades. The remainder (7.0%; $F_{(1258,1311)}$ > 0.05) is among population samples within clades or subclades. Apportionment is similar in analyses restricted to each major clade. For example, the within-population portion of the total pool of variation is 81% for Clade 1 samples and 85% for those of Clade 2 while the among-clade portion is 4% and 10%, respectively. One might expect the highest portion of variation at the clade/subclade level if a high phylogenetic signal were present in the data, which is certainly not the case. Thus, the AMOVA results suggest rather poor phylogenetic signal in the microsatellite dataset, at least based on the mtDNA clade structure.

While a relatively small amount of the total variation in allelic divergence is apparently due to clade effects, there is substantial empirical phylogenetic structure in the 18 microsatellites, and that structure is completely concordant with both mitochondrial and nuclear sequences. We generated a matrix of Fst distance values among all population pairs for the 52 samples for which sample size was at least 5 individuals, using the GDA software (Lewis and Zaykin, 2002) and then constructed a neighbor-joining tree from this matrix, visualizing it with TreeView, version 1.6.6 (Page, 1996) as an unrooted topology (Fig. 14). This analysis includes multiple samples of all mtDNA subclades, except subclade 2E, which is represented by a single population sample, and subclade 1D (*insularis*, from Isla Ángel de la Guarda in the Gulf of California), for which no data are available. As is apparent in Fig. 14, the two mtDNA clades and their member subclades are each completely delineated as unique clusters of samples by the 18 microsatellites.

Moreover, there is clear concordance in the phylogenetic positioning of subclades in this tree relative to the topologies generated by both the mitochondrial *cyt-b* (Fig. 5) and nuclear *Fbg-17* (Fig. 13) trees. In the microsatellite tree, both major clades, the coastal Clade 1 and desert Clade 2, are apparent; the coastal subclades 1B and 1C appear as sisters relative to subclade 1A; and desert subclades 2A and 2B form a sister pair relative to 2C, 2D, and 2E, with subclades 2C and 2D linked relative to subclade 2E.

Fig. 14 thus empirically documents a degree of phylogenetic structure using the microsatellite data that is consistent with that obtained with other molecular data, and thus seems contradictory to the AMOVA results wherein the overwhelming amount of variation is distributed within local populations and not among clades or subclades. However, this tree provides no estimates of the strength of any nodes, which are likely to be low given the typically long branch lengths distributed throughout the tree. Moreover, branch lengths within subclade clusters visually appear to be as great or considerably greater than internal ones linking any pair or other set of subclades. Given this pattern of branch lengths, it is perhaps not surprising that most of the total pool of variation across the microsatellite loci is distributed among population samples rather than among clades or subclades, depending upon how we structured the AMOVA analysis.

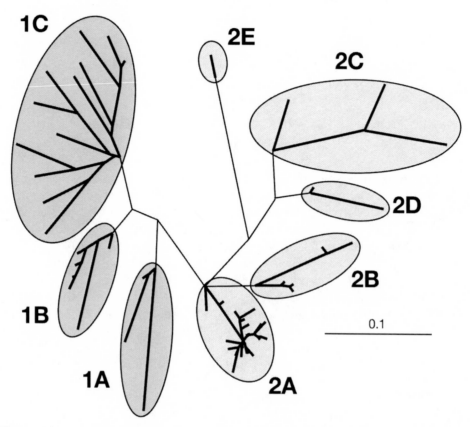

Figure 14. Unrooted neighbor-joining tree of Fst distances among 52 samples of the *Neotoma lepida* group where sample size is < 5. Clusters of samples belonging to each mtDNA *cyt-b* subclades (1A through 2E) are identified by separate ellipses, with Clade 1 and Clade 2 groups separated by different gray tones. Branch lengths are drawn proportional; the scale given in the lower right.

COLORIMETRIC VARIATION AND COLOR PATTERN

RELATIONSHIP OF WOODRAT COLORS TO THE MUNSELL SYSTEM

We performed a principal components analysis (PCA) to compare the multivariate space that includes all 3,379 specimens measured with a set of standard Munsell colors that we chose a priori by comparison to the basic colors observed on these woodrat study skins. Separate analyses were performed for each topographic region of the study skin, using the three trichromatic X, Y, and Z variables. In each case (Figs. 15 and 16), an ellipse that encompasses all study skin measurements is contained within a larger envelope based on the Munsell colors. Note that both dorsal and tail PCA scores are narrowly defined (Fig. 15) and range largely between the Munsell hue, value, and chroma designations of 5YR/6/8 (reddish yellow) to 10YR/2/2 (very dark brown) and from 7.5YR/4/0 (dark gray) to 2.5YR/N2.5/0 (black). The PC scores for both lateral and mid-chest color of the woodrat samples are more broadly distributed (Fig. 16) but still contained within a broader spectrum of color defined by the Munsell colors we chose, ranging largely from 7.5YR/8/6 (reddish yellow) and 7.5YR/6/0 (gray) to 10YR/2/2 (very dark brown) and 2.5YR/2.5/0 (black). Lateral color scores are more centrally distributed within the envelope defined by the Munsell colors than are those for the other three topographic regions of the study skins measured. The PC plots in Figs. 15 and 16 also compare color variation for samples belonging to the coastal (mtDNA Clade 1) with those belonging to the desert molecular clade (mtDNA Clade 2). In all cases, the coastal clade specimens define an ellipse that is smaller and wholly contained within that defined by the desert clade samples; scores for each are significantly different by paired t-tests for both PC-1 and PC-2 axes in each case (student's t ranges from 9.137 to 23.831, with accompanying p-values from 0.0031 to < 0.0001).

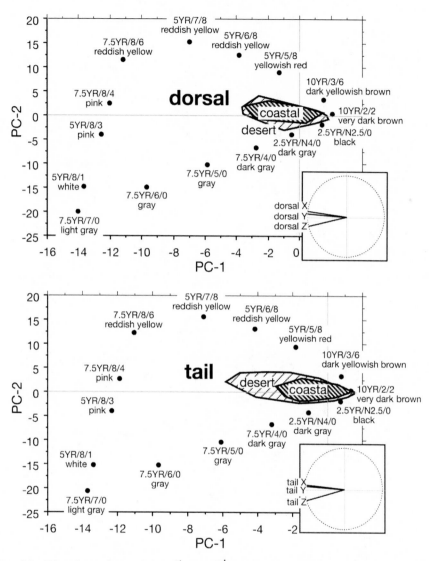

Figure 15. Bivariate plots of the 1st and 2nd principal components analysis of the X, Y, and Z values from all 3, 379 woodrat specimens for which dorsal (above) and tail (below) color was measured. PC-scores for all woodrats are contained within the bold ellipses, which also separate specimens belonging to the coastal from desert mtDNA clades. Measurements of 15 Munsell colors are identified by their Munsell notation of hue, value, and chroma and their corresponding English color names. The insets illustrate the correlation diagram of the X, Y, and Z values for each specimen relative to their respective PC-1 and PC-2 scores.

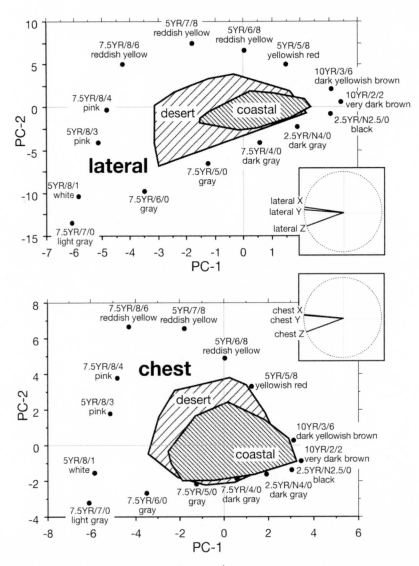

Figure 16. Bivariate plots of the 1st and 2nd principal components analysis of the X, Y, and Z values from all 3, 379 woodrat specimens for which lateral (above) and chest (below) color was measured. PC-scores for all woodrats are contained within the bold ellipses, which also separate specimens belonging to the coastal from desert mtDNA clades. Measurements of 15 Munsell colors are identified by their Munsell notation of hue, value, and chroma and their corresponding English color names. The insets illustrate the correlation diagram of the X, Y, and Z values for each specimen relative to their respective PC-1 and PC-2 scores.

COMPARISONS OF MELANIC POPULATIONS

Melanic individuals with dark, gray-black dorsal pelage and often fulvous-tinged ventral color are found at a number of basalt lava fields throughout the deserts of western North America (Leiberman and Lieberman, 1970). In some cases, this color formed part of the basis for the formal description (e.g., *nevadensis*, 1910; *bensoni*, 1935). Since the localities where melanic individuals are known are of limited aerial extent and widely separated geographically, we asked whether melanic specimens shared a similar set of colorimetric attributes and overall color pattern. Our analysis included 170 individuals from four different basalt flows (Table 8), all of the desert mtDNA Clade 2 and collectively covering a substantial portion of the total range of this phylogeographic group: (1) the lava fields in the Owens Valley, Inyo Co., California (*lepida* and molecular subclade 2A, pooled samples from near Little Lake [locality CA-381] and Big Pine [locality CA-388] on the eastern side of the Sierra Nevada); (2) the lava fields in the Toroweap Valley north of the Grand Canyon, Mohave Co., Arizona (*monstrabilis* and molecular subclade 2B, pooled samples from the floor of Toroweap Valley and nearby Mt. Trumbull [localities AZ-14-17]); (3) the lava fields north of Flagstaff, Coconino Co., Arizona (*devia* and molecular subclade 2C, pooled samples from Wupatki National Monument, Cameron, and Tanner Tank [localities AZ-45-50]); and (4) the Pinacate lava fields in northwestern Sonora, Mexico, and adjacent Yuma Co., Arizona (*bensoni* and molecular subclade 2E, pooled samples from various individual localities within the continuous expanse of basalt extending from Tanque de los Papagos in Sonora north to just across the border in Arizona [localities S-1-3 and AZ-81]).

We divided individuals from each sample into three phenotypic classes: (1) those that are clearly melanic (with a very dark gray to black dorsum, heavily black dorsal tail stripe, and usually strongly buff venter mixed with dark gray); (2) a class we termed "normal," individuals that could not be distinguished from those of non-melanic populations; and (3) "dark" or "intermediate" individuals that are not as dark as true melanics nor as pale as normal individuals. We note that the distribution of the three color morph classes among the four geographic samples differs significantly ($X^2 = 37.120$; df = 6; $p < 0.0001$), which may reflect differences in the aerial extent of a given lava field and/or the inter-dispersion of black basalt and "normal" colored substrates. For example, the proportion of normally pigmented individuals from each sample area is roughly concordant with the degree to which the relevant basalt flows are continuous over larger geographic areas (e.g., the Pinacate lava field, where no "normal" individuals are present) or only intermittently exposures of relatively small size contained within a matrix of normally colored soil types (e.g., the Owens Valley fields, where most individuals are normally colored). Alternatively, some of the variation in phenotypes among the lava flows compared in Table 8 could be due, at least in part, to differences in absolute ages of each basalt field.

In general, the "intermediate" individuals were, indeed, intermediate in their X, Y, and Z values for each of the four topographic areas of the skin that was measured (Fig. 17). Although "normal" individuals are uniformly always significantly different from "intermediate" ones, the latter are not always statistically different from "melanics." The latter observation likely reflects our rather subjective separation of "intermediate" from "melanic" individuals.

Table 8. Samples of melanic populations, separated by designated phenotype.

General locality	Melanic	Intermediate	Normal
Owens Valley	5	16	25
Pinacate	15	4	0
Tanner Tank	16	8	14
Toroweap Valley	14	8	3

We compared the four basalt field samples with melanic individuals in a PCA that included only the trichromatic X-variables from each topographic region of the study skin. The first axis accounts for 70.15% of the total pool of variation with each variable loading nearly equally, as their individual eigenvector range only from 0.7152 (Chest-X) to 0.8939 (Dorsal-X). The mean and 95% confidence limits for each color morph from each of the four sampled populations overlap broadly on PC-1 (Fig. 18), but significant differences among individual pairs of samples do exist. Inter-sample differences are marginally significant for "melanic" individuals by ANOVA ($F_{(3,46)}$ = 3.609, 46, p = 0.0201) and non-significant for "normal" individuals ($F_{(2,46)}$ = 2.811, 45, p = 0.0708). Moreover, comparisons among all combinations of pairs of samples are either non-significant or generally weakly significant, mostly with p-values only 0.05 at best. Consequently, it is unclear if real genetic differences exist in the expression of melanism, in particular, among these samples, as is true for melanic samples of the Rock pocket mouse, *Chaetodipus intermedius* (Hoekstra and Nachman, 2003; Nachman et al., 2003). However, even "melanic" woodrats are not entirely black (as is true for the pocket mice) but express a complex of underlying colors with strongly black or very dark gray overtones. Therefore, it is likely that the differences among the four geographic samples we compare here also reflect subtle variations in other colors expressed and that any molecular genetic analysis of their color will be complex.

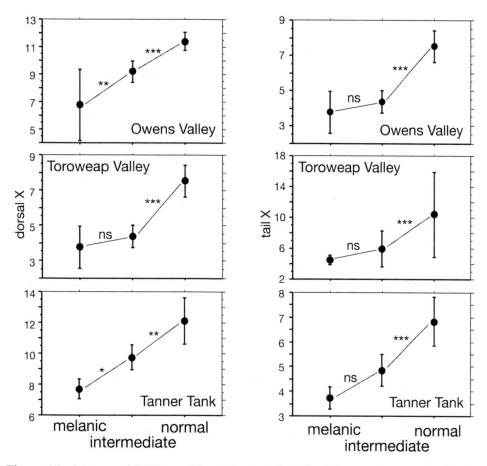

Figure 17. Mean and 95% confidence limits of each of three color morphs for the trichromatic X variable for dorsal (left panel) and tail (right panel) measures, and for three separate geographic basalt flows that contain melanic individuals. The Pinacate sample is not included, because it contained only completely melanic specimens.

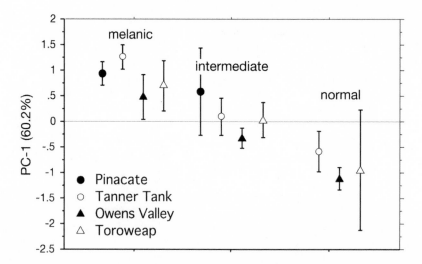

Figure 18. Means and 95% confidence limits of PC-1 scores for "melanic," "intermediate," and "normal" individuals for each of four melanic samples of desert woodrats of mtDNA clade2.

GEOGRAPHY OF COLOR DIFFERENTIATION

The trichromatic X, Y, and Z values are highly intercorrelated for each topographic region of the study skin measured, with Pearson product-moment correlation coefficients always greater than 0.906 (Chest-X versus chest-Z) and p-values based on Fisher's Z-test < 0.0001 in all cases. Moreover, in a PCA that included all individual specimens examined and all variables, the eigenvectors for the three variables for each topographic region are uniform both in magnitude and sign for each extracted PC axis (Table 9), again supporting the high intercorrelations among topographic variables even in multivariate space. As a consequence, we use only the trichromatic X-variable from each topographic site on the study skin in all comparisons among samples in our analyses.

We examined geographic variation in the trichromatic X-variables for each of the four topographic regions of the study skin through the use of both univariate and multivariate analyses. We present here the global patterns in color across the entire sampled range and save the detailed analyses of more restricted geographic areas for the separate transect analyses that we describe in the separate sections, below.

Table 9: Factor coefficients for trichromatic color variables X, Y, and Z for each topographic region of the study skin for the first four PC axes. Eigenvalues and the proportion of variance explained are also given.

Variable	PC-1	PC-2	PC-3	PC-4
Dorsal-X	0.809	0.259	-0.292	0.422
Dorsal-Y	0.814	0.254	-0.293	0.423
Dorsal-Z	0.785	0.204	-0.328	0.441
Tail-X	0.713	0.591	0.260	-0.258
Tail-Y	0.715	0.587	0.267	-0.260
Tail-Z	0.715	0.532	0.327	-0.268
Lateral-X	0.777	-0.346	-0.369	-0.352
Lateral-Y	0.779	-0.348	-0.372	-0.353
Lateral-Z	0.749	-0.347	-0.375	-0.344
Chest-X	0.716	-0.508	0.435	0.137
Chest-Y	0.719	-0.501	0.439	0.146
Chest-Z	0.673	-0.408	0.522	0.196
eigenvalue	6.718	2.186	1.595	1.203
% contribution	56.0	18.2	13.3	10.0

We correlated the trichromatic-X color variables for each topographic region with the latitude and longitude of sample localities. We then repeated the analysis using PCA scores in place of individual measurements. Regardless of whether variables are treated separately or combined in PCA scores for each orthogonal axis, the geographic position of localities as determined by either latitude or longitude explained very little of the variation present. Correlation coefficients were often highly significant ($p = 0.0001$, but R^2 values were uniformly less than 0.13. Clearly, there is no general pattern to color variation in these woodrats that is reflected solely by the geographic position of their populations.

We also examined the relationship between each trichromatic-X color variable and PC scores with the reduced environmental parameters derived from the principal components analysis of 19 bioclimatic variables obtained from the WorldClim database (described above). The first two axes from the bioclimatic PCA explain 73% of the total pool of variation among the 19 original variables. The first PC axis (44.3% explained variation) contrasts cold temperatures (mean temperature of the coldest quarter [loading 0.982], mean temperature of the coldest month [0.971], and mean annual temperature [0.947]) with dry precipitation variables (precipitation in the driest quarter [loading -0.921] and precipitation in the driest month [-0.900]). On the second axis (28.7% explained variation), wet precipitation variables (precipitation in the wettest quarter [0.935], precipitation in the wettest month [0.916], and annual precipitation [0.876]) contrast with warm temperatures (maximum warmest temperature [-0.739], temperature seasonality [-0.637], and mean warmest temperature quarter [-0.586]).

A strong correlation exists between color PC-1 and PC-2 scores and bioclimatic PC-1 and PC-2 scores, although the overall explanatory power for each pair is limited (R^2 = 0.127 or less). Color PC-1 scores, which correspond to overall color tones (Figs. 15 and 16), are significantly correlated with bioclimatic PC-1 (r = 0.256, Z-value = -4.905, p < 0.0001) and PC-2 scores (r = 0.357, Z-value = 11.737, p < 0.0001), indicating that pale animals are typically associated with the driest and warmest habitats and dark animals with the wettest and coldest.

Comparisons between the two mtDNA clades ("desert" and "coastal" groups) or among the subclades within each provide evidence for significant differentiation in nearly all comparisons. ANOVAs that compare the four univariate trichromatic-X variables between both clades or scores for PC axes based on a PCA of those same four variables are all highly significant, with p-values < 0.0001 in all cases. The latter set of observations repeats results from the principal components analyses that include the Munsell colors, described above (Figs. 15 and 16). In a PC analysis comparing color characteristics of each of the eight subclades, all but two comparisons between a subclade belonging to the coastal mtDNA clade and those of the desert clade differ significantly (ANOVA) in both PC-1 and PC-2 scores, with pairwise p-values ranging from 0.0228 to < 0.0001 (Fig. 19). This level of difference extends to comparisons among the individual subclades within each clade. Among the three subclades of the coastal clade only 1B and 1C are quite similar, being barely significantly different on the first PC axis (p = 0.04) and not significantly different on the second axis (p = 0.2231). Within the desert clade, subclades 2A and 2B differ significantly on both axes, but subclades 2C, 2D, and 2E are largely non-significantly different. Clearly, substantial geographic differentiation is present in color characteristics across the full range of desert woodrats.

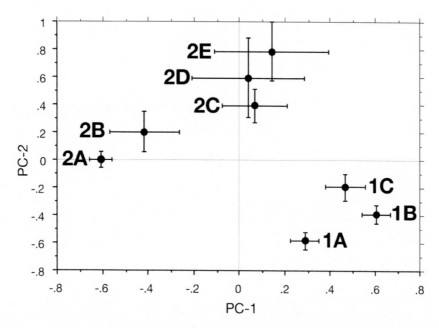

Figure 19. Mean and 95% confidence limits to scores along both the first and second PC axes in a principal components analysis that included all nine mtDNA subclades, 1A-1D of the coastal clade and 2A-2E of the desert clade (see map, Fig. 6).

GEOGRAPHY OF MORPHOLOGICAL VARIATION

MORPHOLOGICAL DIFFERENCES BETWEEN "COASTAL" AND "DESERT" SAMPLES

Grinnell and Swarth (1913, p. 338) detailed a set of morphological differences between woodrats of the *Neotoma lepida* group from coastal California (which they considered to make up the subspecies *N. i. intermedia*) and the interior desert of eastern California (which they considered to be the subspecies *N. i. desertorum*), which we summarize here.

Neotoma i. intermedia

Coloration above dark: blackish mid-dorsally, mixed with clay color, this most pure along the sides and about the face; beneath white, with base of hairs deep plumbeous throughout (occasional examples with hairs on small pectoral patch white to base); ankles dusky; tail black above.

Pelage harsh: hairs relatively stiff and coarse.

Size large: average 13 males, length 336, tail vertebrae 159, hind foot 34.3, ear 30.4.

Tail "long": ratio of tail to body in 13 males, 89%.

Skull large, this involving all features, save that rostrum and nasals are relatively longer and narrower, while audital bullae are actually as well as relatively less inflated.

Neotoma i. desertorum

Coloration above pale: sepia mid-dorsally, mixed with pinkish buff, this color clearest anteriorly and along sides; beneath white, with base of hairs pale plumbeous except on pectoral region and narrow line mid-ventrally which are pure white; ankles white; tail grayish brown above (variable to blackish).

Pelage soft: hairs relatively fine and silky.

Size small: average 10 males, length 288, tail vertebrae 134, hind foot 30.8, ear 28.5.

Tail "short": ratio of tail to body in 10 males, 87%

Skull small, this involving all features, save that rostrum and nasals are relatively shorter and hence blunter, while the audital bullae are distinctly larger, more inflated.

Three years earlier, in his revision of the genus *Neotoma*, E. A. Goldman (1910) had considered these two taxa as separate species, even placing them in different species groups. He considered the two so distinct that he made no specific comparison between them, commenting only on their difference in overall size (p. 16-17). Seventy years later, Mascarello (1978), in his review of the woodrats along both sides of the lower Colorado River, noted that animals from coastal southern California and Baja California (his "Baja type") differed from those of the California and Arizona deserts (his "Western" and "Eastern" types) along multivariate discriminant axes and in the frequencies of three qualitative craniodental characters among the 12 that he scored. His "Western" type lacked accessory mental foramina, had a short posterior (labial) re-entrant angle of M3 that is at right angles to the toothrow, and possessed an M1 with a very distinct V-shaped notch on the anterior loop. Mascarello (1978) also noted substantive difference in the length of the baculum and soft anatomical features of the glans of the glans penis between his "Baja" and his two desert types. We describe these phallic and bacular differences separately, below.

We will discuss the reliability of the characters delineated by Grinnell and Swarth (1913) in a later section that specifically includes the samples they used in their comparisons and will review also those features identified by Mascarello, again using the same set of specimens he examined. In this section we examine the pattern of overall differentiation between *N. intermedia* and *N. desertorum* (= *N. lepida*, following Goldman, 1932), using the distribution maps of these two taxa in Goldman (1910, p. 43 and 77) as the basis for grouping localities. Because each of Goldman's taxa are now included within the single species, *N. lepida*, with multiple subspecies (Hall, 1981), we simply refer to our comparison groups as "coastal" and "desert," geographic units that in the United States are generally divided by the Sierra Nevada, Tehachapi Mts., and Transverse and Peninsular Ranges of California. As is apparent from the description of variation in both the mtDNA cytochrome-b (*cyt-b*) and nuclear β-Fibrinogen-intron 7 (*Fbg-I7*) gene sequences above, Goldman's *N. intermedia* and *N. desertorum* are largely concordant with Clades 1 and 2. We also confine our analyses here to samples from the United States and extreme northern Mexico and so avoid any potential confounding variation along the 1,000 km length of the Baja peninsula in the overall differences we seek to evaluate at this point in our analyses. At this point, thus, our "coastal" group includes three subspecies that are currently recognized, namely *N. l. intermedia*, *N. l. gilva* (with *sola* a synonym), and *N. l. californica* (Hall, 1981). Our "desert" group includes eight taxa currently listed as valid subspecies or species: *N. l. lepida* (with *desertorum* a synonym), *N. l. nevadensis*, *N. l. monstrabilis*, *N. l. auripila* (with *aureotunicata, bensoni, flava*, and *harteri* as synonyms), *N. l. marshalli*, *N. l. grinnelli*, *N. l. sanrafaeli*, and *N. devia* (Hall, 1981; Hoffmeister, 1986; Musser and Carleton, 2005).

As posited by Goldman (1910), "coastal" samples are indeed significantly larger in body size, whether measured by Total Length (TOL) or Head and Body Length (HBL) than "desert" samples (Table 10). This size difference extends to all cranial dimensions except the bullar dimensions BUL and BUW, which are our proxy for overall bullar inflation. Thus, the more comprehensive observations of Grinnell and Swarth (1913, listed above) regarding various size differences between their *intermedia* and *desertorum* are certainly correct, including their recognition that the degree of bullar inflation of the "coastal" form is both absolutely and relatively smaller than that of the "desert" type.

The differences in overall size, in absolute as well as relative length of the tail, and in absolute as well as relative size of the bullae between "coastal" and "desert" samples of the *Neotoma lepida* group are readily apparent in simple bivariate scatterplots combining these characters. For example, in the comparison of Tail Length (TAL) relative to Head-and-Body Length (HBL), the regression lines of each group are significantly different both in slope (ANOVA, $F_{(1,1513)}$ = 18.676, p < 0.0001) and Y-intercept (ANOVA, $F_{(1,1514)}$ = 1588.981, p < 0.0001), as "coastal" animals have longer tails in relation to their head and body lengths than "desert" animals (Fig. 20). Similarly, bullar length (BUL) in relation to cranial length (CIL) is significantly different both in slope (ANOVA, $F_{(1, 1690)}$ = 14.521, p = 0.0001) and Y-intercept (ANOVA, $F_{(1,1691)}$ = 551.896, p < 0.0001), again with "coastal" specimens having smaller bullae but longer skulls (Fig. 21). Although there is some overlap in the measurements of individual variables between the two groups, they are nevertheless readily separable by simple comparisons between individual bullar length and tail length (Fig. 22).

Table 10. Mean and standard error for external and craniodental measurements of "coastal" and "desert" samples of the *Neotoma lepida* group, exclusive of samples from Baja California. Significance level is based on one-way ANOVA.

Variable	"Coastal" N = 515 - 679	Significance level[1]	"Desert" N = 851 - 1025
external			
TOL	324.23±0.702	****	291.82±0.489
HBL	171.01±0.487	****	162.33±0.373
TAL	153.22±0.437	****	129.52±0.345
HF	33.60±0.059	****	30.65±0.046
E	30.30±0.101	****	28.91±0.091
craniodental			
CIL	39.569±0.056	****	37.365±0.048
ZB	21.723±0.036	****	21.723±0.027
IOC	5.513±0.009	****	5.054±0.007
RL	16.274±0.028	****	15.186±0.025
NL	15.803±0.030	****	14.765±0.026
RW	6.578±0.012	****	6.165±0.009
OL	14.336±0.021	****	13.606±0.015
DL	11.395±0.027	****	10.794±0.022
MTRL	8.206±0.012	****	7.908±0.009
IFL	8.751±0.020	****	8.240±0.015
PBL	18.096±0.029	****	17.272±0.025
AW	7.616±0.010	****	6.997±0.008
OCW	9.616±0.012	****	8.939±0.009
MB	17.292±0.020	****	16.718±0.019
BOL	5.924±0.014	****	5.528±0.011
MFL	7.811±0.017	****	7.435±0.016
MFW	2.681±0.008	****	2.327±0.006
ZPW	4.145±0.010	****	4.071±0.008
CD	15.721±0.019	****	15.384±0.015
BUL	6.714±0.009	****	7.195±0.008
BUW	7.062±0.010	****	7.482±0.008

[1] **** = $p < 0.0001$

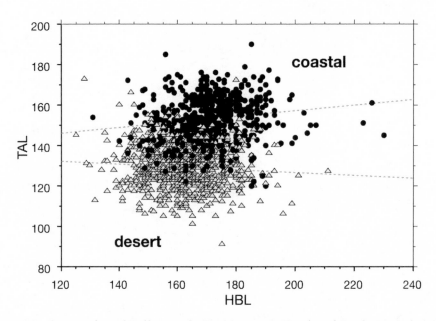

Figure 20. Scatterplot of Tail Length (TAL) versus Head-and-Body Length (HBL) for "coastal" (solid circles) and "desert" (gray-filled triangles) samples of the *Neotoma lepida* group from the United States. Regression lines for each group are: desert: TAL = 140.714 – 0.069 HBL, r = -0.077; coastal: TAL = 130.118 + 0.135 HBL, r = 0.152.

The separation of "coastal" and "desert" individuals in tail length (TAL) results from an actual difference in the number of vertebral elements in the tail. We have available for examination complete skeletons of 25 "coastal" and 57 "desert" individuals, all from localities in southern California in Kern, Riverside, San Bernardino, and Imperial counties. There is a mean of 30.48 elements (range 29-34) in the tail of "coastal" specimens and 25.30 (range 20-29) in "desert" ones. This difference is highly significant (ANOVA: $F_{(1,80)}$ = 120.351, p < 0.0001). Tail length thus differs between "coastal" and "desert" samples of woodrats at least partly as a result of the number of vertebral elements. We did not measure the lengths of individual vertebrae to determine if these differ as well.

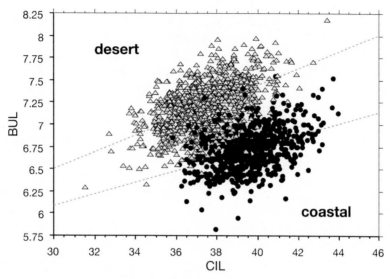

Figure 21. Scatterplot of Bullar Length (BUL) versus Condyloincisive Length (CIL) for the same samples in Fig. 20. Regression lines for each group are: desert: BUL = 3.666 + 0.094 CIL, r = 0.543; coastal = BUL 4.092 + 0.066 CIL, r = 0.410.

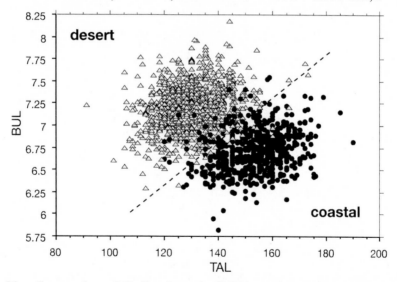

Figure 22. Scatterplot of Bullar Length (BUL) versus Tail Length (TAL) for "coastal" (solid circles) and "desert" (gray-filled triangles) samples of the *Neotoma lepida* group. The dashed line separates 95% of the specimens of each group by the combination of these two variables.

Given the overall size difference between "coastal" and "desert" samples, as well as their separation in combinations of bivariate scatterplots, it is not surprising that the two groups are well separated in multivariate space defined by a PCA (Fig. 23). Based on the 21 log-transformed craniodental variables, both groups are nearly non-overlapping along the first two axes, which combine to explain 69.4% of the total pool of variation. Other axes individually explain no more than 4.6% of the variation. All variables except the two bullar measurements load highly and nearly equally on the first axis, with both BUL and BUW the only highly loading variables on the second axis (Table 11). This difference in the loadings of these contrasting sets of variables is evident in the vector diagram in Fig. 23 (inset) and mirrors the univariate comparisons presented directly above: "coastal" and "desert" samples of the *Neotoma lepida* group differ substantially in overall size, as indexed by the long and positive vectors for all variables except bullar dimensions, with the latter decidedly larger in the smaller bodied "desert" group than in the larger "coastal" form.

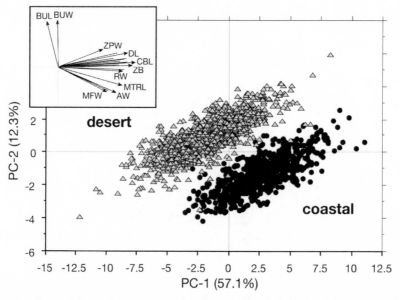

Figure 23. Scatterplot of scores on the 1st and 2nd Principal Components Axes based on the 21 log-transformed craniodental variables, with "coastal" (solid circles) and "desert" (gray-filled triangles) individuals. The percent of the total variance explained by each axis is indicated. The inset box illustrates character vectors along both axes.

Table 11. Principal component factor loadings for 21 log-transformed craniodental variables of "coastal" and "desert" morphological groups of the *Neotoma lepida* group from the United States and Mexico.

Variable	PC-1	PC-2
log CIL	0.969	0.091
log ZB	0.923	0.041
log IOC	0.663	-0.433
log RL	0.905	0.098
log NL	0.886	0.094
log RW	0.805	-0.064
log OL	0.867	-0.033
log DL	0.869	0.264
log MTRL	0.445	-0.287
log IFL	0.814	0.116
log PBL	0.895	0.168
log AW	0.722	-0.444
log OCW	0.795	-0.319
log MB	0.865	0.160
log BOL	0.796	0.045
log MFL	0.726	0.166
log MFW	0.573	-0.412
log ZPW	0.551	0.317
log CD	0.779	0.267
log BUL	-0.157	0.840
log BUW	-0.016	0.862
eigenvalue	11.9785	2.5601
% contribution	57.05	12.29

In our own initial examination of museum specimens of the *Neotoma lepida* group, we also identified several qualitative features in which differences among regional samples were apparent. These characters differentiate samples from nearly all of Baja California and coastal California from those of the interior deserts of northeastern Baja and the US, and thus also the "coastal" versus "desert" morphological groups delineated by both univariate and multivariate analyses of morphometric variables. Mascarello (1978, character 6) identified the first of these characters, the V-shaped notch on the anterior loop of M1, but previous workers to our knowledge have not mentioned the other two. We could not confirm the utility of two additional qualitative characters mentioned by Mascarello as distinctive in the separation of these two global geographic groups (presence or absence of accessory mental foramina and the length of the posterior re-entrant angle of M3).

Anteromedian flexus on anteroloph of M1

This flexus is deeply notched in young individuals (Age 5) of the "coastal" morphological group (Fig. 24, upper row) and remains evident even in older individuals (Age 2-3) following successive wear. The flexus is only weakly developed in the youngest individuals of the "desert" morphological group (Age 5, Fig. 24, lower row) and becomes mostly obliterated with increasing age (age classes 2-3). As clear as this difference is, because the depth and angularity of the flexus decreases with age, care must be taken when placing older individual specimens with regard into either of the two geographic groups.

Given this caveat, however, there is a clear and general relationship between M1 anteroloph marginal shape with respect to the two major mtDNA clades. We illustrate this by arranging holotypes of named forms that we have examined, and for which we have recorded camera lucida drawings of this tooth, on the phylogenetic clade structure (Fig. 25). Note that all coastal Clade 1 holotypes have deep anteromedian flexi, except the holotypes of *insularis* and *pretiosa*, which are old individuals (age class 1) with well-worn teeth. On the other hand, all desert Clade 2 holotypes lack an anteromedian notch.

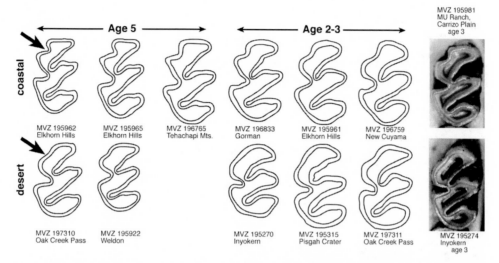

Figure 24. Differences in the depth of the anteromedian flexus of the upper first molar (M1) between "coastal" and "desert" morphological groups of woodrats. This flexus is a deep notch in young aged individuals (Age 5) of the "coastal" form, and remains evident even with extended wear (Age 2-3); the flexus is nearly imperceptible in all age classes of the 'desert' morphological group.

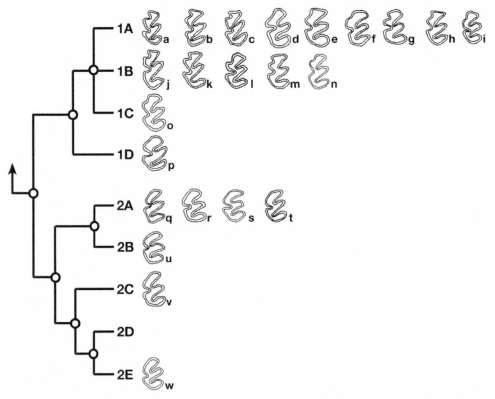

Figure 25. Camera lucida drawing on the first upper molar (M1) of 23 holotypes of the *Neotoma lepida* group, arranged by mtDNA clade, as defined by the Bayesian tree presented in Fig. 5. a = *bryanti*, USNM 186481; b = *nudicauda*, USNM 79073; c = *perpallida*, USNM 79061; d = *pretiosa*, USNM 146123; e = *abbreviata*, MCZ 12260; f = *vicina*, USNM 146803; g = *ravida*, USNM 140692; h = *notia*, USNM 146794; i = *aridicola*, SDNHM 15595; j = *intermedia*, ANSP 8343; k = *gilva*, ANSP 1665; l = *anthonyi*, USNM 137156 (paratype); m = *martinensis*, USNM 81074; n = *egressa*, MVZ 50142; o = *petricola*, MVZ 30202; p = *insularis*, USNM 198405; q = *desertorum*, USNM 25739; r = *bella*, MCZ 5308; s = *marshalli*, USNM 263984; t = *grinnelli*, MVZ 10438; u = *monstrabilis*, USNM 243123; v = *devia*, USNM 226376; w = *flava*, MVZ 62657.

Incisive foramen septum

The structure of the medial septum of the incisive foramen also differs between "coastal" and "desert" samples of the *Neotoma lepida* group. In "coastal" animals, the maxillary spine is shallow and the maxillo-vomerine notch is elongated, resulting in an elongated vacuity (Fig. 26). The opposite conditions characterize "desert" specimens.

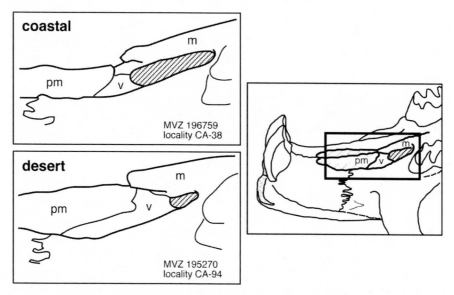

Figure 26. Lateral view of the medial septum of the incisive foramen: the vomerine portion is smaller, resulting in a larger vacuity, in "coastal" as compared to "desert" samples of the *Neotoma lepida* group (see text). **pm**, premaxilla; **v**, vomer; **m**, maxilla.

Position of lacrimal bone with reference to the frontal-maxillary suture

The frontal-maxillary suture intersects the lacrimal at its midpoint or on its anterior half in "coastal" specimens so that contact between the lacrimal and frontal bones is either equal to or longer than contact with the maxillary (Fig. 27, top). In "desert" individuals, the frontal-maxillary suture intersects the lacrimal on its posterior half, resulting in a short contact between the lacrimal and frontal bones and a longer one with the maxillary (Fig. 27, bottom).

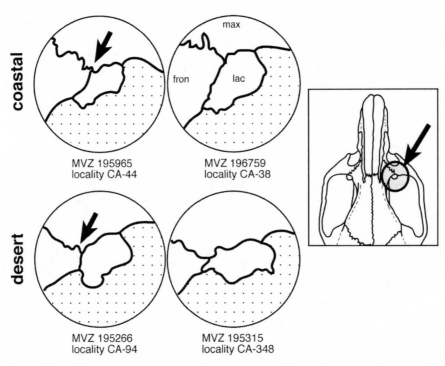

Figure 27. Differences in the position of the frontal-maxillary suture relative to the position of the lacrimal bone in the anterior orbit of individuals of the "coastal" (above) and "desert" (below) morphological groups. **fron**, frontal; **max**, maxillary; **lac**, lacrimal.

The position of the frontal-maxillary suture with respect to the lacrimal bone is also concordant with the mtDNA clade assignments of each holotype that we have examined (Fig. 28). The suture is positioned at the mid-point of the lacrimal bone, or slightly more anteriorly, so that contact with the frontal is longer than that with the maxillary, in those holotypes belonging to Clade 1. In all Clade 2 holotypes, the position of the suture is more posterior so that this bone primarily contacts the maxilla.

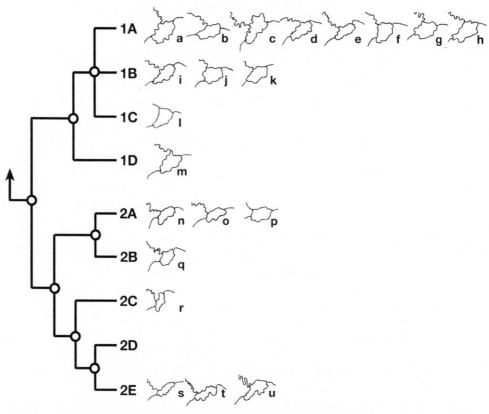

Figure 28. Shape of the lacrimal bone in the holotypes of 17 taxa of the *Neotoma lepida* group, arranged by phylogenetic position within the mtDNA Bayesian tree (Fig. 5). a = *bryanti*, USNM 186481; b = *nudicauda*, USNM 79073; c = *pretiosa*, *USNM 146123*; d = *vicina*, USNM 146803; e = *ravida*, USNM 140692; f = *notia*, USNM 146794; g = *molagrandis*, SDNHM 14065; h = *aridicola*, SDNHM 15595; i = *anthonyi*, USNM 137156 (paratype); j = *martinensis*, USNM 81074; k = *egressa*, MVZ 50142; l = *petricola*, MVZ 30202; m = *insularis*, USNM 198405; n = *desertorum*, USNM 25739; o = *marshalli*, USNM 263984; p = *grinnelli* Hall, MVZ 10438; q = *monstrabilis*, USNM 243123; r = *devia*, USNM 226376; s = *flava*, MVZ 62657; t = *aureotunicata*, SDNHM 10907; u = *harteri*, SDNHM 11462.

Glans penis, including baculum

The glandes of all members of the *Neotoma lepida* group are similar to all woodrats in lacking dorsal lappets, ventral lappets, a urethral process, a dorsal papilla, lateral bacular mounds, a ventral shield, spines on the crater walls, and any corrugation (e.g., Hooper, 1960; Hoffmeister, 1986; Matocq, 2002). They do exhibit the subterminal urinary meatus and crater hood so characteristic of *Neotoma*. In all glandes in this complex, however, the crater hood is apparently elongated relative to that of most or all other species in the genus. We base this conclusion on the fact that, in the other species in the genus, the base of the glans has spines whereas the crater hood lacks them. In this complex, from 40-80% of the distal end of the glans has no spines; we interpret this region as the hood.

With the exception of specimens from Isla Ángel de la Guarda, all well-preserved glandes have an elongate hood extending 75-80% of the length of the glans (Fig. 29). They also sort easily into what we term the coastal morph (= Baja type of Mascarello, 1978) and the desert morph (= western and eastern types of Mascarello, 1978), despite some variation in the desert morph (described below).

Mascarello (1978) stated that the coastal morph curved dorsally but remained straight distally to the tip. Because all of the formalin preserved and many of the rehydrated specimens exhibited a double curve, we believe that this represents the natural condition of all of these glandes; Mascarello used only rehydrated specimens, many of which are badly stretched and greatly over-cleared. We interpret the straight condition of the glandes he used as artifactual. The tip of the coastal morph recurves only slightly and tapers to a split point. The extreme tip of the desert type tightly recurves dorsally in most specimens and ends in a more rounded point, often double, again as described by Mascarello.

Mascarello (1978) recognized differences between the tip of the glans in his eastern and western types (both in our desert morph). He described and figured the eastern type as bifurcated with attenuate tips and the western type as essentially non-bifurcated with a blunt tip. His western specimens came from within the range of our mtDNA subclade 2A and his eastern ones from the range of our subclades 2C, 2D, and 2E; he examined no specimens from the range of subclade 2B. Our specimens came from throughout this entire range and from all five subclades (Fig. 30, compare to Fig. 6). In our analyses, the differences in tip type Mascarello described generally hold in the areas from which he had specimens. Importantly, however, most specimens from subclade 2B exhibit the eastern type of tip, although this subclade is distributed to the north of the Colorado River and is the sister of the western desert subclade 2A (Fig. 5). Both tip types co-occur in our sample from Mokaac Wash, Mojave Co., Arizona (locality AZ-7; two western and five eastern type), and our single specimen from the type locality of *N. l.*

monstrabilis (Ryan, Coconino Co., Arizona; locality AZ-21) is of the western type; all other specimens of subclade 2B from Arizona and Utah north of the Grand Canyon possess the eastern tip type (Fig. 30). Moreover, both western and eastern tip types also co-occur at one locality west of the Colorado River in southeastern California (locality CA-314; Big Maria Mts., Riverside Co.; 4 western and 1 eastern individuals) as well as east of the lower Colorado River in northwestern Sonora (locality S-2; Tanque de los Papagos; 2 eastern and 1 western). Thus, the distribution of the two tip types described by Mascarello is not completely consistent with his use of this character to diagnose *N. lepida* and *N. devia* as species separated by the Colorado River. Moreover, the discordance in the distribution of tip type and both other nonmolecular characters and, importantly, phylogenetic clade structure (Fig. 5 and Matocq et al., 2007) must result either from character convergence in the glans penis or from differential sorting of an ancestral polymorphism during the diversification of the clades now occupying the two sides of the Colorado River (see historical scenario below); it cannot be due to gene flow across the Grand Canyon, as posited by Hoffmeister (1986).

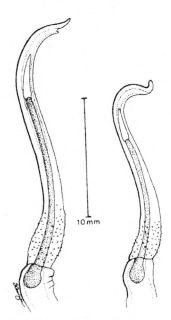

Figure 29. Camera lucida drawing of two glandes of the *Neotoma lepida* group, each with its venter to the right. The "coastal" morph is on the left (LACM 13693; La Zapopita, Baja California, Mexico; locality BCN-20) and the "desert" morph is on the right (LACM 36952; 27.9 mi NE Glamis, Milpitas Wash, Hwy. 78, Imperial Co., California; locality CA-209).

Hoffmeister (1986) recognized five slight variants of our desert morph in Arizona, based mostly on degrees of indentation of the tip to produce the double point, and also disagreed with Mascarello's geographically discrete tip types. Hoffmeister (1986:414) used 18 glandes from Arizona. DGH examined all preserved bacula and glandes in the UIMNH and could score only 16 with respect to details of their morphology. The remainder either consisted only of bacula or had the tip badly damaged or lost. All 16 of these glandes have bifurcated tips, mostly deeply so, as described by Mascarello (1978). Because Hoffmeister did not give the specimen identifications of any of the five tip variants he illustrated in his figure 5.226, we are not sure we saw all specimens upon which he based his observations. The glandes we examined all fit one of his C, D, or E variants, all which belong to Mascarello's eastern type of glans.

The difference between the western and eastern types of Mascarello (1978) is slight and may be compromised by distortion of poorly preserved specimens. This difference is real, however, even if slight and subject to minor overlap and discordant with other characters north of the Grand Canyon in Arizona and Utah (Fig. 30).

Hooper (1960) originally described and figured the glans of *N. lepida*. Based on his figure, the two specimens he examined represent our coastal type.

The baculum of both coastal and desert morphs curves dorsally into the hood but does not protrude into its ventrally curved portion. A large, cartilaginous tip caps the distal end of the baculum and extends into the ventrally curved portion of the hood. This cartilaginous tip appears very flexible as it is quite distorted in some specimens, occasionally bent double on itself. We measured the straight-line length of the baculum to the nearest millimeter but did not measure the cartilaginous tip. The baculum of 31 glandes from coastal animals is longer, averaging 15.4 mm (range = 13-18 mm, standard deviation = 1.58 mm). The baculum of 47 glandes from inland individuals averaged 10.8 mm (range = 8-12 mm, standard deviation = 1.09 mm). Mascarello (1978) obtained similar results. In addition the entire glans of the coastal morph is usually much larger than that of the inland type. The distinctiveness of these two morphs permits their use in field identification of hand-held live animals, by simply rolling back the prepuce to display the glans.

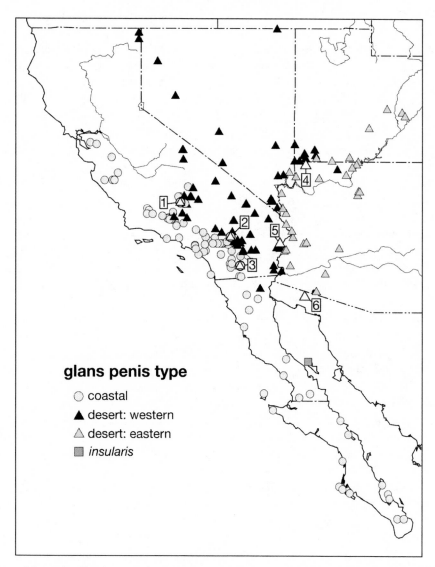

Figure 30. Distribution of the glans penis types described by Mascarello (1978): gray squares = Baja (= our "coastal" group), black circles = western (mtDNA subclade 2A, our "desert" group), and open circles = eastern (mtDNA subclade 2C, 2D, and 2E, our "desert" group), plus square = *insularis* from Isla Ángel de la Guarda described herein. Numbered localities are those where two tip types co-occur: coastal and desert-western types at (1) locality CA-80, (2) locality CA-341, and (3) locality CA-178a; desert-western and desert-eastern types at (4) locality AZ-7, (5) locality CA-314, and (6) locality S-2.

All identifiable glandes from Arizona, Nevada, Utah, and the Mojave Desert and eastern Colorado Desert regions of California are of the desert morph (Fig. 30). All identifiable glandes from coastal California and that part of the Colorado Desert west of the Salton Sea are of the coastal morph. With the exception of one locality from the extreme northeastern part of Baja California (Cerro Prieto, 20 mi. SSE Mexicali; locality BCN-101), all identifiable glandes from the mainland of the peninsula represent the coastal morph; those from Cerro Prieto represent the desert morph. All of the identifiable glandes from the islands off the peninsula also resembled closely the coastal morph except those from Isla Ángel de la Guarda. Hence, the coastal and inland phallic types we identify here are geographically concordant with "coastal" and "desert" craniodental morphological groups and the mtDNA clades 1 and 2 described above.

In the vast majority of the preserved glandes that we deemed unidentifiable, the distal tip was either missing or badly damaged. Individuals with either of these tip types co-occur at three localities (Fig. 30): in Kelso Valley in eastern Kern Co., CA (locality CA-80; 17 coastal and 4 desert), Morongo Valley in San Bernardino Co., CA (locality CA-341; 2 coastal and 34 desert), and near Ocotillo Wells in eastern San Diego Co. (CA-178a; 1 coastal and 1 desert). As we document in the transect analyses below, the first two of these localities are also areas of sympatry and occasional hybridization between the "coastal" and "desert" craniodental morphs.

The glans (Fig. 31) of the five specimens from Isla Ángel de la Guarda does not closely resemble those of any described species in the genus. This glans is relatively short and thick and has a covering of spines over most of its surface; only the distal tip, which we interpret as the hood, lacks these spines. The hood, although more elongate than that of most species in the genus, makes up only 40-50% of the length of the glans and does not have the characteristic tip of either of the other two morphs in the *lepida* group; rather, the tip appears simply collapsed. The baculum is also relatively short and thick and is capped by a large cartilaginous tip. These character states resemble somewhat those of such species as *N. albigula*, *N. floridana*, and *N. mexicana*. In general morphology, the glans of this insular taxon is intermediate between most other species in the genus and the coastal and desert morphs of the *lepida* group described above.

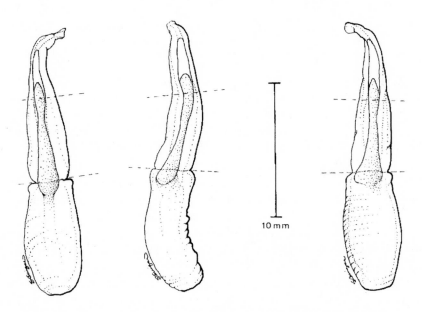

Figure 31. Camera lucida drawing of the glans of SDNHM 19201, from the north end of Isla Ángel de la Guarda, Baja California, Mexico (locality BCN-95). Left, ventral; middle, right lateral; and right, dorsal views. Spines are present in the area between the dashed lines.

The polarity of character change in glans development within the *lepida* group is unclear at the moment, since the shortened glans (and baculum) of animals from Isla Ángel de la Guarda (subclade 1D) is phylogenetically nested within other members of Clade 1 and those of Clade 2 (Fig. 5), all of which have the elongate glans (Figs. 29 and 30). If the shortened glans (and baculum) is considered ancestral for the *lepida* complex, then elongation must have occurred convergently in the other lineages. Similarly, if elongation is viewed as ancestral, then the truncated glans of insularis would be a reversal. In either case, a minimum of two steps is required to derive the glandes types of the *lepida* group from their closest relatives (Edwards and Bradley, 2001; Matocq et al., 2007). Understanding both the mechanism and genetic control of glans development (e.g., Matocq et al., 2007) may provide a means to choose between these alternative evolutionary scenarios.

TRANSITIONS BETWEEN AND WITHIN "COASTAL" AND "DESERT" MORPHOLOGICAL GROUPS

Given the clear set of differences in both qualitative and quantitative morphological variables between the broadly distributed "coastal" and "desert" groups, we undertook a series of more detailed analyses designed to determine the patterns of character variation among local samples over relatively confined geographic regions within the overall range of the *Neotoma lepida* group. We organize these analyses as a series of transects, each incorporating separate, although partially overlapping, geographic regions that encompass areas of transition between the two well-defined morphological groups. We then focus on peninsular Baja California and its associated Gulf and Pacific coast insular populations. We treat this region separately in order to concentrate on population and taxon comparisons along the spine of the peninsula as well as between the various insular taxa and their mainland counterparts. Many of the insular taxa have been traditionally regarded as distinct species, yet have been uniformly included as part of a larger "*Neotoma lepida* group" (e.g., Goldman 1910, 1932; Hall 1981). Finally, we complete our morphological studies by examining Mascarello's (1978) hypothesis of a species-level boundary between desert woodrats separated by the Colorado River in Arizona and California.

The two morphological groups are in contact, or near contact, in four areas in southern California. We examine each of these transition areas as four separate transects (Fig. 32): (1) A Tehachapi Transect—across the Tehachapi Mts. from their northern boundary at the southern end of the Sierra Nevada to their point of contact with the Transverse Ranges in Kern, Ventura, and Los Angeles Cos.; (2) a Cajon Pass Transect—across Cajon Pass between the San Gabriel and San Bernardino Mts., in Riverside and San Bernardino Cos.; (3) a San Gorgonio Pass Transect—across San Gorgonio Pass between the Transverse and Peninsular Ranges in Riverside and San Bernardino Cos., which is the transect originally described by Grinnell and Swarth (1913) that established the current taxonomy of this complex of woodrats (Goldman, 1932); and (4) a San Diego Transect—along the international border in San Diego and Imperial Cos., California, and northern Baja California. Here, we describe patterns of variation in morphological (qualitative external and morphometric craniodental as well as pelage color) characters and place these in a genetic context based on our analyses of mtDNA and nuclear microsatellite loci.

Figure 32. Approximate linear positions of four transects in the analysis of morphological and genetic variation in woodrats of the *Neotoma lepida* group in southern California (see text beyond).

Tehachapi Transect

This transect proceeds from the Caliente Mts. on the west side of the Carrizo Plains east through the foothills and mountains around the southern and southeastern end of the San Joaquin Valley (including Mt. Pinos and Tejon Pass) as far north as Porterville in Tulare Co. and then east across the Tehachapi Mts. and Kern River plateau through the Mojave Desert in San Bernardino Co. Samples encompass localities assigned to three subspecies (Grinnell, 1933; Hall, 1981): *gilva* on the west; *intermedia* from the southern Sierra Nevada; and *lepida* from the eastern slopes of the Tehachapi Mts. and Mojave Desert. The type locality of one formal taxon, *N. desertorum sola* (listed as a synonym of *N. lepida gilva* by Goldman, 1932, and subsequent authors), is contained within this transect. The samples include representatives of the two morphological groups ("coastal" and "desert") defined above, and three mtDNA clades (the coastal subclades 1C and 1B and the desert subclade 2A). These are juxtaposed geographically in a complex and discordant pattern in the middle part of the transect.

    Localities and sample sizes.—To facilitate the analysis of variation across the transect, we organized locality samples into seven geographic samples and placed those specimens from contact or near-contact localities between the two morphological groups into an "unknown" sample. There are five "coastal" samples, all from localities to the west of the Tehachapi Mts. (from the Caliente and Temblor Ranges south along the margins of the San Joaquin Valley and north to the Sierra Nevada foothills of Tulare Co.) and two "desert" samples from the Mojave Desert slopes of the Tehachapi Mts. and Antelope Valley east through the deserts of San Bernardino Co. (Fig. 33). Specimens from the eastern parts of the Kern River Plateau (between Weldon and Onyx), Kelso Valley between the Piute and Scodie Mts., and those from the foothills bordering both sides of the Antelope Valley (Kern and Los Angeles Cos.) made up the "unknown" sample. We list locality numbers (from the Appendix), sample size for each dataset (craniodental [n_m], color [n_c], glandes [n_g], and DNA sequence [n_{DNA}]), and museum catalog numbers for all specimens examined.

    Carrizo (total $n_m = 30$, $n_c = 26$, $n_g = 10$, $n_{DNA} = 29$)
    CALIFORNIA:– SAN LUIS OBISPO CO.: (1) CA-38: $n_m=3$, $n_c=3$, $n_{DNA}=3$; MVZ 196759-196761; (2) CA-39: $n_m=1$; USNM 128812; (3) CA-40: $n_m=6$, $n_c=6$, $n_g = 4$, $n_{DNA}=6$; MVZ 196975-195980; (4) CA-41: $n_m=8$, $n_c=8$, $n_g = 3$, $n_{DNA}=8$; MVZ 196967-195974; (5) CA-42: $n_m=5$ $n_c=2$, $n_{DNA}=5$; MVZ 196754-196758; (6) CA-43: $n_m=1$, $n_c=1$, $n_{DNA}=1$; MVZ 195966; (7) CA-44: $n_m=5$, $n_c=5$, $n_g = 2$, $n_{DNA}=5$; MVZ 195961-195965; (8) CA-45: $n_m=1$, $n_c=1$, $n_g = 1$ $n_{DNA}=1$; MVZ 195981.

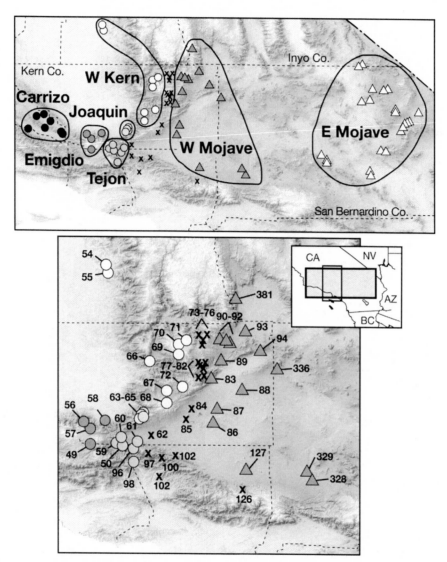

Figure 33. Above – Map of localities included in the Tehachapi Transect (circles = five "coastal" morphological samples; triangles = two "desert" morphological samples; and "✗"s = localities regarded as "unknown" in the morphological analyses [specimens from the eastern end of the Kern River Plateau, Kelso Valley, and margins of the Antelope Valley; see text for explanation]). Below – Localities in the general contact region numbered as in the list of specimens examined (Appendix). Inset – positions of the broader transect and the contact area in southern California.

San Emigdio (total n_m = 14, n_c = 9, n_{DNA} = 7)
        CALIFORNIA:– KERN CO.: (1) CA-56: n_m=1, n_c=1; USNM 31517/43382, USNM 31516 – skin, lectotype of *N. desertorum sola*; (2) CA-57: n_m=3, n_c=3, n_g = 1, n_{DNA}=3; MVZ 198581-198583; (3) CA-58: n_m=1; MVZ 28207; (4) CA-59: n_m=5, n_c=5, n_g = 1, n_{DNA}=4; MVZ 196097-196100, SDNHM 5988. VENTURA CO.: (5) CA-49: n_m=3; MVZ 5331, 5376, 5378; (6) CA-50: n_m=1; MVZ 5333.

Ft. Tejon (total n_m = 46, n_c = 44, n_g = 22, n_{DNA} = 45)
        CALIFORNIA:– KERN CO.: (1) CA-60: n_m=32, n_c=32, n_g = 16, n_{DNA}=32; MVZ 196771-196779, 196809-196821, 200730-200739; (2) CA-61: n_m=1, n_c=2, n_g = 1, n_{DNA}=2; MVZ 196765. LOS ANGELES CO.: (3) CA-96: n_m=6, n_c=5, n_g = 2, n_{DNA}=6; MVZ 196832-196834, 198328-198330; (4) CA-97: n_m=5, n_c=5, n_g = 3, n_{DNA}=5; MVZ 196766, 196835-196836, 198331-198332; (5) CA-98: n_m=2; LACM 55070-55071.

Joaquin Flat (total n_m = 41, n_c = 41, n_g = 16, n_{DNA} = 41)
        CALIFORNIA:– KERN CO.: (1) CA-63: n_m=3, n_c=3, n_{DNA}=3; MVZ 196768-196770; (2) CA-64: n_m=35, n_c=35, n_g = 15, n_{DNA}=35; MVZ 196822-196829, 198584-198596, 200715-200729; (3) CA-65: n_m=3, n_c=3, n_g = 1, n_{DNA}=3; MVZ 196830-196831, 196767.

W Kern River (total n_m = 21, n_c = 23, n_g = 3, n_{DNA} = 16)
        CALIFORNIA:– KERN CO.: (1) CA-66: n_m=3, n_c=5; MVZ 15459-15460, 15462; (2) CA-67: n_m=1, n_c=1; MVZ 60228; (3) CA-68: n_m=5, n_c=5, n_{DNA}=5; MVZ 195912-195216; (4) CA-69: n_m=1, n_c=4; MVZ 15455; (5) CA-70: n_m= 4, n_c=4, n_g = 1, n_{DNA}=4; MVZ 195930-195933; (6) CA-71: n_m=2, n_c=2, n_g = 1, n_{DNA}=2; MVZ 195934-195935; (7) CA-72: n_m=2, n_c=2, n_{DNA}=2; MVZ 197308-197309. TULARE CO.: (8) CA-54: n_m=2; USNM 156651-156652; (9) CA-55: n_m=3, n_{DNA}=3; MVZ 196074-196076; (10) CA-55a; n_g = 1; LACM 63739.

W Mojave (total n_m = 104, n_c = 91, n_g = 25, n_{DNA} = 36)
        CALIFORNIA:– INYO CO.: (1) CA-381: n_m=19, n_c=19, n_g = 9, n_{DNA}=24; MVZ 202459-202483. KERN CO.: (2) CA-83: n_m=21, n_c=11, n_g = 14, n_{DNA}=28; MVZ 199786-199796, 215764-215780; (3) CA-86: n_m=1, n_c=1; MVZ 42465; (4) CA-86a: n_g = 1; CSULB 3015; (5) CA-87; n_m=2, n_c=2; MVZ 26327-26328; (6) CA-87a: n_g = 1; LACM 75421; (7) CA-88: n_c=1; MVZ 103278; (8) CA-89; n_m=10; LACM 75426-75427, 75444-75448, 75451-75453; (9) CA-91: n_m=2, n_c=2, n_g = 1, n_{DNA}=25; MVZ 195264-195265; (10) CA-92: n_m=7, n_c=5, n_g = 3; LACM 63726-63728, MVZ 140500-140502, 143941, 143943-143944, 186336; (11) CA-92A: n_g

= 1; LACM 63721; (12) CA-93: $n_m=1$, $n_c=1$; MVZ 134633; (13) CA-94: $n_m=10$, $n_c=10$, $n_g=2$, $n_{DNA}=10$; MVZ 195266-195275. LOS ANGELES CO.: (14) CA-127: $n_m=1$; MVZ 125887. SAN BERNARDINO CO.: (15) CA-328: $n_m=18$; MVZ 6081, 6084-6092, 6827-6828, 5995, 6006-6007, 6077, 6080; (16) CA-329: $n_m=8$; MVZ 28208, 31434-31439, 31441; (17) CA-329A: $n_g=1$; LACM 29973; (18) CA-335: $n_m=3$, $n_c=3$; MVZ 21035-21037; (19) CA-336: $n_m=1$, $n_c=1$; MVZ 145684.

E Mojave (total $n_m=104$, $n_c=83$, $n_g=14$, $n_{DNA}=39$)
CALIFORNIA:– SAN BERNARDINO CO.: (1) CA-334: $n_m=2$, $n_c=2$; MVZ 65594-65595); (2) CA-346: $n_m=2$, $n_c=2$; MVZ 31425, 31427; (3) CA-347: $n_m=2$, $n_c=3$; MVZ 31431-31433; (4) CA-348: $n_m=7$, $n_c=7$, $n_{DNA}=9$; MVZ 215601-215609); (5) CA-349: $n_m=7$, $n_c=7$, $n_g=6$, $n_{DNA}=7$; MVZ 195313-195319; (6) CA-349a: $n_g=1$; LACM 36954; (7) CA-351: $n_m=1$, $n_c=1$; MVZ 121169; (8) CA-352: $n_m=1$, $n_c=1$, $n_{DNA}=1$; MVZ 195320; (9) CA-353: $n_m=1$, $n_c=3$; MVZ 81957, 93063-93064; CA-353a: $n_g=1$; CSULB 2983; (10) CA-354: $n_m=2$, $n_c=2$; MVZ 196354-196355; (11) CA-355: $n_m=1$; MVZ 81956); (12) CA-356: $n_m=1$, $n_c=1$; MVZ 80250; (13) CA-357: $n_m=19$, $n_c=19$; MVZ 80251-80257, 80259-80270; (14) CA-358: $n_m=1$, $n_c=1$; MVZ 143950; (15) CA-359: $n_m=13$, $n_c=11$; MVZ 80236-80240, 80242-80249; (16) CA-360: $n_m=1$, $n_c=1$; MVZ 81946; (17) CA-361: $n_m=5$, $n_c=6$; MVZ 81950-81955; (18) CA-362: $n_m=2$, $n_c=2$; MVZ 81944-81945); (19) CA-363: $n_m=1$, $n_c=1$; MVZ 81942; (20) CA-364: $n_m=4$, $n_c=4$; MVZ 80230-80233; (21) CA-365: $n_m=1$; MVZ 31418; (22) CA-366: $n_m=5$, $n_c=5$, $n_g=1$, $n_{DNA}=5$; MVZ 195308-195312; (23) CA-367: $n_m=11$, $n_g=4$, $m_{DNA}=17$; MVZ 215580-215596; (24) CA-368: $n_m=1$; MVZ 61182; (25) CA-369: $n_m=1$, $n_c=1$; MVZ 86564; (26) CA-370: $n_m=3$, $n_c=3$; MVZ 86548, 86550, 86552; (27) CA-371: $n_m=1$, $n_c=1$; MVZ 86547; (28) CA-372: $n_m=3$, $n_c=3$; MVZ 86533-86534, 86558; (29) CA-372a: $n_g=1$; CSULB 10541; (30) CA-373: $n_m=3$, $n_c=2$; MVZ 86545, 93060, 93062; (31) CA-374: $n_m=1$, $n_c=1$; MVZ 86546; (32) CA-375; $n_m=1$, $n_c=1$; MVZ 86544.

unknown (total $n_m=123$, $n_c=119$, $n_g=38$, $n_{DNA}=83$)
CALIFORNIA:– KERN CO.: (1) CA-62: $n_m=3$, $n_c=3$, $n_g=1$, $n_{DNA}=3$; MVZ 196762-196763, 196837; (2) CA-73: $n_m=7$, $n_c=19$; MVZ 15467-15470, 15472-15474, 15478-15481; (3) CA-74: $n_m=4$, $n_c=4$, $n_g=2$, $n_{DNA}=4$; MVZ 195919-195922; (4) CA-75: $n_m=7$, $n_c=7$, $n_{DNA}=2$; MVZ 15454, 15483, 15485, 15491, 15494, 195917-195918; (5) CA-76: $n_m=6$, $n_c=6$, $n_g=2$, $n_{DNA}=6$; MVZ 195923-195929; (6) CA-77: $n_m=6$, $n_c=6$, $n_g=2$, $n_{DNA}=6$; MVZ 199797-199802; (7) CA-78: $n_m=10$, $n_c=12$; MVZ 60229-60240; (8) CA-79: $n_m=10$, $n_c=10$, $n_g=3$, $n_{DNA}=10$; MVZ 199772-199781; (9) CA-80: $n_m=33$ $n_c=27$, $n_g=21$, $n_{DNA}=41$; MVZ 202496-202500, 202502-202504, 202507-202517, 202519-202922, 215781-215786, 21796-215803; (10) CA-81: $n_m=4$, $n_c=4$, $n_g=2$, $n_{DNA}=4$; MVZ 199782-199785;

(11) CA-82: n_m=3, n_c=3; MVZ 15506, 60241-60243; (12) CA-84: n_m=7; USNM 136032-136033, 136035-136039; (13) CA-85: n_m=4, n_c=4, n_g = 3, n_{DNA}=4; MVZ 197310-197313); (14) CA-90: n_m=4, n_c=4; MVZ 15457, 15496-15497, 15504. LOS ANGELES CO.: (15) CA-99: n_m=3, n_g = 1, n_{DNA}=1; MVZ 198353-198354, 198579; (16) CA-100: n_m=1, n_{DNA}=1; MVZ 198580; (17) CA-101: n_m=8; MVZ 5370-5373,5383-5384, 6967-6968; (18) CA-102: n_m=2, n_g = 1, n_{DNA}=1; MVZ 198577-198578; (19) CA-126: n_m=1; MVZ 42464; (20) CA-127a: n_g =1; LACM 36953.

Habitat.—Woodrats of the *Neotoma lepida* group that occur in the western half of the Tehachapi Transect live in the more arid habitats of this region, ranging from coastal scrub and chaparral to dry rock outcrops. Here, individuals typically build nests at the base of clumps of Our Lord's Candle (*Yucca whipplei*; Figs. 34 and 35) or in the interstices of rock exposures. Animals in the Mojave Desert construct nests in rock outcrops composed of granite boulders or basalt flows, but also are commonly found on the desert floor in nests constructed at the base of Joshua Tree (*Yucca brevifolia*) and Mojave Yucca (*Yucca schidigera*); see Fig. 36. Contact areas between individuals of the two coastal mtDNA clades, 1B and 1C are in typical dry scrub habitats within the oak woodland (Fig. 37). Where mtDNA clades 1C and 2A are in sympatry, at Joaquin Flat in the Tehachapi Mts. (locality CA-62), the habitat is a complex of granite boulders exposed above open grasslands at the lower edge of the blue oak woodland (Fig. 38). Where "coastal" and "desert" morphology individuals meet in Kelso Valley (locality CA-80), it is a mixture of coastal oak scrub and Mojave Desert Joshua Tree woodland (Fig. 39).

Figure 34. Elkhorn Hills, San Luis Obispo Co. (locality CA-41), looking west across the Carrizo Plain to the Caliente Range; Our Lord's Candle (*Yucca whipplei*) and White Sage (*Salvia apiana*) in the foreground are characteristic of the "coastal" morphological type of the desert woodrat. Photo taken in October 2000.

Figure 35. Nest of a *Neotoma lepida* constructed at the base of an Our Lord's Candle, Elkhorn Plain Ecological Reserve, Elkhorn Hills, San Luis Obispo Co. (locality CA-41). Photo taken in October 2000.

Figure 36. Desert woodrat house constructed at the base of a Mojave Yucca at Halloran Spring, San Bernardino Co. (locality CA-366). Photo taken in July 2000.

Figure 37. Open hillside 1.5 mi SE Ft. Tejon (locality CA-60) with California Juniper, Our Lord's Candle, and White Sage. Representatives of mtDNA clades 1B and 1C co-occur here. Photo taken in May 2001.

Figure 38. Joaquin Flat, Tehachapi Mts. (locality CA-64) where individuals of mtDNA clades 1C and 2A co-occur. Photo looking west to San Joaquin Valley; taken in March 2001.

Figure 39. Western margin of Kelso Valley (locality CA-80), where individuals of the "coastal" and "desert" morphological groups co-occur. Habitat is a mixture of interior California woodland and Mojave desert scrub. Photo taken in October 2003.

Morphometric differentiation.— Descriptive statistics for all variables for both groups are given in Table 12. Each of the 4 external and 21 craniodental characters exhibit significant differences among the seven samples that make up the Tehachapi Transect, based on one-way ANOVAs. In comparisons between each pair of geographically adjacent samples, there are no significant differences between the two samples of the "desert" group (W and E Mojave, Fig. 33). Alternatively, there are 11 significant character differences between the Joaquin and W Kern samples of the "coastal" group that extend northwards along the slopes of the Tehachapi, Breckenridge, and Greenhorn mountains that border the southeastern margin of the San Joaquin Valley, but no other adjacent pair of "coastal" groups differ by more than 4 variables. In contrast, all 25 external and craniodental characters are highly significantly ($p < 0.001$) different in comparisons between the "coastal" Joaquin or W Kern samples and the W Mojave sample of the "desert" group. These observations are completely consistent with the assignment of individuals to either "coastal" or "desert" morphological groups by more global set of univariate and multivariate comparisons presented above.

Because the "coastal" and "desert" morphological groups are readily separable by most univariate mensural characters, it is not at all surprising that these groups are also well defined by both PCA and CVA analyses. The first two PC axes explain 67.3 percent of the total pool of variation and are the only axes where mean scores of the "coastal" and "desert" morphological groups are significantly different ($p < 0.0001$ in both cases; ANOVA, $F_{(1,273)} = 219.142$ and $F_{(1,273)} = 230.979$, respectively, in the comparison between "coastal" and "desert" group PC-1 and PC-2 scores). Thus, subsequent components, each accounting for no more than 4.5 percent of the total variance, provide no additional insights. As with the global comparison between "coastal" and "desert" morphological groups, above, all variables except BUL and BUW load positively and reasonably uniformly on PC-1 (Table 13) while the two bullar measurements are most important on PC-2. Not surprisingly, therefore, the two morphological groups do not overlap on the combination of both axes, and the "unknown" individuals largely fall into either the "coastal" or "desert" groups (Fig. 40).

Table 12. External and cranial measurements of adult (age classes 1-5) specimens of the "coastal" and "desert" samples of *Neotoma lepida* along the Tehachapi Transect (from San Luis Obispo Co. east through San Bernardino Co.; Fig. 33). Sample mean, standard deviation, sample size, and range are given for each pooled sample.

Variable	Carrizo	Emigdio	Ft. Tejon	Joaquin Flat	W Kern	W Mojave	E Mojave
external							
TOL	334.7±3.545	318.3±3.723	325.9±2.271	333.9±2.296	319.4±3.216	293.2±2.066	293.3±1.791
	22	13	36	31	22	63	71
	307-363	291-336	305-368	313-375	282-347	255-330	261-334
TAL	159.6±1.976	149.7±2.329	154.1±1.49	162.7±1.476	153.7±1.48	129.9±1.264	128.8±1.201
	22	14	36	31	22	63	71
	143-174	135-162	133-177	149-190	138-166	110-154	110-160
HF	33.8±0.265	32.7±0.377	33.73±0.175	34.1±0.174	34.0±0.247	31.6±0.155	30.6±0.175
	25	15	40	36	22	65	73
	31-36	31-36	30-35	32-36	32-36.5	28-34	28-36
E	33.21±0.255	31.3±1.048	30.9±0.155	32.0±0.174	31.5±0.535	30.0±0.407	27.9±0.284
	24	8	40	36	20	45	73
	31-36	25-34	29-33	30-34	25-35	23-34	22-33
craniodental							
CIL	39.73±0.335	39.22±0.277	39.54±0.198	39.73±0.193	38.45±0.289	37.69±0.157	37.22±0.172
	25	15	40	36	20	67	73
	36.89-43.00	36.88-40.95	37.21-42.83	37.47-42.74	36.52-41.25	35.04-40.95	34.33-41.9
ZB	22.25±0.449	21.68±0.179	22.02±0.118	21.86±0.117	21.15±0.151	20.77±0.086	20.35±0.098
	25	15	40	36	20	67	73
	20.65-32.10	20.63-22.80	19.99-23.68	20.44-23.17	20.28-22.78	19.07-22.73	18.64-22.42

Table 12 (continued)

Variable	Carrizo	Emigdio	Ft. Tejon	Joaquin Flat	W Kern	W Mojave	E Mojave
IOC	5.33±0.037	5.41±0.062	5.59±0.036	5.44±0.029	5.51±0.045	5.05±0.023	5.06±0.020
	25	15	40	36	21	67	73
	5.02-5.71	5.00-5.78	5.07-6.09	4.92-5.84	5.15-5.95	4.67-5.55	4.74-5.49
RL	16.47±0.148	16.18±0.15	16.44±0.098	16.58±0.085	16.00±0.171	15.37±0.080	15.17±0.077
	25	15	40	36	21	67	73
	15.37-18.15	15.17-17.26	15.34-18.15	15.63-17.96	14.85-17.74	13.90-17.12	13.74-17.1
NL	15.91±0.164	15.53±0.193	15.68±0.088	15.79±0.088	15.51±0.153	14.94±0.081	14.87±0.077
	25	15	40	36	22	67	73
	14.79-17.74	14.45-17.56	14.66-17.1	14.80-17.75	14.48-17.14	13.67-16.89	13.38-16.24
RW	6.32±0.045	6.37±0.076	6.59±0.039	6.44±0.039	6.57±0.0780	6.25±0.027	6.22±0.028
	25	15	40	36	22	67	73
	5.83-6.74	5.80-6.96	6.06-7.22	5.99-7.05	6.02-7.78	5.58-6.73	5.80-6.80
OL	14.35±0.113	14.08±0.077	14.19±0.082	14.18±0.057	14.12±0.079	13.69±0.054	13.49±0.056
	25	15	40	36	21	67	73
	13.21-15.3	13.62-14.69	13.23-15.37	13.27-14.90	13.7-14.92	12.73-15.01	12.67-14.85
D	11.37±0.16	11.22±0.117	11.38±0.089	11.52±0.098	11.05±0.145	10.88±0.071	10.80±0.081
	25	15	40	36	22	67	73
	10.01-12.90	10.16-11.77	10.16-12.87	10.17-13.29	10.11-12.52	9.80-12.40	9.44-13.14
MTRL	8.40±0.04	8.31±0.068	8.25±0.048	8.30±0.036	8.16±0.042	7.99±0.033	7.89±0.031
	25	15	40	36	22	67	73
	7.95-8.73	7.79-8.89	7.50-8.81	7.73-8.65	7.71-8.56	7.26-8.59	7.22-8.44
IFL	8.79±0.106	8.78±0.086	8.87±0.071	8.95±0.068	8.58±0.103	8.30±0.047	8.27±0.052
	25	15	40	36	22	67	73
	7.92-9.83	7.98-9.28	7.93-9.90	7.50-9.77	7.72-9.38	7.44-9.36	7.25-9.59

Table 12 (continued)

Variable	Carrizo	Emigdio	Ft. Tejon	Joaquin Flat	W Kern	W Mojave	E Mojave
PBL	18.12±0.177	17.88±0.169	17.99±0.102	18.23±0.092	17.66±0.140	17.35±0.074	17.22±0.090
	25	15	40	36	22	67	73
	16.58-19.77	16.23-18.91	16.75-19.89	17.16-19.78	16.82-18.99	15.80-18.73	15.57-19.5
AW	7.66±0.052	7.48±0.065	7.58±0.037	7.79±0.046	7.69±0.054	7.00±0.028	6.96±0.031
	25	15	40	36	22	67	73
	7.16-8.12	7.15-8.17	7.11-8.40	6.70-8.35	7.13-8.29	6.46-7.56	6.42-7.89
OCW	9.73±0.072	9.49±0.079	9.57±0.040	9.56±0.043	9.52±0.061	9.11±0.037	8.94±0.030
	25	15	40	36	22	67	73
	9.01-10.51	9.03-10.21	9.07-10.31	9.07-10.27	8.64-9.95	8.52-10.03	8.39-9.53
MB	17.51±0.101	17.21±0.096	17.43±0.061	17.59±0.076	17.14±0.093	17.00±0.055	16.73±0.058
	25	15	40	36	21	67	73
	16.64-18.37	16.53-17.72	16.75-18.19	16.43-18.41	16.57-18.26	16.06-18.26	15.54-18.13
BOL	6.02±0.076	5.80±0.073	6.12±0.054	6.04±0.046	5.87±0.091	5.54±0.040	5.47±0.040
	25	15	40	36	22	67	73
	5.33-6.77	5.29-6.19	5.54-7.29	5.53-6.74	5.33-6.77	4.95-6.49	4.70-6.33
MFL	7.85±0.079	7.73±0.089	7.82±0.048	7.91±0.048	7.65±0.084	7.67±0.048	7.72±0.044
	25	15	40	36	21	67	73
	7.10-8.43	6.76-8.17	7.29-8.55	7.37-8.50	7.07-8.47	7.17-8.84	6.82-8.47
MFW	2.76±0.036	2.67±0.049	2.61±0.033	2.78±0.040	2.81±0.046	2.33±0.025	2.36±0.024
	25	15	40	36	21	67	73
	2.35-3.02	2.31-2.91	2.19-3.05	2.32-3.34	2.47-3.19	1.94-2.80	1.93-3.00
ZPW	4.12±0.046	4.01±0.057	4.10±0.034	4.29±0.039	4.10±0.069	4.20±0.029	4.06±0.025
	25	15	40	36	21	67	73
	3.65-4.52	3.66-4.36	3.68-4.66	3.93-4.89	3.71-4.85	3.79-4.81	3.48-4.66

Table 12 (continued)

Variable	Carrizo	Emigdio	Ft. Tejon	Joaquin Flat	W Kern	W Mojave	E Mojave
CD	15.47±0.094	15.41±0.102	15.59±0.077	15.62±0.05	15.26±0.078	15.63±0.047	15.37±0.05
	25	15	40	36	21	67	73
	14.76-16.39	14.87-16.36	14.92-16.99	15.09-16.27	14.76-16.01	14.52-16.75	14.56-16.57
BUL	6.98±0.043	6.69±0.059	6.71±0.029	6.72±0.029	6.65±0.051	7.22±0.034	7.19±0.032
	25	15	40	36	21	67	73
	6.48-7.34	6.35-7.10	6.14-7.19	6.36-7.13	6.38-7.25	6.54-7.95	6.62-7.88
BUW	7.11±0.046	7.10±0.05	6.92±0.034	7.05±0.033	7.04±0.045	7.46±0.022	7.43±0.030
	25	15	40	36	21	67	73
	6.66-7.49	6.76-7.54	6.43-7.44	6.49-7.47	6.61-7.47	7.00-7.94	6.97-8.03

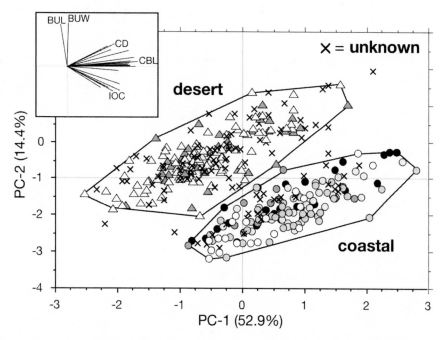

Figure 40. Scatterplot of individual scores on the first two principal components axes. Circles identify individuals with a "coastal" morphology, those that occur from the Carrizo Plains to the eastern margins of the Tehachapi Mts.; triangles are specimens of the "desert" morphology, from localities in the Mojave Desert east of the Tehachapi Mts. Filling of both circles and triangles is keyed to the geographic samples mapped in Fig. 33; "✘"'s indicate the "unknown" specimens from the eastern part of the Kern River plateau, vicinity of Kelso Valley, and the foothills bordering the western Antelope Valley. The inset box illustrates character vectors along both axes, which contrasts the highly positive character vectors for all variables exclusive of those of the bulla (BUL and BUW) on the 1st axis with the strongly positive bullar dimensions on the 2nd.

Table 13. Principal component factor loadings and standardized coefficients from the canonical variates analysis for log-transformed cranial variables of the "coastal" and "desert" morphological groups of the Tehachapi Transect.

Variable	PC-1	PC-2	CAN-1
log CIL	0.963	0.096	-0.64295
log ZB	0.873	-0.004	-0.03584
log IOC	0.623	-0.491	-0.27138
log RL	0.928	-0.005	-0.50191
log NL	0.845	0.070	0.11719
log RW	0.695	-0.085	0.08183
log OL	0.855	0.035	-0.21198
log DL	0.837	0.318	0.10197
log MTRL	0.455	-0.416	-0.14759
log IFL	0.820	0.043	-0.02809
log PBL	0.903	0.177	0.32255
log AW	0.713	-0.474	-0.31724
log OCW	0.717	-0.346	-0.26005
log MB	0.860	0.114	-0.24735
log BOL	0.851	0.009	-0.23318
log MFL	0.592	0.335	0.32997
log MFW	0.553	-0.419	-0.31280
log ZPW	0.541	0.403	0.20371
log CD	0.634	0.446	0.49721
log BUL	-0.095	0.828	0.43829
log BUW	0.011	0.866	0.64572
eigenvalue	11.111	3.013	11.0044
% contribution	52.91	14.35	100.00

Plots of PC scores and mean longitudinal position of grouped samples illustrates the shift in both overall size (as indexed by PC-1) and bullar size (PC-2 scores) across the transect (Fig. 41). No pairs of the five "coastal" or two "desert" samples differ significantly in either PC-1 or PC-2 scores. However, as noted above for individual craniodental variables, highly significant differences are present for both sets of scores in comparisons between the eastern-most "coastal" samples (Joaquin and W Kern) and the western-most "desert" sample (W Mojave). Thus, there is a clear and sharp step in character transition, measured by either univariate or multivariate means, along the western margins of the Mojave Desert, the general area of contact between "coastal" and "desert" morphotypes of the *Neotoma lepida* group. The characters that exhibit these sharp transitions are the same as those identified in our global comparisons among all samples of these woodrats in the US.

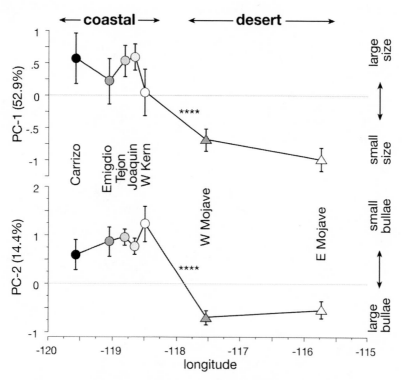

Figure 41. Means (and 95% confidence limits) of PC-1 (above) and PC-2 (below) scores across the Tehachapi Transect, ordered from west to east by the mean longitudinal position of the seven grouped samples. Grouped samples are identified by name and by symbols, as above. Significant differences (**** = p < 0.0001) are present only between eastern-most "coastal" and western-most "desert" samples.

We also performed a canonical analysis with the "coastal" and "desert" morphological samples as pre-defined groups, with the results completely concordant with both univariate and PCA analyses. The variables logCIL and logBUW weigh most heavily, and in opposite directions, on the single CAN axis obtained in the analysis (Table 13). There is nearly complete separation of the two groups, with only three of the 279 individuals of the predefined "coastal" and "desert" groups overlapping (Fig. 42). Both groups are significantly different ($F_{(21,254)} = 133.101$, $p < 0.0001$; mean squared Mahalanobis distance = 44.038). Despite the overlap of three individuals, the a posteriori classification of each specimen relative to its respective a priori group is 100%, with posterior probabilities of the 138 "coastal" individuals of membership to their group always above 0.995 and, in all but three cases, above 0.999. All but one of the 140 "desert" individuals have posterior probabilities of membership to their group of 0.995 or above, and the one that is below this threshold has a probability of 0.925.

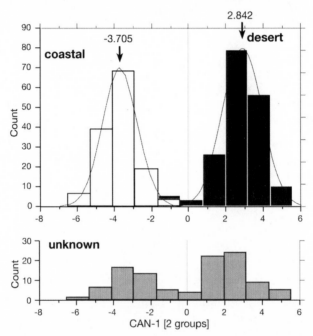

Figure 42. Histograms of canonical variate scores: Top – scores for the "coastal" and "desert" pre-defined morphological groups (see Fig. 33). Mean CAN-1 scores are given for each group. Bottom – distribution of scores for individuals grouped as "unknown" from the contact region in the eastern part of the Kern River Plateau, Kelso Valley, and foothill margins of the Antelope Valley (Fig. 33).

The a posteriori scores of the "unknown" specimens are bi-modally distributed, with each peak similar in position to the means of the two pre-defined morphological groups (Fig. 42). Similarly, and with only two exceptions, each "unknown" specimen is unambiguously associated with either the pre-defined "coastal" or "desert" groups, always with posterior probabilities > 0.98. The two exceptional "unknown" specimens with intermediate posterior probabilities are MVZ 60233 (from Kelso Valley [near locality CA-80]), which is almost exactly intermediate between the two morphological groups (probabilities of membership to "coastal" individuals of 0.562 and to the "desert" group, 0.438), and MVZ 196763 (from Pescadero Creek, on the south side of the Tehachapi Mts. [locality CA-62]), which is more similar to "coastal" animals, with a probability of 0.822 to that group and 0.178 to the "desert" group. The morphological intermediacy of these two individuals suggests that limited hybridization may occur at the points of contact in Kelso Valley and along the southwestern margins of the Tehachapi Mts., a possibility we examine in greater detail below using a suite of molecular microsatellite markers. For comparison to other transects described below, we illustrate the degree of morphological separation of the two pre-defined groups as well as all "unknown" individuals with a scatterplot of their posterior probabilities of group membership and scores on the single canonical axis (Fig. 43). This nicely illustrates the intermediate positions of the single specimens from Kelso Valley (MVZ 60233) and Pescadero Creek (MVZ 196763) relative to the otherwise widely separable pre-defined groups and strong assignments of all other "unknown" specimens to one or the other of those two groups.

With the exception of the two intermediate specimens, all others from areas of contact or near contact between the "coastal" and "desert" morphological groups of the Tehachapi Transect segregate clearly into one morphological group or the other (Fig. 44). However, individuals of both morphological types do co-occur at several specific localities, especially in the vicinity of Kelso Valley on the eastern side of the Piute Mts. (NW Kelso Valley [locality CA-78; one "coastal" and eight "desert" individuals in addition to the single intermediate specimen], Whitney Well [locality CA-80; 12 "coastal" and one "desert"], Schoolhouse Well [locality CA-81; two "coastal" and one "desert"], and Sorrell's Ranch [CA-82; one "coastal" and one "desert"]). Of those specimens collected at locality CA-78, the single "coastal" animal and a "desert" individual were trapped at the same nest on successive nights in November of 1933 (field notes of D. S. MacKay; MVZ archives). The habitat along the western margins of Kelso Valley grades sharply from coastal scrub/woodland vegetation to western Mojave desert scrub (Fig. 39). It is exactly at the ecotone between these vegetation types where "coastal" and "desert" morphological types of woodrats are found in syntopy. At Pescadero Creek (locality CA-62), near the southwestern end of the Tehachapi Mts., the single intermediate specimen was trapped with a "desert" individual. This locality is the western-most patch of Joshua Tree, a diagnostic component of Mojave desert scrub, along the southern margins of the mountain range, juxtaposed with the granite boulder outcrops and coastal scrub habitat typical of "coastal" samples (Fig. 37 and 38). The two morphological types of woodrats thus marginally overlap geographically and in habitat along the eastern fringes of the Tehachapi and Piute mountains, and it is likely that an occasional hybrid individual is produced here, as suggested by the two morphologically intermediate individuals. Further evidence is provided by the analysis of molecular markers we present below.

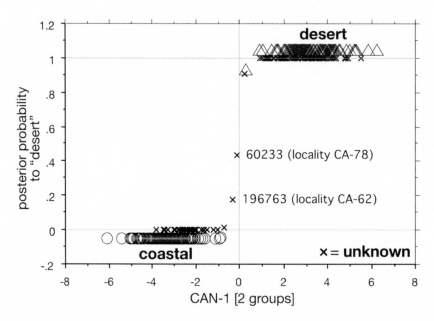

Figure 43. Plot of the posterior probability of membership to the "desert" morphological group from the Mojave Desert for each specimen examined in Tehachapi Transect relative to the score of that individual on the first CAN axis. Points for both pre-defined groups are deliberately offset from the "0" and "1" lines for ease in comparing the distribution of each group and the "unknown" individuals. Note that all individuals of both pre-defined groups have very high posterior probabilities to their respective groups, while individuals considered as "unknown" include a large number also belonging to one or the other of these two groups along with two specimens (MVZ 60233 and MVZ 196763) that are morphologically intermediate, at least as suggested by their intermediate posterior probabilities.

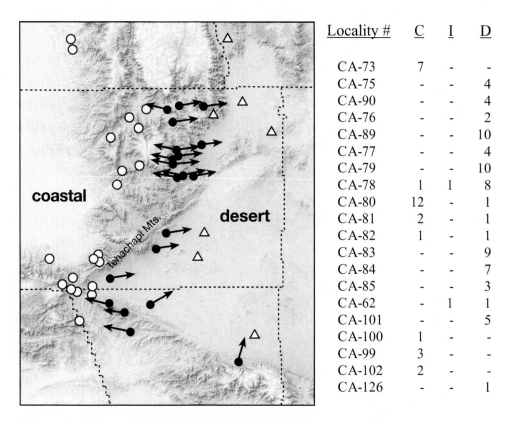

Locality #	C	I	D
CA-73	7	-	-
CA-75	-	-	4
CA-90	-	-	4
CA-76	-	-	2
CA-89	-	-	10
CA-77	-	-	4
CA-79	-	-	10
CA-78	1	1	8
CA-80	12	-	1
CA-81	2	-	1
CA-82	1	-	1
CA-83	-	-	9
CA-84	-	-	7
CA-85	-	-	3
CA-62	-	1	1
CA-101	-	-	5
CA-100	1	-	-
CA-99	3	-	-
CA-102	2	-	-
CA-126	-	-	1

Figure 44. Left – map of the region of contact between "coastal" (open circles) and "desert" (open triangles) morphological groups in the general region of the Tehachapi Mts., and the assignments of "unknown" individuals (solid circles) to either of these two groups (directional arrows). Right – table of assignments of each "unknown" locality to "coastal" (C), "desert" (D), or intermediate (I), based on posteriori probabilities (see text).

Color variation.—We organized samples for colorimetric analysis along the Tehachapi Transect into the same seven geographic groupings used for craniodental variation, including the same "unknown" localities. Based on our global analysis of color variation and the relationship among color traits for all sampled individuals (see Colorimetric Analysis, above), we limit our analysis to the trichromatic X-coefficients for the dorsal, tail, lateral, and chest regions of the study skin. There is a significant relationship between an individual's X-value for all four topographic regions of the skin, with correlation coefficients ranging from relatively weak (Tail-X versus Chest-X; r = 0.133, Z-value = 2.574; p = 0.01) to quite strong (Dorsal-X versus Lateral-X; r = 0.509, Z-value = 10.837; p < 0.0001). Thus, while certainly not perfect, the color of all parts of the external fur of these woodrats is related in a general way and changes in one region of the skin are reflected by similar changes in others.

Each of the four X-coefficients exhibits highly significant differentiation among these samples, but variation across the transect is complex (Table 14; one-way ANOVA, $F_{(7,368)}$, p < 0.0001 in all cases). Color is pale in the western-most sample (Carrizo sample), becomes progressively darker in a steep cline around the southern end of the San Joaquin Valley (San Emigdio) and into in the foothills to the immediate east (Tejon, Joaquin, and W Kern samples), and then becomes markedly paler again in the shift to the two samples from the Mojave Desert (W Mojave and E Mojave). The geographically adjacent pairs of San Emigdio-Ft. Tejon and Ft. Tejon-Joaquin Flat differ significantly for a single colorimetric variable (Dorsal-X and Chest-X, respectively; ANOVA, Fisher's PLSD post-hoc test, p < 0.01 in each case), and the two Mojave Desert samples also differ in a single variable (Dorsal-X, p = 0.0018). However, in the comparisons between adjacent "coastal" and "desert" samples (Joaquin Flat or W Kern River versus W Mojave), all four variables exhibit highly significant differences with 0.001 > p < 0.0001. The Carrizo sample is nearly as pale as either of those of the Mojave Desert.

We used principal components analysis to summarize colorimetric variation along the transect, with the four X-coefficients as the included variables (Table 14). The first axis is the only one with an eigenvalue greater than 1; it explains 48.2% of the total pool of variation present in the sample (Table 15). All four trichromatic X variables load highly and evenly on this axis, and all four are significantly (p < 0.0001 in all cases) and negatively correlated with their respective PC-scores (r-values range from -0.555 [PC-1 versus Chest-X, Z-value = -12.083] to -0.816 [PC-1 versus Dorsal-X, Z-value = -22.085]). Thus, variation along PC-1 expresses primarily the degree of darkness (positive PC-1 scores) or paleness (negative PC-1 scores) in individuals across all four topographic regions of the skin. Separation of samples on the second axis is due primarily to reciprocal differences in Tail-X and Chest-X, with the former becoming paler from the coast to the desert but the latter expressing the opposite trend; that is, paler on the coast to darker inland. These colorimetric trends along the Tehachapi Transect are illustrated by the bivariate relationship of PC-1 scores and latitude (Fig. 45).

Table 14. Descriptive statistics for the colorimetric X-measurement for the four regions of the woodrat study skins. Means ± one standard error, sample sizes, and ranges are given for each of five pooled geographic samples along the Tehachapi Transect (see text for the rationale behind and membership in each group).

Sample	Dorsal-X	Tail-X	Lateral-X	Chest-X
Carrizo	11.46±0.29	11.86±0.39	29.35±0.76	49.29±1.25
	26	26	26	26
	8.4–14.3	7.6-17.1	23.2-37.1	38.1-62.1
San Emigdio	10.65±0.75	9.99±0.76	26.20±1.71	44.25±2.22
	9	9	9	9
	6.1-14.5	6.5-14.0	15.1-30.4	35.8-58.2
Ft. Tejon	8.50±0.17	8.13±0.25	23.00±0.43	46.26±1.29
	43	43	43	43
	6.4-11.1	5.3-13.2	18.0-28.5	9.5-60.7
Joaquin Flat	7.91±0.18	7.33±0.27	21.61±0.52	42.01±0.79
	41	27	27	27
	5.1-10.2	4.5-11.9	12.9-29.3	30.2-59.0
W Kern River	9.15±0.25	8.87±0.40	21.53±0.73	43.31±0.79
	25	25	25	25
	7.31 – 11.29	5.6-14.8	12/9-29.1	35.2–50.9
W Mojave	13.96±0.52	9.92±0.14	31.15±0.60	48.08±0.99
	26	26	26	26
	8.3-19.8	4.8-15.8	25.2-37.9	38.2-61.8
E Mojave	12.39±0.24	8.80±0.26	32.05±0.47	46.94±0.71
	91	91	91	91
	6.6-16.9	4.2-14.8	21.3-43.8	29.1-60.0

Table 15. Principal component eigenvalues and factor loadings of colorimetric variables from all samples of the Tehachapi Transect.

Variable	PC-1	PC-2	PC-3
Dorsal-X	0.816	-0.103	-0.238
Tail-X	0.597	-0.613	0.496
Lateral-X	0.772	0.083	-0.460
Chest-X	0.555	0.695	0.457
eigenvalue	1.926	0.876	0.723
% contribution	48.2	21.6	21.6

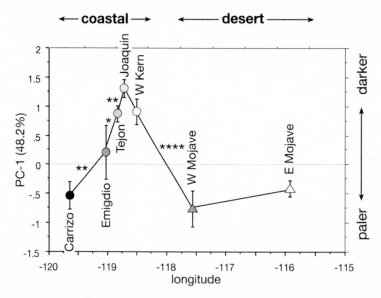

Figure 45. Means and 95% confidence limits for variation in the first principal component axis (PC-1), comparing the seven geographically grouped samples across the Tehachapi Transect. Significance levels between adjacent samples, based on ANOVA (Fisher's PLSD post-hoc tests), are indicated: $* = p < 0.05$, $** = p < 0.01$, and $**** = p < 0.0001$).

We performed a canonical analysis to determine the degree to which the colorimetric variables can identify the group membership of individuals, particularly those from the contact areas in the Kern River plateau, Kelso Valley, and margins of the Antelope Valley. We excluded the pale Carrizo and San Emigdio samples and grouped the remaining samples as either "coastal" or "desert" based on their craniodental characters. These two groups are strongly different (Mahalanobis $D^2 = 11.658$, $F_{(4,221)} = 160.824$, $p < 0.0001$) and correct classification of individuals to group is nearly perfect (nine of 226 [4%] are misclassified). However, a general overlap of CAN scores compromises the ability of color alone to allocate any individual to one group or the other. We illustrate this overlap in a plot of the posterior probability of membership in the "desert" group relative to an individual's CAN score (Fig. 46), a diagram that contrasts with the strong separation of these same individuals based on craniodental variables (Fig. 43). Moreover, a large number of the "unknown" individuals exhibit intermediate posterior probabilities (between 0.2 and 0.8 relative to the "desert" group) so that their individual assignments are ambiguous at best. Hence, while there is a very good ability to distinguish color between coastal and desert samples, there appears to be sufficient background clinal variation likely due to increasing aridity from west to east to obfuscate the color assignments of any individual specimen to one group or the other, particularly in the transitional area across the Tehachapi Mts. and the Piute and Scodie Mts. immediately to the north.

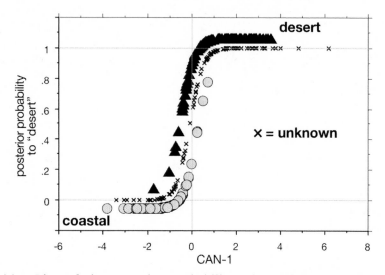

Figure 46. Plot of the posterior probability of membership in the "desert" colorimetric group for each specimen examined in the Tehachapi Transect (excluding the Carrizo and San Emigdio samples) relative to the score of that individual on the single CAN axis. Points for both pre-defined groups are deliberately offset from the "0" and "1" lines for ease in comparing the distribution of each group and the "unknown" individuals.

Morphological – mtDNA clade concordance.—Specimens belonging to three mtDNA clades are found among the samples of the Tehachapi Transect (Fig. 47). Individuals with haplotypes of the coastal 1C subclade are present at all localities from the vicinity of the Carrizo Plains east along the southern margins of the San Joaquin Valley, across Tejon Pass, and into the western margins of the Tehachapi Mts. (localities CA-38, CA-40-CA-45, CA-57, CA-59-CA-60, CA-64, and CA-97). Individuals belonging to subclade 1B occur at four localities in the vicinity of Tejon Pass (localities CA-60-CA-61 and CA--CA-97), co-occurring with subclade 1C individuals at Ft. Tejon (locality CA-60) and east of Gorman (locality CA-97). Specimens with haplotypes of the desert subclade 2A are found at the same localities as those of the coastal subclade 1C at Joaquin Flat, on the western slope of the Tehachapi Mts. (locality 21) as well as at two localities along the southwestern margins of the Antelope Valley (east of Three Points [locality CA-100] and near Red Mountain [locality CA-102]). Otherwise, this subclade is distributed from the western foothills of the Sierra Nevada and Tehachapi Mts. east throughout the Mojave Desert. The distribution of subclade 2A individuals, in particular, is discordant with the morphological group membership of these same specimens (compare Fig. 47 to Figs. 33 and 44). As noted above, all individuals in the areas of contact between "coastal" and "desert" morphologies (the eastern end of the Kern River Plateau and Kelso Valley) have haplotypes of the desert subclade 2A. We thus performed separate morphometric analyses grouping individuals by their mtDNA clade, to determine more explicitly their morphological relationships relative to clade membership.

Individuals of subclades 1C and 1B overlap on the first two principal components axes (Fig. 48) and are distributed within the bivariate space occupied by "coastal" morphology individuals (compare to Fig. 40). Scores on both the 1st and 2nd PC axes cannot discriminate these two groups (PC-1 – ANOVA, $F_{(1,95)}$ = 0.004, p = 0.9693; PC-2 –ANOVA, $F_{(1,95)}$ = 2.785, p = 0.0984). On the other hand, specimens belonging to subclade 2A are broadly distributed across the scatterplot, with some individuals overlapping the distributions of both subclades 1C and 1B and others occupying the multivariate space of the "desert" morphological group (compare to Fig. 40).

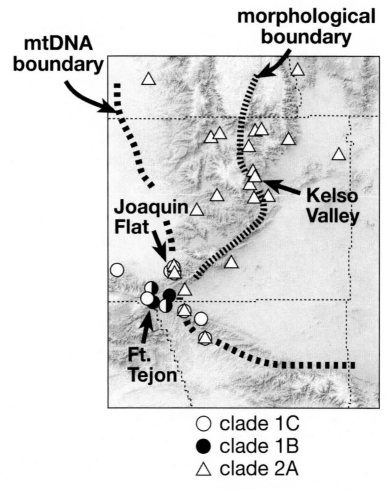

Figure 47.   Sample localities for haplotypes of three mtDNA clades from the contact region of the Tehachapi Transect.   Open circles identify individuals with haplotypes of the coastal subclade 1C, solid circles are those with haplotypes of the coastal subclade 1B, and open triangles are those with haplotypes of the desert subclade 2A.   Overlapping symbols indicate areas of clade co-occurrence (clades 1B and 1C, Ft. Tejon [CA-60]; clades 1C and 2A, Joaquin Flat [CA-64]).   The general area of Kelso Valley is also identified (see text for further details).   The approximate positions of the boundaries between the morphological "coastal" and "desert" groups and the mtDNA clades 1 and 2 are also indicated.

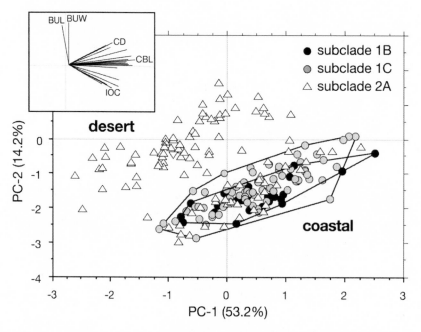

Figure 48. Scatterplot of individual scores on the first two principal components axes for those specimens of the Tehachapi Transect that were sequenced for their mtDNA. Note that specimens of the coastal subclades 1B (black circles) and 1C (gray circles) overlap in their respective distributions, while some individuals of the desert subclade 2A (open triangles) overlap with those of the coastal subclades and others occupy their own multivariate space. Compare this distribution to that depicted in Fig. 40, where samples are organized by their morphological group membership.

Results of a canonical variates analysis, where specimens are pre-grouped by their respective mtDNA clades, are completely concordant with the PCA, both in the morphological distinction of "coastal" and "desert" animals and in emphasizing the discordance between morphology and mtDNA across the Tehachapi Mts. (Fig. 47). The first canonical axis explains 88.2% of the variation. Specimens of the two coastal subclades 1B and 1C exhibit indistinguishable unimodal distributions (Fig. 49; mean CAN -1 scores -1.950 and -1.229, respectively; ANOVA, $F_{(1,95)} = 3.554$, p = 0.0625). CAN-1 scores of individuals of subclade 2A, however, when placed into their "coastal" and "desert" morphological groups, are bimodally distributed with their means significantly different (Fig. 49; mean CAN-1 scores 0.038 and 1.941, respectively; ANOVA,

$F_{(1,61)}$ = 97.665, p < 0.0001). Interestingly, the mean CAN-1 score of those subclade 2A specimens that have the "coastal" morphology is intermediate between the mean scores of either subclade 1B or 1C and that of subclade 2A individuals with the "desert" morphology. Such a skew in CAN scores suggests that a residual expression at the nuclear gene level controlling morphology remains of the past event(s) that produced the original discordance in morphology and mtDNA haplotypes across the Tehachapi Mts. The discordance between an individual's morphology and mtDNA haplotype is emphasized by those specimens at Joaquin Flat (locality CA-64) on the west side of the Tehachapi Mts. where both subclade 1C and 2A individuals were found. These individuals all have high (> 0.9) posterior probabilities to the "coastal" morphological group.

Of those males for which we examined the glans penis, all individuals of subclade 1B or 1C possess the coastal tip morphology (Figs. 29 and 30). Moreover, coincident with the canonical analyses, males of subclade 2A that have "coastal" craniodental morphologies all possess the coastal tip type while all individuals classified as "desert" by their craniodental morphology have a desert tip type. Thus, there is complete concordance in all morphological traits, craniodental and phallic, across the transect, even if there is discordance between mtDNA subclade assignments and morphology in the eastern Tehachapi Mts. and Kern River Plateau. This correspondence in craniodental and phallic characters extends to males from the contact localities in Kelso Valley (e.g., locality CA-80) where both "coastal" and "desert" morphological types of woodrats co-occur and hybridize on occasion (see immediately below).

Morphology, mtDNA, and nuclear gene markers.—The above analyses document two important aspects of character change along the Tehachapi Transect. First, morphologically intermediate individuals do occur at contact points between the "coastal" and "desert" groups, suggesting that limited hybridization takes place. Second, the discordance between patterns of relationship suggested by individual morphologies and mtDNA haplotypes throughout the Tehachapi Mts., Kern River Plateau, and western foothills of the southern Sierra Nevada indicates that genetic interaction also occurred between these two groups at some time in the past. We address scenarios for the widespread distribution of the "desert" mtDNA subclade 2A within the background of "coastal" morphology in a later section. Here, we determine the relationship between an individual's morphology and its mtDNA haplotype using 18 nuclear microsatellite loci. We examine this relationship along the entire transect but especially at two contact localities (Joaquin Flat and Kelso Valley) where individuals of different mtDNA clade and/or morphology co-occur.

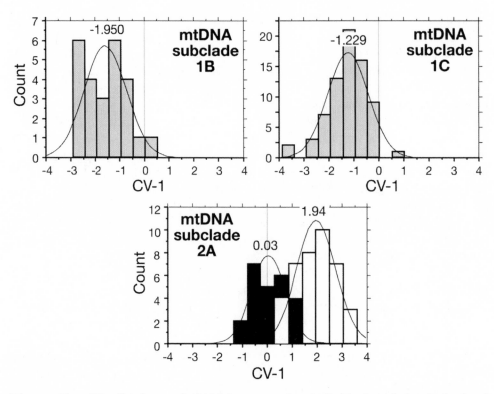

Figure 49.   Distributions of CAN-1 scores for individuals of the Tehachapi Transect of known mtDNA haplotype clade membership, the coastal 1B (upper left) and 1C (upper right) subclades, and the desert 2A subclade (lower). Individuals of subclade 2A are grouped by their pre-defined morphological assignments into the "coastal" (gray bars) or "desert" (black bars) groups. Mean CAN-1 scores for each group are given above the distribution curves.

Table 16 provides a general summary of allelic variation in the 18 microsatellite loci we examined. We provide data for the same set of pooled samples from the western side of the transect that we used for the morphological analyses, above. Initial analyses, including the summary data in Table 16, involve only the set of "coastal" and "desert" samples. We then use two assignment test methods (Pritchard et al., 2000; Anderson and Thompson, 2002) to align individuals of the various "unknown" morphological samples with respect to each of these two geographic groupings. Both methods construct posterior probability assignments to the "parental" types, but without either class being specified a priori. Anderson and Thompson's (2002) method has the added advantage of assigning individuals to specific hybrid classes (e.g., F1 versus F2 or backcrosses to either parent).

There is no significant correlation between sample size and either the number of alleles (Z-test, r = 0.186, p = 0.5517) or gene diversity among individuals per sample (r = -0.356, p = 0.2390) across the transect. Moreover, while Fis values, which measure the proportional reduction in heterozygosity within subpopulations due to inbreeding, vary among samples by a factor of 10, none are significantly different from 0 (Table 16). The lack of significant Fis values generalizes the conclusions of Matocq and Lacey (2004) on the related species *Neotoma macrotis* that woodrat populations are not typically characterized by spatial clustering of related males and/or females, and thus the likelihood of mating by close relatives is limited. Moreover, because the majority of our samples were taken at one time and typically from a single rock outcrop or other limited area, spatial clustering of kin is unlikely for populations of the *Neotoma lepida* group.

However, overall allelic diversity, whether measured across loci within individuals or among individuals, does differ in comparison between the five "coastal" and eight "desert" samples. For both the mean numbers of alleles per sample (5.49 versus 8.08, t = -3.0275, p = 0.0115) and mean gene diversity (0.6346 versus 0.7982, t = -9.1654, p < 0.0001), "coastal" samples contain less diversity than do those of the "desert" group.

There is also a substantial shift in allelic presence and/or frequency from "coastal" morphological samples on the west to "desert" morphological samples on the east (raw data not shown). Given the large number of alleles at each locus and the number of loci, we summarize pairwise genic similarity among samples with Wright's fixation index (Fst), estimated using the method of Weir and Cockerham (1984) as implemented in the GDA software program (Lewis and Zaykin, 2002). Figure 50 illustrates the geographic placement of the 13 samples examined and a neighbor-joining tree, based on the pairwise Fst matrix, that illustrates relationships among these samples. The tree is drawn with branch lengths proportional to the measured distances between the pairs of samples. Two important elements deserve note in the pattern of differentiation exhibited along the transect. First, there is significantly greater differentiation among "coastal" samples (mean Fst = 0.0761 ± 0.0154 standard error) than among those of the "desert" group (mean Fst = 0.0152 ± 0.0021; t-value = 5.8807, p < 0.0001). And, second, there is sharp differentiation in the comparisons between "coastal" and "desert" samples, with a mean Fst of 0.2434 (0.0006 standard error), significantly higher than that of within either "coastal" (Fisher's PLSD critical difference = 0.018, p < 0.0001) or "desert" samples (critical difference = 0.014, p < 0.0001). Hence, although coastal samples exhibit substantial differentiation among them, differences between "coastal" and "desert" groups are considerably higher, more than 3 times in comparisons between means, with differentiation among the desert samples virtually non-existent. Furthermore, the shift in allelic differentiation along the Tehachapi Transect corresponds geographically to the morphological transition area, namely along the eastern margins of the Tehachapi Mts. and western versus eastern parts of the Kern River plateau, not to the mtDNA clade boundary (compare Fig. 50 to Fig. 47).

Table 16. Measures of diversity in 18 microsatellite loci for 13 samples (5 of the "coastal" and 8 of the "desert" morphological groups; see Fig. 50) of the Tehachapi Transect. Samples are identified by their mtDNA subclade membership and locality number(s), if pooled (see Appendix).

Sample (clade, locality number)	Mean N	Mean # alleles	Gene diversity	H_o	H_e	F_{is}[1]
1C - Carrizo (CA-38-45)	28.4	6.00	0.656	0.653	0.612	0.064
1C - San Emigdio (CA-57, 59)	6.7	4.17	0.595	0.602	0.581	0.037
1B/1C - Ft. Tejon (CA-60)	31.6	5.17	0.611	0.611	0.583	0.046
1C/2A - Joaquin Flat (CA-64)	41.2	6.44	0.659	0.657	0.654	0.004
2A - W Kern (CA-55, 68, 70-72)	15.3	5.67	0.652	0.649	0.577	0.015
2A - Oak Creek Pass (CA-85)	3.9	4.50	0.779	0.768	0.713	0.083
2A - Hoffman Summit(CA-83)	25.6	10.44	0.811	0.811	0.793	0.022
2A - E Kern (CA-74-76)	11.9	7.28	0.802	0.791	0.771	0.026
2A - Freeman Canyon (CA-92)	21.6	8.61	0.765	0.762	0.758	0.006
2A - Inyokern (CA-94)	8.9	7.72	0.799	0.801	0.807	0.007
2A - Halloran Spr (CA-367)	15.4	8.50	0.762	0.761	0.742	0.026
2A - Pisgah (CA-348)	8.6	8.06	0.863	0.826	0.764	0.079
2A - Little Lake (CA-381)	24.6	9.56	0.805	0.803	0.772	0.040

[1] not significantly different from 0, based on bootstrapping over loci with 1000 repetitions

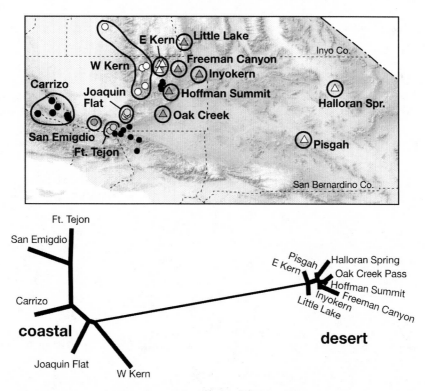

Figure. 50. Above – map of five "coastal" (circles), eight "desert" (triangles), and the "unknown" samples (black dots) of the Tehachapi Transect analyzed for variation in 18 microsatellite loci (redrawn from Fig. 33). Below – neighbor-joining tree of relationships among these 13 "coastal" and "desert" samples, based on a matrix of pairwise Fst distances. Branch lengths are drawn proportional to the distance among groups.

We also used the microsatellite to ask if there is evidence of genetic admixture within and among any of our samples, specifically with reference to individuals from the "unknown" localities relative to the "coastal" and "desert" samples (Fig. 50). We used the program STRUCTURE (Pritchard and Wen, 2003), as detailed in the Methods section. The analysis was run at k-values ranging from 2 to 8 with the resulting individual assignments highly correlated (r > 0.997 in all comparisons) regardless of the value used. Probabilities we report are based on k = 2. The number of loci used (18) falls well within the estimated 12-24 required for the efficient detection of hybrid individuals (Vähä and Primmer, 2006).

Individuals of all 5 "coastal" and 8 "desert" samples were assigned to their respective groups with posterior probabilities greater than 0.996 in each case (Table 17). The average assignment probability within the "coastal" samples is 0.997 (0.0049 standard error, range 0.941 to 0.999); that of the "desert" samples is 0.995 (0.00064 standard error, range 0.956-0.999). Moreover, most individuals of the "unknown" samples were likewise assigned to either the "coastal" or "desert" groups with equally high probabilities, typically to the source population geographically closest (Fig. 51). For example, 19 of 22 specimens from the "unknown" group of localities from the desert slopes along the western margins of the Antelope Valley in Kern and Los Angeles Cos. (localities CA-61, CA-62, CA-96, CA-97, CA-99, CA-100, and CA-102) were assigned to the "coastal" group with probabilities greater than 0.996. Of the remaining 3 individuals, one (MVZ 196762, from Pescadero Creek on the south slope of the Tehachapi Mts.; locality CA-62), belonged to the "desert" group (p = 0.991) and two (both from the Pescadero Creek sample [MVZ 196763 and 196837]) had somewhat intermediate probabilities, although close to the "desert" group (p to "coastal" = 0.140 and 0.247, respectively; Fig. 51). On the other hand, while 51 of the 62 individuals from the set of four localities in the vicinity of Kelso Valley (localities CA-77, CA-79, CA-80, and CA-81) sort strongly to either "coastal" or "desert" samples for the most part, 11 specimens exhibit intermediate assignment probabilities (Table 17, Fig. 51).

Table 17. Probabilities of assignment for 340 individuals, based on 18 microsatellite loci, along the Tehachapi Transect. The number assigned to "coastal" or "desert" groups (probabilities > 0.95) or arbitrarily classified as "intermediate" (probabilities 0.95 to 0.05 to either "coastal" or "desert" are given.

Sample	Probability to "coastal" > 0.95	Intermediate	Probability to "desert" > 0.95
"coastal" samples	126	0	0
"desert" samples	0	0	130
western AntelopeValley			
Gorman (CA-96, CA-97)	11	0	0
Cement Plant (CA-61)	2	0	0
Three Points (CA-99, CA-100)	4	0	0
Red Mountain. (CA-102)	2	0	0
Pescadero Creek (CA-62)	0	2*	1
Kelso Valley			
Harris Grade (CA-77)	0	2**	4
St John Mine (CA-79)	0	0	10
Whitney Well (CA-80)	27	10***	6
Schoolhouse Well (CA-81)	4	0	0

*	probabilities to "desert" of 0.860 and 0.753.
**	probabilities to "desert" of 0.816 and 0.745.
***	probabilities to "coastal" or "desert" range from 0.556 to 0.884.

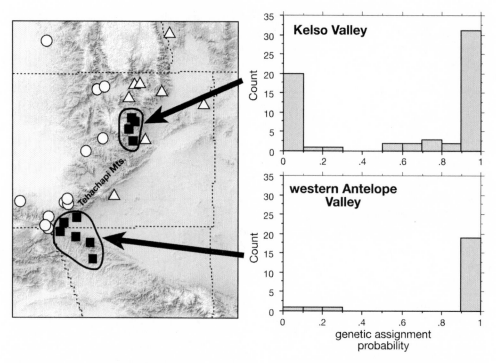

Figure 51. Map of contact region along the margins of the Tehachapi Mts., Kern Co. In samples identified by open circles and open triangles, all individuals are assigned to the "coastal" or "desert" groups, respectively, with probabilities > 0.99. The "unknown" samples (black squares; identified in Figs. 33 and 50) are divided into two geographic regions, for which histograms of probability assignments to the "coastal" group are illustrated on the right. Note the presence of individuals with intermediate probabilities, particularly those from Kelso Valley.

Because of genetically intermediate individuals within the collective "unknown" pool of samples, we investigated the likelihood of occasional hybridization further using Anderson and Thompson's (2002) model-based NEWHYBRID program. Their approach is similar to that used in STRUCTURE but calculates Bayesian posterior probabilities of membership in parental groups as well as potential F1, F2, and backcross combinations to either parent based on an individual's combination of multi-locus genotypes. In our analysis, the parental groups again correspond to the "coastal" and "desert" groupings used in the morphological analysis, above. The posterior probabilities calculated for each individual from either "coastal' or "desert" samples are highly and positively correlated with the same value obtained from the STRUCTURE analysis, including those from the "unknown" samples assigned to either of these two groups (r = 0.892, $F_{(1,128)}$ = 497.607, p < 0.0001). However, the NEWHYBRID analysis assigns the 14 individuals with intermediate probabilities in the STRUCTURE analysis (Table 15) as backcross individuals, either to the "coastal" (8 of 14, all from Whitney Well, Kelso Valley [locality CA-80]) or "desert" groups (5 of 14, one from Whitney Well and two each from Harris Grade [locality CA-77] and Pescadero Creek [locality CA-62]). One individual from Whitney Well is a possible F1 hybrid, with a probability of assignment to this class of 0.792 and to backcross to "coastal" of 0.192). The STRUCTURE assignment of this individual (MVZ 202503) was 0.567 to "coastal" and 0.433 to "desert".

Several points are worth emphasizing with regard to the assignment tests of group membership based on the microsatellite data. First, hybridization between "coastal" and "desert" groups does occur, even if limited both in number of examples and in geographic extent. Importantly, all hybrid class individuals were found at localities of actual contact, or near contact, between the morphologically "coastal" and "desert" groups (Figs. 47 and 51). These points of contact occur where coastal woodland and scrub vegetation communities interdigitate with those of the western Mohave Desert (Fig. 39). Moreover, genetic interaction is not limited to the production of F1 individuals, as only one potential F1 was found among the 14 hybrid-class individuals defined by their microsatellite assignments (Table 17). The identification of backcross individuals means that F1 individuals are sufficiently fertile so that introgression in either direction is possible. Genic assignments are thus in complete accord with morphological analyses in supporting limited genetic interaction between both groups of woodrats. Second, there is an apparent asymmetry in hybridization towards the "coastal" group, since the number of backcrosses to that category is eight times that to the "desert" parental group, at least at the one locality where both types of woodrats co-occur (Whitney Well in Kelso Valley). Such an asymmetry suggests either that coastal males or desert females may be less discriminatory with regard to mating than the opposite sex within their respective groups. The direction of asymmetry may also explain why the desert mtDNA has spread widely through those populations of the "coastal" form that occur throughout the Tehachapi Mts. and western parts of the Kern River plateau immediately west and north of the contact area. However, the process(es) by which the desert mtDNA lineage has become fixed in all of these populations remains to be identified.

Taxonomic considerations.—The Tehachapi Transect includes the type locality of one formal taxon of the *Neotoma lepida* group: *Neotoma desertorum sola*, from San Emigdio at the southern end of the San Joaquin Valley, Kern Co.

This taxon has been considered a junior synonym of *N. l. gilva* by all authors since Goldman's 1932 revision (for example, Grinnell, 1933; Hall, 1981). Jones and Fisher (1973) noted that the holotype (USNM 31516/43381) was composite, with the skull actually a specimen *of N. fuscipes macrotis* (now = *N. macrotis macrotis*; see Matocq, 2002) and designated the skin (USNM 31516) as the lectotype. By colorimetric measurements, this specimen is somewhat intermediate between our very pale Coastal-w samples from the nearby Carrizo Plains and the darker Coastal-e individuals from the mountains adjacent to Tejon Pass (Fig. 45). However, color characteristics of the lectotype also fall within the range of "desert" samples and, in a canonical analysis with the five grouped localities used as pre-defined reference samples, the posterior probability of assignment is highest to the Desert-e group (0.3608) and next highest to the Coastal-e group (0.1587). Given that color itself varies in a clinal fashion along the entire transect (Fig. 45) and is a poor predictor of group membership for individual specimens in general (Fig. 46), the limited data are inadequate to incontrovertibly place the lectotype within either the "coastal" or "desert" morphological groups. However, because the type locality is well within the range of the coastal group and the color characteristics are not inconsistent with placement there, we see no disagreement with the historical consideration of *sola* Merriam as a synonym of the coastal taxon *gilva* Rhoads (following Goldman, 1932; Grinnell, 1933; Jones and Fisher, 1973; Hall, 1981).

Cajon Pass Transect

This transect runs north from the Riverside-Beaumont area in western Riverside Co. in southern California across Cajon Pass to Hesperia and the Victorville – Oro Grande region along the upper Mojave River in San Bernardino Co. (Fig. 52). We grouped individual localities through this region into seven pooled groups, six of which we view as reference samples and one we group as an "unknown" in the analyses that follow. We list locality numbers (from the Appendix), sample size for each dataset (craniodental [n_m], color [n_c], glandes [n_g], and DNA sequence [n_{DNA}]), and museum catalog numbers for all specimens examined.

Beaumont (n_m=21, n_g = 1, n_{DNA}=17)
CALIFORNIA:– RIVERSIDE CO.: (1) CA-220: n_m=2; MVZ 90673, 90720; (2) CA-221: n_m=5; MVZ 88525-88529; (3) CA-222: n_m=14, n_g = 1, n_{DNA}=17; MVZ 196101-196114.

Riverside (n_m=8)
CALIFORNIA:– RIVERSIDE CO.: (1) CA-214: n_m=2; MVZ 3410-3411; (2) CA-215: n_m=5; USNM 93983, 93986, 93989, 93994-93995; (3) CA-216: n_m=1; MVZ 2534.

Reche Canyon (n_m=28)
CALIFORNIA:– SAN BERNARDINO CO.: (1) CA-315: n_m=6; MVZ 24499-24504; (2) CA-316: n_m=10; MVZ 2668-2669, 2672-2673, 2677-2678, 2681, 2683-2685; (3) CA-317: n_m=9; USNM 127985-127988, 127990-127994; (4) CA-318: n_m=1; SDNHM 16015; (5) CA-320: n_m=1; MVZ 77229; (6) CA-321: n_m=1; MVZ 77227.

Cajon Wash (n_m=6)
CALIFORNIA:– SAN BERNARDINO CO.: (1) CA-319: n_m=6; MVZ 2590, 2592-2593, 2595-2596, 2598.

Lone Pine Canyon (n_m=5, n_g = 1, n_{DNA}=5)
CALIFORNIA:– SAN BERNARDINO CO.: (1) CA-324: n_m=4, n_g = 1, n_{DNA}=4; MVZ 198657-198660; (2) CA-325: n_m=1, n_{DNA}=1; MVZ 198661.

Hesperia (n_m=20)
CALIFORNIA:– SAN BERNARDINO CO.: (1) CA-326: n_m=4; LACM 31718, 31733-31735; (2) CA-327: n_m=16; UCLA 947, 948, 958-961, 965-966, 976-977, 979-980, 982; SDNHM 1030-1031; USNM 42981.

Victorville (n_m=26)
CALIFORNIA:– SAN BERNARDINO CO.: (1) CA-328: n_m=18; MVZ 5995, 6006-6007, 6075, 6077, 6080-6081, 6084-6092, 6828; (2) CA-329: n_m=8; MVZ 28208, 31434-31439, 31441.

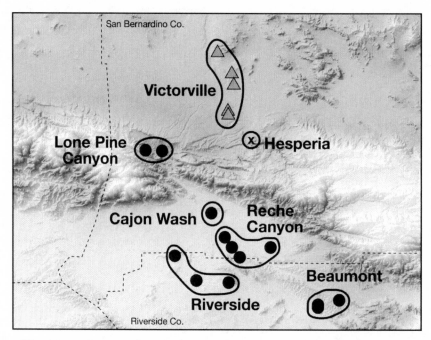

Figure 52. Map of grouped and individual localities of samples examined in the Cajon Pass Transect. Circles are samples of the "coastal" morphological group; triangles are those of the "desert" morphological group; and the single 'ⅹ' marks those specimens from the vicinity of Hesperia, to the north of Cajon Pass, where both "coastal" and "desert" morphological types are present and that are considered as "unknown" in the canonical and discriminant analyses.

MtDNA sequence data are available for only three of these localities, and all belong to the 'coastal' subclade 1B: Lamb Canyon (CA-222, part of the Beaumont sample) and Lone Pine Canyon and Mormon Rocks (CA-324 and CA-325, both included in the Lone Pine Canyon sample). The Hesperia sample includes individuals of both the "coastal" and "desert" morphological groups based on their qualitative craniodental features, which is why we regard specimens of this sample as "unknown" in multivariate analyses.

Multivariate analyses.—We analyzed specimens by both PCA, which makes no a priori assumptions about group membership, and CVA, where a priori groups conform to the grouped localities and their respective inclusive localities, listed above. In both cases, only "adult" animals of age classes 1-5 were considered.

The first two PCA axes explained 61% of the total pool of variation (48 and 13%, respectively), with no more than 5% explicable by any subsequent axis. Individuals assorted into two major groupings along these two axes (Fig. 53), one that encompassed all "coastal" morphology animals (those from the grouped localities south of Cajon Pass) and a second that included all individuals from the Victorville area. Specimens from the general area around Hesperia (indicated by 'x' in Figs. 52 and 53) clearly fell into either of these two groupings, four with the "coastal" group and 12 with the "desert" group. In each case, these individuals were placed within the group where their qualitative morphological features also placed them. The sole individual from Deep Creek, southeast of Hesperia, is of the "coastal" morphology, but at Hesperia, members of both morphological groups are present (e.g., UCLA 949 and 982 are "coastal", all others are "desert"). The latter thus represents an apparent area of true sympatry, although the actual locality where these animals were trapped in 1930 is both unknown and now likely completely obliterated by the recent urban expansion in this area.

We further investigated the likelihood of two morphological types of woodrats in the Hesperia sample using CVA on the log-transformed craniodental variables. We treated the 5 "coastal" localities south of Cajon Pass and the single "desert" locality (Fig. 51) as a priori defined groups and the individuals from Hesperia as "unknown." We then classified each "unknown" into one of 6 a priori groups by their posterior probabilities. The first two axes explained 92% of the variation; these are the only two axes that had eigenvalues > 1.0 and significance levels of $p < 0.001$.

CAN-1 separates two non-overlapping clusters of samples, one with all five "coastal" morphological groups and a second comprising the single "desert" group (Fig. 54). The various pooled samples of the "coastal" type are partially separated on CAN-2, but the differences among them are small. As with the PCA analysis (Fig. 53), 12 "unknown" individuals from Hesperia fall within or near the "desert" group. Four individuals fall within the "coastal" cluster of localities; these are the same identified by the PCA to have "coastal" morphology (LACM 31733 from Deep Creek southeast of Hesperia [locality CA-326], UCLA 949 and 982 from Hesperia [CA-327], and LACM 31735 from the Camp Mojave site near Hesperia [CA-327]).

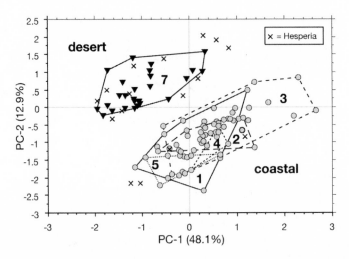

Figure 53. Scatterplot of individual scores on the first two principal components axes. Lines enclose individuals of each pooled locality (Fig. 51). Samples 1-5 (circles) are the "coastal" morphological group; sample 7 (triangles) is a "desert" morphological group. Individuals from Hesperia are indicated by '✕' (see text).

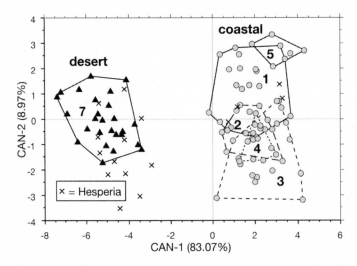

Figure 54. Scatterplot of the first two canonical axes for a priori locality groupings in the Cajon Pass Transect (as defined in Fig. 52). Samples 1-5, identified by open circles, are of the "coastal" morphological group and sample 7 (solid triangles) is of the "desert" morphological group. Individuals from general locality 6, the vicinity of Hesperia, are indicated by '✕'; their scores were based on the discriminant function equation generated by including only localities 1-5 and 7.

Fig. 55 illustrates the probabilities of group membership for each specimen examined. Because there are five "coastal" morphology general localities and only a single "desert" one (Victorville), we use the posterior probability to this grouped locality as our index of how well each specimen "fits" within its prospective group. Probabilities of membership in the Victorville sample are always greater than 0.999, if the individual was classified as belonging to this group, or no greater than 0.001, if it was classified into one of the five "coastal" groups. The assignment of each "unknown" individual is completely unambiguous to either the "coastal" or "desert" morphological groups.

Two points are apparent in these analyses. First, individuals of both morphological groups co-occur in the vicinity of Hesperia. Second, judging from both the placement of individuals in both PCA and CVA plots as well as their posterior probabilities, there is no evidence of morphological intermediacy that might suggest hybridization between the two groups in the vicinity of Hesperia.

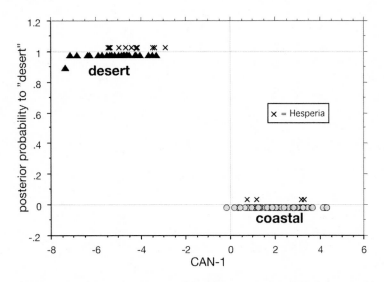

Figure 55. Plot of the posterior probability of membership to the "desert" morphological group for each specimen examined in the Cajon Pass transect relative to the score of that individual on the first CAN axis. Points for both pre-defined groups are deliberately offset from the "0" and "1" lines for ease in comparing the distribution of each group and the "unknown" individuals. Note that all individuals, including those from the area of sympatry near Hesperia, have posterior probabilities relative to the "desert" morphological type of either 1.0 or 0.0, with no evidence of intermediacy.

San Gorgonio Pass Transect

The transition of woodrat morphologies across San Gorgonio Pass is the historical basis for the current systematics of the *Neotoma lepida* group (Grinnell and Swarth, 1913; Goldman, 1932). Four of the currently recognized subspecies (Hall, 1981) are arrayed across this area, from west (*intermedia* and then *gilva*) to east (*lepida* and finally *grinnelli*). Grinnell and Swarth (1913), in their monograph on the birds and mammals of the San Jacinto Mts., documented character shifts between the coastal *Neotoma intermedia* and the desert *Neotoma desertorum*, interpreted this transition as evidence of interbreeding between these two species, and recommended that they be submerged into a single species. Goldman (1932) in his revision of the entire complex accepted their decision, although he recognized that the name *desertorum* was a junior synonym of *Neotoma lepida*, which had been described a year earlier.

Because of the historical significance of the transitional area across San Gorgonio Pass, we focused on details of this transition in our morphometric analyses. Our recent sampling has also emphasized genetic markers of individuals and population samples through the region. Fortunately, quite large samples are available in museum collections from a substantial number of localities from coastal Orange Co. east to the Colorado River in southern California. These permit us to document the nature of variation across this transect area and to match details of morphological features of individual specimens from key localities to their genetic attributes, using both the mitochondrial and nuclear DNA markers. In total, we examined 707 specimens from 136 separate localities.

Localities and sample sizes.—This transect runs west to east from coastal Orange Co. and eastern Los Angeles Co. through the length of Riverside Co. and the southern fringes of San Bernardino Co., and extending across San Gorgonio Pass to the Coachella Valley and east as far as the Colorado River basin in eastern California (Fig. 56). The transect includes samples that we can readily place into the general "coastal" and "desert" morphological groups. We group individual localities into six geographic units linearly arrayed from west to east along the transect, with three subdivisions within both the "coastal" and "desert" morphological groups. We group all localities in the central portion of this transect (generally from near Cabezon northeast to Morongo Valley and southeast through Palm Springs to the Santa Rosa Mts.) into an "unknown" category, because this is the transitional area of contact between "coastal" and "desert" types of woodrats (Grinnell and Swarth, 1913). These somewhat arbitrary subdivisions allow us to examine variation within both morphological groups as well as to document separately details of the transition between these two groups from San Gorgonio

Pass through the Coachella Valley. We are also able to determine more effectively the degree of intermediacy of any specimens from "unknown" localities by assignment via discriminant analysis to pre-defined reference samples, such as the pooled morphological "coastal" and "desert" groups. Resulting Mahalanobis distances, or the posterior probabilities of membership to either "coastal" or "desert" groups, also permit us to examine concordance between an individual's genetic (based on both mitochondrial and nuclear markers) and morphological characters, including both craniodental and color morphometrics data.

We group individual localities into pooled samples (Fig. 56) and list these by number, as in the Appendix, along with sample size for craniodental morphometric (n_m), colorimetric (n_c), glans penis (n_g), and molecular samples (n_{DNA}), with respective museum catalog numbers.

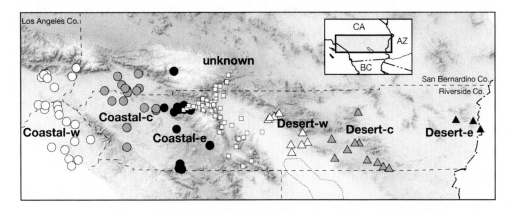

Figure 56. Map of localities included in the San Gorgonio Pass Transect. Circles identify localities from which all individuals examined possessed the "coastal" qualitative morphological characters; triangles represent those localities where all individuals are of the "desert" morphological type; and squares identify those localities that are regarded as "unknown" in all analyses (see text for explanation). Both "coastal" and "desert" groups are subdivided into three, somewhat arbitrary geographic subgroups linearly arrayed from west to east. These include the four recognized subspecies of this region of California (open circles = Coastal-w [*intermedia*], gray circles = Coastal-c [transition between *intermedia* and *gilva*], solid circles = Coastal-e [*gilva*], open triangles = Desert-w [*lepida*], gray triangles = Desert-c [transition between *lepida* and *grinnelli*], and solid triangles = Desert-e [*grinnelli*]). Inset – Generalized position of San Gorgonio Pass Transect in southern California and adjacent states.

Coastal-w (total $n_m = 76$, $n_c = 15$, $n_g = 5$, $n_{DNA} = 4$)
    CALIFORNIA:– LOS ANGELES CO.: (1) CA-118: $n_m = 3$; LACM 88273-88274, 88276; (2) CA-120: $n_m = 3$, $n_c = 1$; LACM 20635, 20637, 21234, MVZ 9059; (3) CA-121: $n_m = 2$; LACM 29962, 44974; (4) CA-122: $n_m = 1$; LACM 20616; (5) CA-123: $n_m = 1$ $n_c = 1$; MVZ 25557; (6) CA-124: $n_m = 2$; LACM 49628, 96062; (7) CA-125: $n_m = 1$; MVZ 65593. ORANGE CO.: (8) CA-128: $n_m = 1$; LACM 29940; (9) CA-129: $n_m = 1$; LACM 31729; (10) CA-130: $n_m = 1$; LACM 29938; (11) CA-131: $n_m = 1$; LACM 29939; (12) CA-132: $n_m = 2$; LACM 29960, 44969; (13) CA-133: $n_m = 7$, $n_c = 8$; MVZ 2359-2366; (14) CA-133a: $n_g = 1$; CSULB 2945; (15) CA-134: $n_m = 3$; LACM 29933, 29935, 29937; (16) CA-134a-b: $n_g = 2$; CSULB 2580, 3132; (17) CA-135: $n_m = 1$; LACM 29951; (18) CA-136: $n_m = 12$, $n_c = 5$; MVZ 2342-2346; LACM 29950, 29952, 29954-29957, 29959; (19) CA-137: $n_m = 2$; LACM 29947-29948; (20) CA-138: $n_m = 2$; LACM 29941-29942; (21) CA-139: $n_m = 24$; LACM 44061-44064, 44066=44068, 44073-44074, 44076-44077, 44079, 44082-44086, 44088-44089, 44091, 44117, 44126-44127; (22) CA-140: $n_m = 1$; LACM 44970; (23) CA-134: $n_m = 1$; LACM 44972; (24) CA-141: $n_m = 2$; USNM 149849, 149851; (25) CA-142: $n_m = 4$, $n_g = 2$, $n_{DNA} = 4$; MVZ 197375-197378.

Coastal-w (total $n_m = 76$, $n_c = 15$, $n_g = 5$, $n_{DNA} = 4$)
    CALIFORNIA:– LOS ANGELES CO.: (1) CA-118: $n_m = 3$; LACM 88273-88274, 88276; (2) CA-120: $n_m = 3$, $n_c = 1$; LACM 20635, 20637, 21234, MVZ 9059; (3) CA-121: $n_m = 2$; LACM 29962, 44974; (4) CA-122: $n_m = 1$; LACM 20616; (5) CA-123: $n_m = 1$ $n_c = 1$; MVZ 25557; (6) CA-124: $n_m = 2$; LACM 49628, 96062; (7) CA-125: $n_m = 1$; MVZ 65593. ORANGE CO.: (8) CA-128: $n_m = 1$; LACM 29940; (9) CA-129: $n_m = 1$; LACM 31729; (10) CA-130: $n_m = 1$; LACM 29938; (11) CA-131: $n_m = 1$; LACM 29939; (12) CA-132: $n_m = 2$; LACM 29960, 44969; (13) CA-133: $n_m = 7$, $n_c = 8$; MVZ 2359-2366; (14) CA-133a: $n_g = 1$; CSULB 2945; (15) CA-134: $n_m = 3$; LACM 29933, 29935, 29937; (16) CA-134a-b: $n_g = 2$; CSULB 2580, 3132; (17) CA-135: $n_m = 1$; LACM 29951; (18) CA-136: $n_m = 12$, $n_c = 5$; MVZ 2342-2346; LACM 29950, 29952, 29954-29957, 29959; (19) CA-137: $n_m = 2$; LACM 29947-29948; (20) CA-138: $n_m = 2$; LACM 29941-29942; (21) CA-139: $n_m = 24$; LACM 44061-44064, 44066=44068, 44073-44074, 44076-44077, 44079, 44082-44086, 44088-44089, 44091, 44117, 44126-44127; (22) CA-140: $n_m = 1$; LACM 44970; (23) CA-134: $n_m = 1$; LACM 44972; (24) CA-141: $n_m = 2$; USNM 149849, 149851; (25) CA-142: $n_m = 4$, $n_g = 2$, $n_{DNA} = 4$; MVZ 197375-197378.

Coastal-c (total $n_m = 57$, $n_c = 72$, $n_g = 1$, $n_{DNA} = 2$)

CALIFORNIA:– RIVERSIDE CO.: (1) CA-214: $n_m = 2$, $n_c = 6$; MVZ 2434-2437, 3410-3411; (2) CA-215: $n_m = 5$, $n_c = 7$; USNM 93982-93984, 93986, 93989, 93994-93995; (3) CA-216: $n_m = 1$, $n_c = 3$; MVZ 2534-2536; (4) CA-217: $n_m = 1$; SDNHM 6644; (5) CA-217a: $n_g = 1$; CSULB 10244; (6) CA-218: $n_m = 1$; USNM 70039; (7) CA-219: $n_m = 2$, $n_c = 1$, $n_g = 1$; MVZ 121585-121586; (8) CA-220: $n_m = 2$, $n_c = 2$; MVZ 90673, 90720); (9) CA-221: $n_m = 5$, $n_c = 5$; MVZ 88525-88529.   SAN BERNARDINO CO.: (10) CA-315: $n_m = 6$, $n_c = 6$; MVZ 24499-24504; (11) CA-316: $n_m = 10$, $n_c = 19$; MVZ 2663, 2666-2669, 2673-2679, 2681-2687; (12) CA-317: $n_m = 9$, $n_c = 9$; USNM 127985-127988, 127990-127994); (13) CA-318: $n_m = 1$; SDNHM 16015; (14) CA-319: $n_m = 6$, $n_c = 9$; MVZ 2590-2598; (15) CA-319a; $n_g = 1$; CSULB 7404; (16) CA-320: $n_m = 1$, $n_c = 1$; MVZ 77229; (17) CA-321: $n_m = 1$, $n_c = 2$; MVZ 77227-77228; (18) CA-322: $n_m = 2$, $n_c = 2$, $n_g = 1$, $n_{DNA} = 2$; MVZ 196052-196053; (19) CA-323: $n_m = 3$; USNM 94019-94021.

Coastal-e (total $n_m = 82$, $n_c = 83$, $n_g = 21$, $n_{DNA} = 31$)

CALIFORNIA:– RIVERSIDE CO.: (1) CA-222: $n_m = 14$ $n_c = 14$, $n_g = 10$, $n_{DNA} = 14$; MVZ 196101-196114; (2) CA-223: $n_m = 7$ $n_c = 1$; MVZ 89850-89852, 89854-89855, 89857-89858; (3) CA-224: $n_m = 7$ $n_c = 12$; MVZ 2289-2297, 2300-2301; (4) CA-225: $n_m = 7$, $n_c = 4$; MVZ 89865-89971; (5) CA-226: $n_m = 3$, $n_c = 3$; MVZ 89861-89862, 89864; (6) CA-228: $n_m = 2$, $n_c = 2$; MVZ 84462-84463; (7) CA-229: $n_m = 15$ $n_c = 24$, $n_g = 2$; ANSP 1665 [holotype of *gilva* Rhoads], MVZ 1424-1447; (8) CA-232: $n_m = 13$, $n_c = 13$, $n_g = 8$, $n_{DNA} = 13$; MVZ 196119-196131; (9) CA-275: $n_m = 1$ $n_c = 2$; MVZ 1871-1872; (10) CA-276: $n_m = 2$, $n_c = 3$, $n_g = 1$; MVZ 123545-123547; (11) CA-277: $n_m = 1$; MVZ 123544; (12) CA-278: $n_m = 11$ $n_c = 4$, $n_{DNA} = 4$; MVZ 198349-198352; SDNHM 1767, 1769, 1774-1777.   SAN BERNARDINO CO.: (13) CA-330: $n_m = 1$, $n_{DNA} = 1$; MVZ 198678; (14) CA-331: $n_m = 1$; MVZ 6833.

Desert-w (total $n_m = 31$, $n_c = 10$, $n_g = 24$, $n_{DNA} = 22$)

CALIFORNIA:– RIVERSIDE CO.: (1) CA-289: $n_c = 2$; MVZ 125888-125889; (2) CA-291: : $n_m = 8$, $n_g = 10$, $n_{DNA} = 21$; MVZ 215710-215730; (3) CA-291a-f; $n_g = 7$; CSULB 5286-5287, 5300, 5303, 5306, 5309; (4) CA-292: $n_m = 2$; LACM 29923-29924; (5) CA-293: $n_m = 1$; LACM 29925; (6) CA-294: $n_m = 3$, $n_c = 4$; MVZ 104034-104037; (7) CA-295: $n_m = 1$, $n_c = 1$, $n_g = 1$, $n_{DNA} = 1$; MVZ 199815; (8) CA-296: $n_m = 12$, $n_c = 2$; MVZ 149326, 149331, 149333, 149337, 149340-149342, 149344-149347, 149349; (9) CA-296a; $n_g = 6$; CSULB 10476-10481; (10) CA-297: $n_m = 1$, $n_c = 1$; MVZ 64817; (11) CA-298: $n_m = 1$; LACM 75543; (12) CA-299: $n_m = 2$; LACM 61849-61850.

Desert-c (total $n_m = 35$, $n_c = 3$, $n_g = 20$, $n_{DNA} = 41$)

CALIFORNIA:– RIVERSIDE CO.: (1) CA-300: $n_m = 15$, $n_g = 19$,, $n_{DNA} = 40$; MVZ 215659-215701; (2) CA-301: $n_c = 1$; MVZ 84765; (3) CA-302: $n_m = 1$, $n_c = 1$; MVZ 84764; (4) CA-303: $n_m = 4$; LACM 75470, 75473-75474, 75476; (5) CA-304: $n_m = 1$, $n_c = 1$, $n_g = 1$, $n_{DNA} = 1$; MVZ 199816; (6) CA-305: $n_m = 2$; LACM 75462, 75464; (7) CA-306: $n_m = 1$; MVZ 104033; (8) CA-307: $n_m = 3$; LACM 75491, 75485, 75487; (9) CA-308: $n_m = 3$; LACM 7550, 75507, 75509; (10) CA-309: $n_m = 3$; LACM 75521, 75523, 75526; (11) CA-310: $n_m = 2$; LACM 91642, 91656.

Desert –e (total $n_m = 7$, $n_c = 11$, $n_g = 6$, $n_{DNA} = 9$)

CALIFORNIA:– RIVERSIDE CO.: (1) CA-312: $n_m = 5$, $n_c = 4$, $n_g = 1$; MVZ 149261-149264, 149266 ; (2) CA-313: $n_m = 1$, $n_c = 1$, $n_{DNA} = 1$; MVZ 199817; (3) CA-314: $n_m = 1$, $n_g = 5$, $n_{DNA} = 8$; MVZ 215702-215709; (4) CA-315: $n_c = 1$; MVZ 10427. SAN BERNARDINO CO.: (5) CA-376: $n_c = 5$; MVZ 20974, 20976-20978, 20980.

unknown (total $n_m = 360$, $n_c = 244$, $n_g = 93$, $n_{DNA} = 163$)

CALIFORNIA:– RIVERSIDE CO.: (1) CA-227: $n_m = 1$, $n_c = 1$; MVZ 84464; (2) CA-229: $n_m = 9$, $n_c = 11$; MVZ 89873-89879, 89881-89883, 89887; (3) CA-230: $n_m = 6$, $n_c = 6$, $n_g = 3$, $n_{DNA} = 6$; MVZ 196132-196137; (4) CA-231: $n_m = 3$, $n_c = 1$; MVZ 90682, SDNHM 1650, 1658; (5) CA-232: $n_m = 10$, $n_c = 16$; MVZ 1333-1338, 1341, 1344-1347, 1350-1353, 1524; (6) CA-233: $n_m = 1$; SDNHM 159; (7) CA-234: $n_m = 4$, $n_c = 2$; LACM 85176, 85178-85179, MVZ 84465-84466; (8) CA-235: $n_m = 3$, $n_c = 3$; MVZ 90234-90236; (9) CA-236: $n_m = 1$, $n_c = 1$; MVZ 80688; (10) CA-237: $n_c = 1$; MVZ 84468; (11) CA-238: $n_m = 7$, $n_c = 7$; MVZ 90237-90242, 99967; (12) CA-239: $n_c = 1$; MVZ 84467; (13) CA-240: $n_m = 1$; LACM 29893; (14) CA-241: $n_m = 6$, $n_c = 10$; MVZ 1514-1516, 1525-1529, 1531; (15) CA-242: $n_m = 1$; MVZ 77230; (16) CA-243: $n_m = 5$; LACM 29902, 29904-29907; (17) CA-246: $n_m = 17$; LACM 85180-85186, 85188-85191, 85194-85199; (18) CA-247: $n_m = 5$, $n_g = 3$, $n_{DNA} = 5$; MVZ 206794-206798; (19) CA-248: $n_m = 1$, $n_c = 1$, $n_{DNA} = 1$; MVZ 196144; (20) CA-249: $n_m = 3$, $n_c = 6$; MCZ 5302; USNM 53979, 150589; (21) CA-250: $n_m = 2$, $n_c = 6$; MVZ 1518, 1520-22, 1533-1535; (22) CA-252: $n_m = 15$, $n_g = 8$, $n_{DNA} = 15$; MVZ 206830-206844; (23) CA-253: $n_m = 33$, $n_c = 35$; LACM 29909-29910, MVZ 84469-84477, 88533-88548, 90243-90248, 90251-90253; (24) CA-254: $n_m = 3$, $n_c = 3$; MVZ 88530-88532; (25) CA-255: $n_m = 6$, $n_c = 6$; MVZ 85136, 85139-85140, 88549-88551; (26) CA-256: $n_m = 3$, $n_c = 3$, $n_g = 1$, $n_{DNA} = 3$; MVZ 196138-196140); (27) CA-257: $n_m = 17$, $n_c = 16$; MVZ 90254-90269, 90723; (28) CA-258: $n_m = 1$, $n_c = 1$; MVZ 90270; (29) CA-259: $n_m = 1$, $n_c = 1$; MVZ 39972; (109) CA-260: $n_m = 1$, $n_c = 1$; MVZ 39971; (30) CA-261:

$n_m = 3$, $n_c = 3$, $n_g = 3$, $n_{DNA} = 3$; MVZ 196141-196143; (31) CA-262: $n_m = 3$; LACM 85171-85713; (32) CA-263: $n_m = 23$, $n_g = 9$, $n_{DNA} = 23$; MVZ 206799-206821; (33) CA-264: $n_m = 4$; LACM 29917-29920; (34) CA-265: $n_m = 5$, $n_g = 3$, $n_{DNA} = 5$; MVZ 206845-206849; (35) CA-266: $n_m = 8$, $n_g = 6$, $n_{DNA} = 8$; MVZ 206822-206929; (36) CA-267/268: $n_m = 22$, $n_c = 7$; LACM 1446, 1447, 3316, 20496, 21355; MCZ 5308 [holotype of *bella* Bangs]; MVZ 23920-23921, 55179, 62646-62650; UCLA 994, 1012-1013, 1026, 1446-1447, 7167, 7191, 7212; USNM 44985; (37) CA-269; $n_g = 1$; CSULB 1553; (38) CA-270: $n_m = 4$; LACM 29886-29889; (39) CA-271: $n_m = 3$; LACM 29891-29892, 29922; (40) CA-271a-b; $n_g = 5$; CSULB 11807-11809, 11816-11817; (41) CA-272: $n_m = 2$, $n_c = 3$; MVZ 2065-2067, 39970; (42) CA-273: $n_m = 1$, $n_c = 1$; MVZ 47523; (43) CA-274: $n_m = 3$, $n_c = 3$; MVZ 1950, 1954, 1956; (44) CA-279: $n_m = 6$, $n_c = 12$; MVZ 2056, 90674-9080, 90683-90684, 90721-90722; (45) CA-280: $n_m = 4$, $n_c = 5$; MVZ 39963, 39965-39969; (46) CA-281: $n_m = 7$, $n_c = 7$, $n_g = 1$, $n_{DNA} = 7$; MVZ 196145-196151; (47) CA-282: $n_m = 1$; MVZ 80689; (48) CA-283: $n_m = 1$; LACM 29894; (49) CA-285: $n_m = 3$, $n_c = 2$; MVZ 1951, 1957; LACM 90371; (50) CA-286: $n_m = 2$; LACM 20726-20727; (51) CA-287: $n_m = 1$, $n_c = 1$; MVZ 186337; (52) CA-290: $n_m = 1$; LACM 1503; (53) CA-292: $n_m = 1$; UCLA 9550; (54) CA-290a-b; $n_g = 9$; CSULB 11182, 11841-11848. SAN BERNARDINO CO.: (55) CA-338: $n_m = 7$, $n_c = 7$ $n_{DNA} = 7$; MVZ 197174-197178, 198333-198334; (56) CA-339: $n_m = 7$; LACM 1976, 1982, 21261-21262, 22720-22721, 22771; (57) CA-340: $n_m = 6$, $n_c = 6$, $n_g = 2$, $n_{DNA} = 6$; MVZ 198365-198370; (58) CA-341: $n_m = 62$, $n_c = 46$, $n_g = 39$, $n_{DNA} = 70$; MVZ 195321-195325, 198355-198364, 199804-199810, 202523-202546, 215731-215754, USNM 151295; (59) CA-342: $n_m = 4$, $n_c = 4$ $n_{DNA} = 4$; MVZ 199811-199814.

Habitat.—Grinnell and Swarth (1913) detail the shift in habitat from the Pacific side of this transect to the vicinity of Palm Springs and the Santa Rosa Mts., describing the vegetation characteristics at each locality they visited (pg. 201-214) and placing each locality within Merriam's Life Zone concept (pg. 215-217, plates 6 and 7). They also describe details of microhabitats where they caught woodrats. Our own trapping experiences across this same area are identical to the observations made by Grinnell and Swarth nearly a century earlier, except that increasing human development has radically altered the natural landscape throughout the region, most notably on the eastern slope of San Gorgonio Pass and in the Palm Springs area along the northeastern margins of the San Jacinto and Santa Rosa Mts. Many areas where woodrats were present a half-century or more ago are now devoid of any vestige of natural habitat and woodrats have been locally extirpated as a result.

In general, woodrats were most commonly found by all collectors, from Grinnell's days to the present, in the extensive rock outcrops that border the flank of the San Jacinto and Santa Rosa Mts. (Figs. 57 and 58) or in Creosote Bush (*Larrea tridentata*) and Mojave Yucca communities, commonly also associated with small rock exposures, on the flat lands of the northern Coachella Valley (Fig. 59). Animals exhibiting the "coastal" morphology (subspecies *intermedia* and *gilva*) were almost exclusively found in rocky outcrops, where piles of stick debris and fecal pellets are often evident in crevices or on flat rock surfaces. In contrast, the "desert" morphological type of woodrat (subspecies *lepida*) was most commonly associated with Creosote Bush and Mojave Yucca, within or outside of rocky exposures, at least in the transition area from Cabezon to Desert Hot Spring or Palm Springs. East of the Coachella Valley, from the Indio and Mecca Hills to the Colorado River, woodrats of this morphological type (subspecies *lepida* or *grinnelli*) again are found more commonly in rocky exposures (Fig. 60). Finally, in the transition between Colorado and Mojave Desert through Morongo Valley, woodrats make characteristic stick nests in rocky outcrops and at the base of both Mojave Yucca and Joshua Tree in otherwise dense Catclaw (*Acacia greggii*), Creosote Bush, and *Opuntia* sp. desert scrub vegetation (Fig. 61).

Figure 57. Granite boulder slopes of Lamb Canyon, south of Beaumont on Hwy. 79, Riverside Co., California (locality CA-222); habitat of *Neotoma lepida gilva*. Photo taken in December 2000.

Figure 58. Granite boulder and desert vegetation at Piñon Flat, Santa Rosa Mts., Riverside Co., California (locality CA-281). Habitat of *Neotoma lepida gilva* on the eastern margins of its range, where animals make stick nests both at the base of junipers or yucca or in crevices among exposed boulders. Photo taken in December 2000.

Figure 59. Creosote Bush desert east of Whitewater Hill, Riverside Co., California (locality CA-252), an area of overlap between "coastal" and "desert" morphological groups of woodrats. Photo taken in April 2004.

Figure 60. Woodrat nest constructed almost exclusively of rock chips on rocky outcrop, Red Cloud Wash, west slope of the Chuckwalla Mts., Riverside Co., California (locality CA-304). Photo taken in October 2002.

Figure 61. Mojave Yucca and Joshua Tree habitat at the eastern end of Morongo Valley, San Bernardino Co., California (locality CA-341) where "coastal" and "desert" morphological and molecular types of woodrats were taken in adjacent traps. Photo taken in March 2005.

Morphometric differentiation.—Pairwise comparisons of the 21 craniodental variables among the three "desert" pooled localities are uniformly non-significant in all but two of 42 cases (Table 18), and these two are significant at only $p < 0.05$. We conclude that samples from the eastern half of Riverside Co. are morphologically uniform from the eastern margins of the Coachella Valley to the Colorado River. Differentiation among the samples of the "coastal" morphological type is greater, however, with slightly more than half of all pairwise comparisons exhibiting some level of significance (22 of 42, Table 18). The eastern-most locality of the "coastal" group (Coastal-e, samples from the vicinity of Banning, the type locality of *Neotoma intermedia gilva* [locality CA-229]) differs more strongly from Coastal-c than Coastal-c differs from Coastal-w. The largest degree of difference is between the eastern sample of the "coastal" group (Coastal-e) and the western sample of the "desert" group (Desert-w). In this comparison, all 21 variables are significantly different, most (16 of 21) at $p < 0.001$ or 0.0001 (Table 18). Because each set of pooled localities is relatively uniform within its respective morphological unit, but strongly separable from other units, we combined samples into "coastal" and "desert" morphological groups, and provide standard descriptive statistics (mean, standard error, sample size, and range) for all external and craniodental variables for these two groups in Table 19.

The differences in tail length and bullar dimensions described above in the global comparison between the "coastal" and "desert" morphological groups are evident in the San Gorgonio Pass Transect. We illustrate the significant shift in character means from the "coastal" set of samples to those of the "desert" group noted in Table 19, for two univariate variables and scores for the first principal components axis in Fig. 62. Here, the similarity among the three "desert" pooled localities is evident in all comparisons, as is either the statistical uniqueness of the Coastal-e samples relative to the other two "coastal" pooled localities (CIL, Fig. 62, upper left; PC-1, Fig. 62, bottom) or uniformity (BUL, Fig. 62, upper right). Thus, although there is a general clinal shift from one end of the transect to the other, a substantial step occurs in that cline between the eastern-most "coastal" and western-most "desert" samples. This clinal pattern from Coastal-w to our Desert-w samples was apparent to Grinnell and Swarth (1913) in their early analyses (described in greater detail below).

Table 18. Results of MANOVA comparisons for 21 craniodental variables between geographically adjacent grouped localities for the "coastal" [w = Coastal-w, c = Coastal-c, and e = Coastal-e pooled samples] and "desert" [w = Desert-w, c = Desert-c, and e = Desert-e pooled samples] morphological groups in the San Gorgonio Pass Transect, as well as between the "coastal" and "desert" groups.

Variable	"coastal" w <-> c	"coastal" c <-> e	"coastal" vs "desert"	"desert" w <-> c	"desert" c <-> e
CIL	ns	***	****	ns	ns
ZB	ns	**	****	ns	ns
IOC	*	ns	****	ns	ns
RL	ns	***	****	ns	ns
NL	ns	***	****	ns	ns
RW	ns	**	***	ns	*
OL	***	***	****	ns	ns
DL	**	**	*	ns	ns
MTRL	***	ns	****	ns	ns
IFL	ns	ns	****	ns	ns
PBL	*	**	***	ns	ns
AW	*	ns	****	*	ns
OCW	**	ns	****	ns	ns
BOL	ns	ns	****	ns	ns
MFL	ns	ns	**	ns	ns
MB	***	*	*	ns	ns
MFW	ns	ns	****	ns	ns
ZPW	ns	ns	**	ns	ns
CD	*	***	**	ns	ns
BUL	ns	ns	****	ns	ns
BUW	ns	*	****	ns	*

ns = non-significant ($p > 0.05$), * = $p < 0.05$, ** = $p < 0.01$, *** = $p < 0.001$, and **** = $p < 0.0001$.

Table 19. External and craniodental measurements of adult (age classes 1-5) specimens of the "coastal" and "desert" samples of *Neotoma lepida* across the San Gorgonio Pass Transect (from Orange Co. east through Riverside Co.; see Fig. 56). Mean, standard deviation, sample size, and range are provided.

Variable	"coastal"	"desert"		"coastal"	"desert"
external					
TOL	325.5 ± 1.26	291.2 ± 2.93	HF	34.0 ± 0.12	30.3 ± 0.18
	183	41		195	49
	279-387	253 - 334		27-38	27 - 32
TAL	153.5 ± 0.79	132.2 ± 1.5	E	30.2 ± 0.18	29.6 ± 0.36
	183	41		124	49
	122-185	111 - 155		26-35	21 - 35
craniodental					
CIL	39.87 ± 0.10	37.46 ± 0.23	AW	7.64 ± 0.020	6.98 ± 0.06
	214	48		214	49
	36.71 − 45.17	34.23 − 40.87		7.02 − 9.18	6.03 − 8.73
ZB	21.83 ± 0.07	20.41 ± 0.13	OCW	9.57 ± 0.02	8.92 ± 0.04
	214	48		214	49
	19.30 − 26.51	18.50 − 23.06		8.82 − 10.71	8.13 − 9.47
IOC	5.52 ± 0.02	5.05 ± 0.03	MB	17.03 ± 0.04	16.82 ± 0.09
	214	49		214	49
	5.01 − 6.21	4.59 − 5.58		15.88 − 18.95	15.14 − 18.16
RL	16.38 ± 0.05	15.19 ± 0.11	BOL	5.94 ± 0.03	5.42 ± 0.04
	214	49		214	49
	13.26 − 18.93	13.39 − 17.45		4.86 − 7.02	4.72 − 6.00
NL	16.04 ± 0.06	14.88 ±0.13	MFL	7.95 ± 0.03	7.55 ± 0.06
	214	49		214	49
	13.94 − 18.56	12.48 − 17.03		6.64 − 9.38	6.49 − 8.24
RW	6.71 ± 0.02	6.30 ± 0.04	MFW	2.71 ± 0.02	2.327 ± 0.03
	214	49		214	49
	5.94 − 7.69	5.51 − 7.02		2.05 − 3.57	1.83 − 2.72
OL	14.49 ± 0.04	13.66 ± 0.08	ZPW	4.17 ± 0.02	4.08 ± 0.04
	214	48		214	49
	13.00 − 15.97	12.52 − 14.92		3.58 − 5.12	3.41 − 4.58
DL	11.468 ±.05	10.81 ± 0.11	CD	15.96 ± 0.04	15.51 ± 0.07
	214	49		214	48
	9.76 − 13.47	9.5 − 12.40		14.76 − 18.00	14.66 − 17.04
MTRL	8.24 ± 0.02	7.94 ± 0.05	BUL	6.78 ± 0.02	7.28 ± 0.04
	214	49		214	49
	7.06 − 9.08	7.19 − 8.58		5.82 − 7.52	6.70 − 7.92

Table 19 (continued)

Variable	"coastal"	"desert"		"coastal"	"desert"
IFL	8.92 ± 0.03	8.35 ± 0.07	BUW	7.17 ± 0.02	7.49 ± 0.04
	214	49		214	49
	7.78 – 10.80	7.55 – 9.39		6.47 – 7.99	6.62 – 8.33
PBL	18.11 ± 0.05	17.29 ± 0.12			
	214	49			
	15.68 – 21.09	15.76 – 18.94			

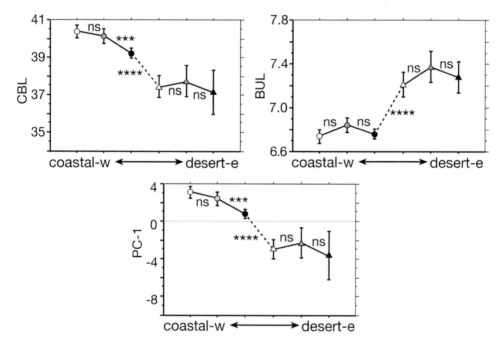

Figure 62. Mean and 95% confidence intervals for two variables (above) and PC-1 scores (bottom) across the west-to-east general localities of Coastal-w to Desert-e for. Symbols identifying samples are as in Fig. 56. Significance levels between the San Gorgonio Pass Transect geographically adjacent grouped localities along the transect are indicated: ns = non-significant (p > 0.05), *** = p < 0.001, **** = p < 0.0001 (based on ANOVA, using Fisher's PLSD post-hoc test).

In a principal components analysis that included all pooled samples of the San Gorgonio Pass Transect, the first PC axis represents general size, since factor loadings are both generally even and positive for all variables and individual PC-1 scores are positively correlated with individual craniodental variables (Table 20). For example, the correlation between Condyloincisive Length (CIL) and individual PC-1 scores is highly significant (r = 0.976, Z-value = 49.385, p < 0.0001). Only six variables (Interorbital Constriction [IOC], Molar Toothrow Length [MTRL], Mesopterygoid Fossa Length [MFL], Zygomatic Plate Width [ZPW], Bullar Length (BUL), and Bullar Width [BUW]) have correlation coefficients below 0.800, and only for BUL is the r-value non-significant (r = 0.002, Z-value = 0.056, p = 0.9552). There is also a highly significant relationship between PC-1 scores and longitude, although with much scatter (adjusted R^2 = 0.262; $F_{(1,541)}$ = 141.313, p < 0.0001). Animals generally get smaller from west to east along the transect when all individuals are included. A shift to smaller size is also generally seen among the three "coastal" grouped localities (the slope of the relationship is 0.710; r = 0.349, $F_{(1,211)}$ = 29.198, p < 0.0001) but not for the "desert" grouped localities (slope = 0.141, r = 0.080, $F_{(1,46)}$ = 0.299, p = 0.5870; Fig. 62, bottom). Not surprisingly, therefore, a multivariate perspective of craniodental variation from west to east mirrors that generally observed for individual characters (Fig. 62, top), including the substantial step in the cline in the middle part of the transect between the Coastal-e and Desert-w samples.

Two non-overlapping groups are apparent in a bivariate plot of the first two PC axes (Fig. 63), when localities are identified a posteriori as "coastal" or "desert." These axes explain 66.2% of the variation (PC-1, 54.5%; PC-2, 11.7%; Table 20). We use this simplified two group structure, rather than showing the three separate grouped localities within each, because most of the univariate character variation is distributed as differences between the "coastal" and "desert" groups with only minimal differences among any of the subgroups within each (above). This pattern of separate groups defined by the PCA is the same as that exhibited across both the Tehachapi and Cajon Pass transects, above (compare Fig. 63 to Figs. 40 and 53). Moreover, the same pattern of character vectors, with bullar dimensions contrasting all others, is apparent in each transect analysis as well (compare insets in Figs. 63 and 40). Individuals in the "unknown" pool are widely scattered across the diagram, with a substantial number falling between the two group clusters rather than being divided equally within each. This pattern contrasts with that apparent across the Cajon Pass Transect (Fig. 53), where all "unknown" individuals fall essentially within the ellipses of either pre-defined group, but is similar to that present across the Tehachapi Transect (Fig. 40). The two morphological groups, "coastal" and "desert," are significantly separated on both PC-1 and PC-2 axes, but not on PC-3 (for PC-1 scores, $F_{(1,260)}$ = 119.995, p < 0.0001; for PC-2 scores, $F_{(1,260)}$ = 211.723, p < 0.0001; for PC-3 scores, $F_{(1,260)}$ = 0.001, p = 0.9777).

Table 20.   Principal component eigenvectors and standardized coefficients for canonical variables for log-transformed cranial characters of the "coastal" and "desert" morphological groups of the San Gorgonio Pass Transect.

Variable	PC-1	PC-2	CAN-1
log CIL	0.972	0.043	-0.91478
log ZB	0.905	0.033	0.73565
log IOC	0.626	-0.462	0.61385
log RL	0.909	0.035	0.56024
log NL	0.890	0.037	0.27989
log RW	0.808	-0.069	0.22891
log OL	0.843	-0.039	-0.17047
log DL	0.874	0.221	-0.54322
log MTRL	0.286	-0.291	0.11392
log IFL	0.784	-0.035	0.23579
log PBL	0.893	0.107	-0.00281
log AW	0.693	-0.426	0.24703
log OCW	0.721	-0.279	0.41529
log MB	0.819	0.257	-0.27773
log BOL	0.796	-0.037	0.10121
log MFL	0.694	0.132	0.16323
log MFW	0.527	-0.393	0.42995
log ZPW	0.521	0.344	0.06533
log CD	0.816	0.160	-0.58933
log BUL	0.001	0.857	-0.65452
log BUW	0.215	0.846	-0.42903
eigenvalue	11.444	2.463	6.136
% contribution	54.5	11.7	100.0

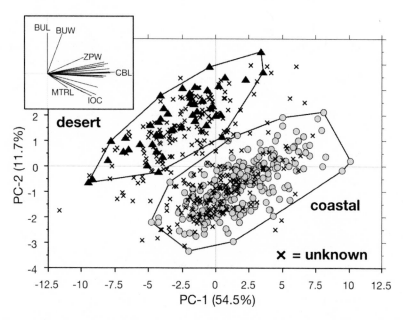

Figure 63. Scatterplot of individual scores on the first two principal components axes. Ellipses enclose individuals clustered with respect to their morphological group membership (Fig. 56). Circles are individuals with a "coastal" morphology and that occur from Banning west through Orange Co.; triangles are specimens of the "desert" morphology from the Indio and Mecca hills east to the Colorado River; "✗s" are the "unknown" specimens from intervening localities (near Whitewater northeast to Morongo Valley and southeast through Palm Springs to the Santa Rosa Mts.). The inset box illustrates character vectors along both axes, which contrast the highly positive vectors for all variables except those of the bulla (BUL and BUW) on the 1st axis with the strongly positive bullar dimensions on the 2nd.

We investigated further the apparent morphological intermediacy of so many individuals in the San Gorgonio Pass Transect through a canonical variates analysis, with the two a priori defined "coastal" and "desert" morphological groups of the transect serving as reference samples. Standardized coefficients for the single canonical axis are provided in Table 20. The two morphological groups separate at a highly significant level (Mahalanobis D^2 = 32.2787, $F_{(15,245)}$ = 76.7949, p < 0.0001), with all 286 specimens correctly classified into their pre-defined groups. The separation of "coastal" and "desert" morphological groups is readily apparent in a histogram of individual scores on the single CAN axis (Fig.

64). Both of the pre-defined groups are unimodal, and their individual distributions are non-overlapping. Not surprisingly, therefore, individuals of the "coastal" group have posterior probabilities of membership in that group ranging from a low of 0.984 to 1.0; those of the "desert" group exhibit a similar range of posterior probabilities of membership in that group of 0.965 to 1.0. The distribution of the "unknown" individuals, those from geographically intervening localities between the pre-defined groups (Fig. 56), is distinctly bimodal, with peaks that largely overlap with those of the "coastal" and "desert" groups. Note, however, that the "unknown" peak that corresponds to that of the "desert" group is shifted slightly to the right, closer in position to the "coastal" peak. No such shift is apparent in the "coastal-unknown" peak, which is directly beneath that of the pre-defined "coastal" group (Fig. 64).

　　　　We explored the intermediacy of the cluster of "unknown" individuals further by examining the distribution of posterior probabilities of group assignment for each individual relative to the pre-defined "desert" morphological group (Fig. 65). The high posterior probabilities (all near 1.0) of both the "coastal" and "desert" individuals to membership in their own groups are readily apparent. Although the two pre-defined groups are completely separated with high individual probabilities, 20 individuals of the "unknown" group exhibit intermediate probabilities (defined as between 0.9 and 0.1), including four that are close to equal in probability of membership to either reference group. This pattern is different than that seen in the more local Cajon Pass Transect (Fig. 55), where no morphologically intermediate individuals were observed, or that seen in Tehachapi Transect (Fig. 43), where only three intermediate individuals were identified. There thus appears to be both a quantitative and qualitative difference in the pattern of "intermediacy" across San Gorgonio Pass compared to that observed at other areas of geographic contact between "coastal" and "desert" morphological groups.

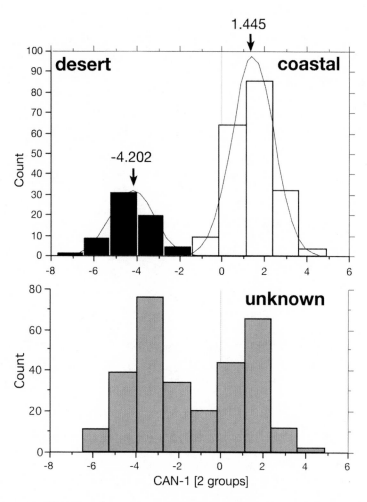

Figure. 64. Histograms of scores on the single canonical variates axis. Upper – scores for the "coastal" and "desert" pre-defined morphological groups (respectively, from Banning west through Orange Co., and from the Indio Hills east to the Colorado River; see Fig. 56). Mean CAN-1 scores are given for each group. Bottom – distribution of scores for individuals labeled as "unknown," from those geographically intervening localities near Whitewater northeast through Desert Hot Springs to Morongo Valley and southeast through Palm Springs to the Santa Rosa Mts.

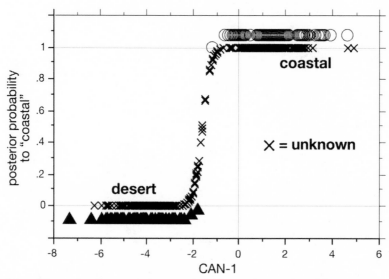

Figure 65. Plot of the posterior probability of membership to the "desert" morphological group (from the Indio Hills east to the Colorado River) for each specimen examined in the San Gorgonio Pass Transect relative to the score of that individual on the first CAN axis. Points for both pre-defined groups are deliberately offset from the "0" and "1" lines for ease in comparing the distribution of each group and the "unknown" individuals.

The "coastal" and "desert" morphological groups are sympatric at seven localities (Fig. 66), based on both posterior probability assignments and qualitative characters. From north to south, these include: 1.2 mi E Pioneertown (locality CA-342), east end Morongo Valley (CA-341), west end Morongo Valley (CA-338), edge San Gorgonio River (CA-246), 0.5 mi N & 4.4 mi W Desert Hot Springs (CA-263), Blaisdell Canyon (CA-257), and Palm Springs (locality CA-267). Intermediate individuals (posterior probabilities between 0.1 and 0.9) are present at 10 localities, including each locality where individuals with "coastal" and "desert" morphologies co-occur as well as at Whitewater (locality CA-249), 2.6 mi E Whitewater (CA-252), and Tahquitz Canyon (CA-280). The probability of assignment of individuals from these localities is unbiased because each was included as an "unknown" and not as part of the pre-defined "coastal" or "desert" groups.

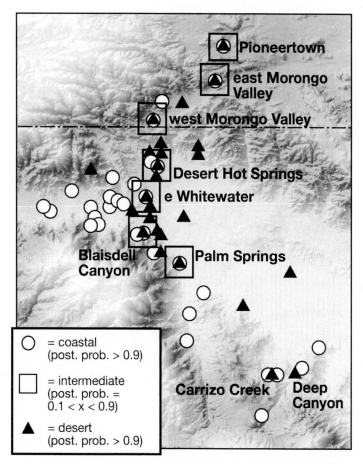

Figure 66. Morphological group assignments (based on posterior probabilities of membership of individual specimens to reference samples) of woodrat samples from San Gorgonio Pass and the western Coachella Valley. Open circles are localities of the "coastal" morphometric group (posterior probabilities > 0.9); solid triangles are localities of the "desert" morphometric group (posterior probabilities > 0.9); open squares are localities where individuals of intermediate morphology were found (posterior probabilities between 0.1 and 0.9 to either the "coastal" or "desert" groups). Overlapping symbols and names identify localities where individuals of "coastal," "desert," and/or intermediate morphologies co-occur.

Sympatry extends to the actual interspersion of nests occupied by both morphological types of woodrats at some localities, thus providing the opportunity for occasional interbreeding. At locality CA-341 (east end of Morongo Valley), we mapped the distribution of trap sites for 46 woodrats collected from 2002 to 2005 by GPS (Fig. 67). Ten of these rats are "coastal" and 33 "desert" in their morphology. However, three rats have intermediate posterior probabilities (p ranges from 0.714 to 0.857 to the "desert" reference sample), suggesting that interbreeding does occasionally occur. The mixture of mtDNA haplotypes in all three morphologically defined groups supports occasional hybridization: 37% of the morphologically defined "coastal" (three of 10) and "desert" (13 of 33) individuals have the "wrong" mtDNA haplotype. We examine the relationship between the morphological and genetic assignments of individual specimens, based on three different sets of markers, in greater detail below.

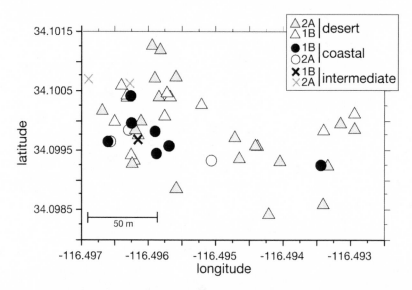

Figure. 67. Individual trap sites of woodrats collected over three trapping sessions from 2002 through 2005 in Morongo Valley (locality CA-341), mapped by GPS. Triangles are trap sites of individuals of the "desert" morphology; circles are sites of the "coastal" morphology; and "Xs" are sites of intermediate morphology. Different shading indicates the mtDNA clade haplotype of each individual.

Color variation.—We organized our samples of colorimetric variables into the same six grouped localities of the San Gorgonio Pass Transect that we used in the analysis of craniodental variables. As we did for the Tehachapi Transect, we restricted our analysis to the trichromatic X-variables for the four topographic regions of the study skin because of the very high correlations between X, Y, and Z coefficients ($r > 0.930$ and $p < 0.0001$ in all comparisons). Consistent with the pattern observed for the Tehachapi Transect, the individual X-coefficients for each topographic region of the study skin are inter-correlated, with all correlation coefficients highly significant although ranging widely (from $r = 0.241$, [Dorsal-X versus Chest-X, Z-value $= 3.325$, $p = 0.0009$] to $r = 0.570$ [Dorsal-X versus Lateral-X, Z-value $= 8.761$, $p < 0.0001$]). These correlations add further support to the general observation from all colorimetric analyses that change in the color of one part of the skin is generally reflected by similar change in all other regions.

Samples on the western half of the transect (the "coastal" samples) have lower X-coefficients for each region of the study skin (i.e., are darker) than those in the eastern portion (the "desert" samples; Table 21). Moreover, there is statistical uniformity in the samples from each global set of "coastal" and "desert" regional samples for each variable ($p \gg 0.05$ in all cases), although the two groups themselves are highly significantly different, whether the comparison is made between the geographically adjacent Coastal-e and Desert-w grouped localities or between pooled "coastal" and "desert" geographic units (ANOVA, $F_{(1,184)}$ ranges from 26.456, $p < 0.0001$ for Chest-X to 11.693, $p < 0.0001$ for Lateral-X). The shift from darker individuals to paler ones is abrupt geographically, occurring between Banning and Cabezon - Whitewater on the western margins of San Gorgonio Pass.

We summarize colorimetric variation across the transect with a principal components analysis. The first axis is the only one with an eigenvalue greater than 1.0; it explains 65.1% of the total pool of variation (Table 22). All four X-coefficients load equally on the first axis, and all four are significantly ($p < 0.0001$) and negatively correlated with individual scores (r-values range from -0.641 [PC-1 versus Chest-X, Z-value $= -10.989$] to -0.882 [PC-1 versus Lateral-X, Z-value $= -15.555$]). As with the Tehachapi Transect, therefore, PC-1 scores reflect the overall degree of darkness to paleness over the entire study skin, from the dorsum to the venter.

Table 21. Descriptive statistics for the colorimetric X-measurement for the four regions of the woodrat study skins. Means ± one standard error, sample sizes, and ranges are given for each of six pooled geographic samples along the San Gorgonio Pass Transect (see text for the rationale behind and membership in each group).

Sample	Dorsal-X	Tail-X	Lateral-X	Chest-X
Coastal-w	8.98±0.39	6.096±0.54	20.29±0.56	36.54±1.16
	16	16	16	16
	6.3–10.9	3.1–11.6	15.7–24.3	27.4–46.6
Coastal-c	9.28±0.16	6.32±0.27	20.43±0.31	37.71±0.63
	70	70	70	70
	6.5–12.6	3.2–15.7	13.4–27.8	25.0–50.3
Coastal-e	8.89±0.18	7.06±0.23	21.25±0.30	40.50±0.61
	84	84	25	84
	5.1-12.9	2.9-12.9	13.8–31.0	26.8–52.1
Desert-w	14.59±057	10.32±1.32	32.18±1.78	46.61±1.98
	9	9	9	9
	11.7–17.6	5.6–18.9	24.9–37.9	36.9–54.9
Desert-c	12.73±0.90	11.32±0.50	31.55±1.79	48.36±3.83
	2	2	2	2
	11.8–13.6	10.8–11.8	29.8–33.3	44.5–52.2
Desert-e	13.76±1.08	12.56±1.26	32.19±1.70	46.80±2.40
	5	5	5	5
	10.6–16.5	9.6–16.8	28.4–28.1	42.26–56.1

Table 22. Principal component eigenvalues and factor loadings of colorimetric variables from all samples of the San Gorgonio Pass Transect.

Variable	PC-1	PC-2	PC-3
Dorsal-X	0.867	-0.341	-0.144
Tail-X	0.814	-0.432	0.342
Lateral-X	0.882	0.210	-0.346
Chest-X	0.641	0.720	0.238
eigenvalue	2.603	0.866	0.314
% contribution	65.1	21.6	7.8

In a pattern similar to that of the Tehachapi Transect, there is a significant, negative relationship between color PC-1 scores and longitude ($r = -0.701$, $F_{(1,191)} = 184.28$, $p < 0.0001$; Fig. 68), providing further documentation that specimens become paler as localities transition between the coast and desert in southern California. However, the pattern of color change along the San Gorgonio Pass Transect is not a gradual clinal shift from west to east, but one with a sharp step at the mid-point and uniform samples to the west and east (Fig. 69). The three "coastal" samples do become slightly paler from west to east (Coastal-e is significantly paler than Coastal-c; $p = 0.05$), but the three "desert" samples are indistinguishable from one another. The difference between these two sets of geographically positioned samples is, however, highly significant ($p < 0.0001$; Fig. 69). Thus, the transition from dark coastal animals to pale desert ones is both abrupt and geographically narrow, beginning immediately east of latitude 116.65^0W (the vicinity of Whitewater on the eastern edge of San Gorgonio Pass) and ending at latitude 116.09^0W (vicinity of Pinyon Wells in the Little San Bernardino Mts.), a linear distance of approximately 25 miles. We will examine this transition in greater detail in the next section.

Unlike the pattern of color variation along the Tehachapi Transect, where color characteristics discriminate populations relatively poorly compared to craniodental characters (despite significant differences in color pattern among pooled samples), the color transition along the San Gorgonio Pass Transect is sharply defined and more useful in distinguishing individuals of the "coastal" and "desert" morphological groups. In a canonical analysis of color variables, the two reference groups are highly significantly different (Mahalanobis $D^2 = 24.7441$, $F_{(4,181)} = 83.8385$, $p < 0.00001$), and all individuals are correctly classified to their respective groups. The non-overlap and wide separation of reference individuals is readily apparent in the scatterplot of their CAN-1 scores and posterior probabilities (Fig. 70) and contrasts sharply with the wide overlap among individuals of the Tehachapi Transect (compare to Fig. 46). The pooled group of "unknown" individuals, those from localities between Cabezon and Palm Springs, however, do exhibit the full range of canonical scores and posterior probabilities spanning that between the two reference samples. A considerable number of these specimens exhibit intermediate probabilities of membership between the coastal or desert reference groups. The extent to which this intermediacy is due to selection for paleness as habitats become more xeric from west to east or to genetic interactions between coast and desert morphotypes where they meet through this region remains to be determined.

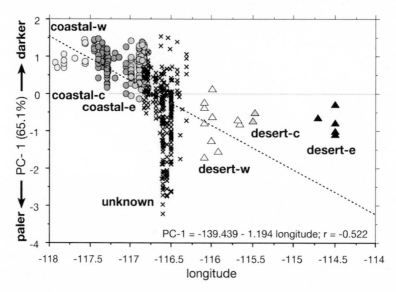

Figure 68. Regression of colorimetric scores on the first principal components axis and longitude along the San Gorgonio Pass Transect. Specimens are separated into geographic groups arranged from west to east (see Fig. 56).

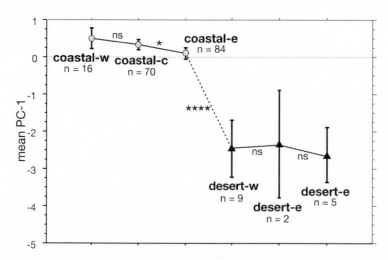

Figure 69. Means and 95% confidence limits of colorimetric PC-1 scores for the six geographic samples along the San Gorgonio Pass Transect. Samples are arranged from west (Coastal-w) to east (Desert-e). Significance levels between adjacent samples (based on ANOVA, Fisher's PLSD post-hoc tests): ns = non-significant; * $p < 0.05$; **** $p < 0.0001$).

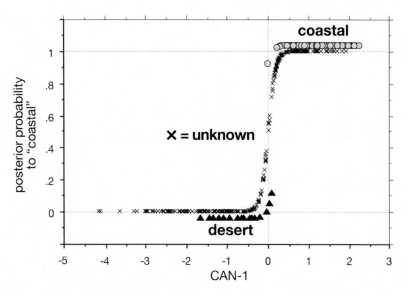

Figure 70. Plot of the posterior probability of membership to the "coastal" colorimetric group for each specimen examined in the San Gorgonio Pass Transect relative to the score of that individual on the single CAN axis. Points for both pre-defined groups are deliberately offset from the "0" and "1" lines for ease in comparing the distribution of each group and the "unknown" individuals. Note that there is no overlap in the distribution of individuals of the a priori defined "coastal" and "desert" groups, although a number of specimens are characterized by intermediate posterior probabilities. On the other hand, individuals in the "unknown" group are widely distributed in their posterior probabilities.

The Grinnell-Swarth transect.—Our analyses include the set of specimens originally collected and examined by Grinnell and Swarth (1913) in their study of the transition across San Gorgonio Pass. These are the specimens that formed the basis for their conclusion that the coastal and desert morphological taxa (*N. intermedia* and *N. desertorum* at that time) were members of a single species:

> It is shown by the forgoing array of facts that, in the white-footed woodrats of the San Jacinto area, there are two diverse types on the remoter parts of the opposite sides of the mountains, namely, *intermedia* on the Pacific side, most typically represented by specimens from Kenworthy, and *desertorum* at the desert base, as illustrated at Whitewater, Palm Springs, and perhaps Dos Palmas. The point of emphasis is that our material, as interpreted by us, would seem to

establish **complete intergradation** [emphasis ours] between the extreme
types names. In all respects as enumerated, we find transition through
various intermediate degrees of difference from one extreme to the other.
This is not in accordance with currently accepted notions as regards the
relationship between *intermedia* and *desertorum* (see Goldman, 1910);
but were we without recourse to previous literature, we should
unhesitatingly place one form as a geographic race of the other without
considering any explanation of our position as called for (p. 345).

At any rate, systematically the proper thing is to employ the
trinomial..., and since the name *intermedia* was first proposed in this
group, it takes precedence in specific combination... (p. 347).

Grinnell and Swarth (1913) argued that the "...extensive intergradation..."
(p. 345) between *N. intermedia* and *N. desertorum* resulted from differential
selection (what they termed "...the direct action of environment..." [p. 346]) due
to the sharp transition in habitat across the transect. They distinguished this from
hybridization, as they noted that samples from separate localities included
individuals with mixtures of characters and that "...in no instance were both typical
intermedia and *desertorum* found in the same locality with examples of
intermediate nature..." (p. 346). They nevertheless admitted that there may not be
"...any intrinsic difference between the results of long-continued hybridization and
'intergradation' " (p. 346), by which they meant that repeated hybridization would
also yield samples where all individuals were mixtures of characters that diagnosed
parental forms outside of the area of contact.

Some of our results support the conclusions of Grinnell and Swarth while
others remain in stark contrast, regardless of whether analyses include all currently
available samples or are restricted solely to the samples available to those authors.

The Grinnell-Swarth samples are from (west to east) Vallevista (locality
CA-242), Banning (CA-229), Cabezon (CA-232), Snow Creek (CA-241),
Whitewater (CA-250), and Dos Palmas Spring (CA-274). We treated the latter
four localities as "unknown" in all craniodental multivariate analyses, but we
included the former two in the Coastal-e grouped locality. Importantly, specimens
from all of the Grinnell-Swarth localities are assigned to the "coastal"
morphological group with very high posterior probabilities (all at a probability
between 0.993 and 1.0). Hence, the clearly delineated transition of size from large
to small across these localities documented by Grinnell and Swarth (1913: Fig. A,
p. 340-341) apparently involves only variation within the "coastal" morphological
type. That is, Grinnell and Swarth do not appear to have collected, and thus to
have included, specimens of "true" *N. desertorum* (= *N. lepida*) in their analyses.
This observation is fully consistent with the mtDNA sequence data for modern
samples through this same region (discussed below). Notably, for example, the

transition in overall size, as referenced by Total Length (TOL) for the samples available to Grinnell and Swarth and obtained by us from many of the same localities (and which all possess mtDNA haplotypes of the coastal Clade 1B, rather than the desert Clade 2A) exhibit the same overall clinal trend. The only difference in our respective temporal samples is the pair from the Santa Rosa Mts. The Grinnell-Swarth sample is noticeably smaller in average body size than ours (Fig. 71). All of their Santa Rosa individuals, however, exhibit the morphological characteristics of the "coastal" group, including small bullae, deep anteromedian flexus on M1, and centrally positioned lacrimal with respect to the fronto-maxillary suture. Each of their specimens also has posterior probabilities that strongly place them within the "coastal" morphological reference group. We conclude, therefore, that the gradation in characters observed by Grinnell and Swarth in their samples apparently did not involve any hybridization with true *N. desertorum*. The character trends that Grinnell and Swarth documented were apparently not due to either hybridization or intergradation, as they posited. These trends are best explained as differential selective responses to the sharply changing environmental condition across San Gorgonio Pass to the Coachella Valley floor.

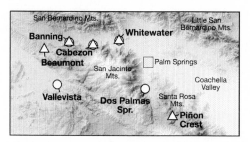 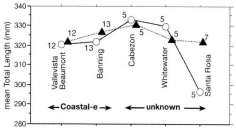

Figure 71. Top – map of San Gorgonio Pass (Banning to Whitewater), and San Jacinto Mts., and the Coachella Valley. Open circles = Grinnell and Swarth (1913) localities; open triangles = nearest recent sample to the Grinnell and Swarth historical locality. Bottom – Mean Total Length for Grinnell and Swarth (open circles) and recent (solid triangles) samples of woodrats across San Gorgonio Pass. Numbers adjacent to the symbols are sample sizes. The pooled sample to which each locality belongs (Fig. 56) is indicated.

Grinnell and Swarth, in their analyses of character change from the coastal *N. intermedia* to the desert *N. desertorum*, also mention a shift in color from "...above dark: blackish mid-dorsally, mixed with clay color..." to "...above pale: sepia mid-dorsally, mixed with pinkish buff..." (1913, p. 338: diagnosis). They noted that specimens from Banning, the type locality of *N. intermedia gilva* (locality CA-229), are paler than those of true *N. intermedia* further to the west, while others from nearby (Vallevista, CA-224) are as pale as those from the eastern desert. These observations are in accord with our analyses of larger series from more localities in vicinity of Banning, as these samples are significantly paler than those further to the west along the transect (Figs. 68 and 69). Moreover, if analyses are restricted to the same set of specimens used by Grinnell and Swarth, color (as indexed by PC-1 scores) does become paler from west to east along their transect (r = -0.497, $F_{(1,43)}$ = 13.745, p = 0.0006). However, the shift to paler coloration is not as sharp as the difference between pooled "coastal" and "desert" samples (excluding the "unknown" individuals). For example, when the Grinnell and Swarth specimens are plotted against latitude (as in Fig. 72), but distinguished from all other specimens, there is only the most minimal degree of overlap in their palest specimens, as indexed by PC-1 scores, with those belonging to our "desert" samples. We conclude, therefore, that the colorimetric data reinforce our craniodental morphometric analyses, and both support the hypothesis that the sample available to Grinnell and Swarth apparently did not include any true *N. desertorum* individuals, only those of what we identify here as the "coastal" group.

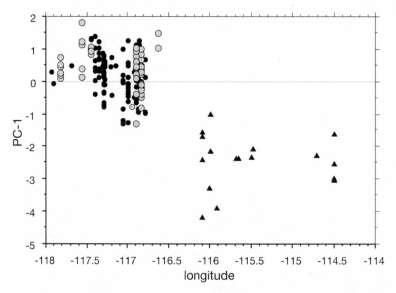

Figure 72. Scatterplot of colorimetric PC-1 scores against the longitudinal position of localities for those specimens included in Grinnell and Swarth's (1913) study (gray filled large circles) and more recently collected specimens from the "coastal" (solid smaller circles) and "desert" (solid triangles) morphological groups.

Although the original Grinnell and Swarth samples likely did not include any individuals of the "desert" morphology characteristic of true *N. desertorum*, the combined samples available to us certainly do. As we document above, there are seven localities in the transitional area between Cabezon and the floor of the Coachella Valley where individuals assigned to both morphological groups, "coastal" and "desert," are present (Fig. 66). Moreover, most of these localities are also sites where individuals with "intermediate" morphology are found. Some of these localities were sampled in the first decades of the 20th century; we trapped at other localities 80-90 years later and the rest were visited in the intervening time period. Hence, the contact between "coastal" and "desert" types of woodrats through this area has been present for at least the last century. The combination of localities with both "pure" parents and intermediates also suggests that occasional hybridization does characterize these points of contact. We detail in a separate section the genetic evidence for hybridization, using a suite of DNA markers.

The transect passes through a complex area, from the confines of San Gorgonio Pass formed by the San Bernardino Mts. to the north and the San Jacinto Mts. to the south, and then extends north through the very narrow Morongo Valley between the San Bernardino and Little San Bernardino Mts. as well as spreading out through the northern Coachella Valley as far east as the Colorado River and south along the margins of the San Jacinto and Santa Rosa mountains. This is an area that has been made even more complex by extensive urbanization, especially in the decades following World War II. As a consequence, many localities from which woodrats were obtained in this general area a century or less ago are now devoid of natural habitat, and woodrats are no longer present. It is unclear how this urbanization and fragmentation of habitat has affected the distribution of woodrats (other than local extirpation) or the nature of the interaction between individuals of both morphometric types over time. Equally important, however, is the imprecision of locality designation in pre-GPS days, which results in uncertainty as to the precise site where specimens were actually obtained. Nevertheless, and given these two caveats (an inability to revisit some localities and the geographic imprecision of many), sympatry or near sympatry of "coastal" and "desert" morphotypes is apparent at several localities from the San Gorgonio River southwest of Whitewater south and east along the western edge of the Coachella Valley to Tahquitz Canyon and Cottonwood Spring in the Santa Rosa Mts., including Palm Springs itself (Fig. 66).

Morphological – molecular concordance.—The San Gorgonio Pass Transect includes haplotypes of only two of the *cyt-b* clades: the coastal subclade 1B and the desert subclade 2A (Fig. 73). The former is distributed from Dana Point (locality CA-142) on the coast of Orange Co. east across San Gorgonio Pass northeast through Morongo Valley and southeast along the margins of the San Jacinto and Santa Rosa mountains in the vicinity of Palm Springs. Individuals with haplotypes of this subclade co-occur with those of the desert subclade 2A in Morongo Valley (localities CA-338 and CA-341) and are otherwise in close proximity throughout the Creosote Bush desert between Whitewater and Desert Hot Springs on the eastern slope of San Gorgonio Pass where coastal and desert vegetation zones meet (Fig. 66). Although few samples are available, desert subclade 2A individuals are distributed throughout the entire eastern portion of Riverside Co. east of the Coachella Valley and San Bernardino Co. north of the San Bernardino Mts. in the Mojave Desert.

In contrast to the Tehachapi Transect, there is complete concordance between those samples placed in pre-defined morphological groups based on qualitative craniodental characters and their respective mtDNA clade memberships. This concordance includes the morphology of the glans penis, for those specimens where this structure was available for examination. All "coastal" morphological samples (localities CA-142, CA-222, CA-230, CA-232, CA-278, CA-324, and CA-325) have haplotypes of the coastal subclade 1B, and "desert" morphological samples (localities CA-291, CA-295, CA-300, CA-304, CA-313, CA-314, CA-332, and CA-333) are characterized by the desert subclade 2A haplotypes, whether the morphometric assessments are based on the principal components or canonical variates analyses. In each case, for example, posterior probabilities of group membership in the "correct" reference sample are 0.965 or higher (see above).

For those localities coded as "unknown" in the morphometric analyses (Fig. 56), all specimens from along the edge of the San Jacinto and Santa Rosa mountains, from Cabezon to Piñon Crest (localities CA-256, CA-261, CA-281, and CA-287), have both the coastal haplotype 1B and posterior probabilities > 0.9997 of belonging to the "coastal" morphological group. In contrast, for those "unknown" localities where sympatry or near-sympatry between "coastal" and "desert" morphological groups occurs (the Whitewater to Desert Hot Springs area [localities CA-241 to CA-255 and CA-262 to CA-266] and Morongo Valley [localities CA-338 to CA-342; Fig. 73), discordance between an individual's morphological assignment by discriminant analysis and its mtDNA haplotype is apparent (Table 23). Two of 22 individuals (9%) from near Desert Hot Springs (one from locality CA-263 and the second from locality CA-266) have the "wrong" haplotype relative to their predicted morphologies. This mismatch is even greater in Morongo Valley where 16 of 43 (37%) are discordant. Moreover, in the Morongo Valley samples, six individuals had intermediate morphologies (three each with posterior probabilities between 0.1 and 0.9 to either "coastal" or "desert" groups). These morphologically intermediate individuals have a haplotype characteristic of either subclade 1B or 2A (Fig. 67). These observations provide further evidence of likely occasional hybridization between coastal and desert woodrats in the areas where they are in contact. As we did above with the Tehachapi Transect, we address this issue in greater detail next using genotypic probability assignments from 18 microsatellite loci.

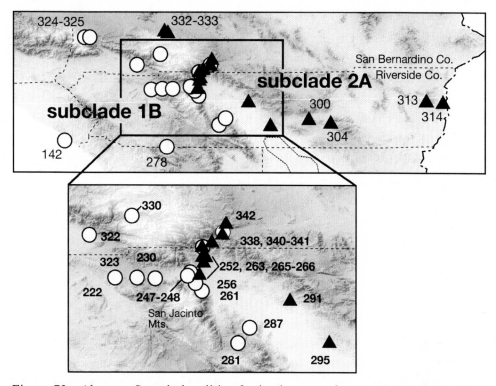

Figure 73. Above – Sample localities for haplotypes of two mtDNA clades along the San Gorgonio Pass Transect. Open circles identify individuals with haplotypes of the coastal subclade 1B and solid triangles indicate those with haplotypes of the desert subclade 2A. Overlapping circles and triangles identify areas where haplotypes of both subclades co-occur. Those localities treated as "unknown" in the morphometric analyses (see Fig. 56) are enclosed in the box. Bottom – Detailed map of sample localities of the "unknown" morphometric samples (Fig. 56) for which mtDNA subclade haplotype is known. Open circles are localities where individuals have mtDNA subclade 1B haplotypes; solid triangles are those with subclade 2A haplotypes; overlapping symbols indicate localities where individuals of both haplotype subclades co-occur. Localities are numbered as in the Appendix.

Table 23. Distribution of specimens from the Whitewater to Desert Hot Springs and Morongo Valley areas along the San Gorgonio Pass Transect relative to their morphological group and mtDNA haplotype clade membership. Morphologically "intermediate" individuals are those with posterior probabilities between 0.1 and 0.9 of assignment to either "coastal" or "desert" morphological groups.

Whitewater – Desert Hot Springs
    (localities CA-247, CA-248, CA-252, CA-263, CA-264, CA-265)

	morphological assignment		
mtDNA clade	coastal	intermediate	desert
1B	3	0	1
2A	1	0	17

Morongo Valley
    (localities CA-338, CA-340, CA-341, CA-342)

	morphological assignment		
mtDNA clade	coastal	intermediate	desert
1B	7	1	13
2A	3	2	20

Morphology, mtDNA, and nuclear gene markers.— In this section, we detail further the degree of admixture between the "coastal" and "desert" groups in the contact region based on genotypic assignments derived from the 18 microsatellite loci. We use data from nine populations outside of the contact zone that we defined above as "parental" samples in both assignment test analyses. These include 5 samples of the "coastal" morphological and mtDNA group on the west side of the transect and 6 of the "desert" group to the north and east (Fig. 74, map). The two groups of reference samples are internally homogeneous yet strongly differentiated, as illustrated by the neighbor-joining tree (Fig. 74, bottom) based on an Fst matrix estimated by the method of Weir and Cockerham (1984), as implemented in GDA (Lewis and Zaykin, 2002). The average Fst value within the "coastal" group of five samples is 0.036, slightly but significantly greater than the mean of 0.022 for the six "desert" samples (ANOVA, Fisher's PLSD, p = 0.0106), but the difference between these groups of samples is an order of magnitude higher (mean Fst = 0.2198, comparison to either "coastal" or "desert" internal samples, ANOVA, Fisher's PLSD, p < 0.0001). The pattern where "coastal" samples are somewhat more differentiated among themselves then are "desert" samples is similar to what we observed for the Tehachapi transect.

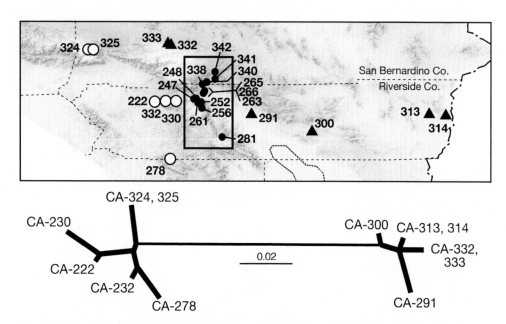

Figure 74. Above – The San Gorgonio Pass Transect with localities used in the microsatellite analyses mapped and numbered individually (see Appendix). Dark gray circles indicate samples of the "coastal" group (as defined by both morphology and mtDNA sequences); pale gray triangles identify those of the "desert" group. The rectangle marks all contact zone localities. Below -- neighbor-joining tree of relationships among the five "coastal" and six "desert" samples, based on a matrix of pairwise Fst distances. Branch lengths are drawn proportional, with the scale provided in the middle.

We provide summaries of allelic variation at the 18 microsatellite loci for these 9 non-contact zone populations in Table 24. The "desert" samples are significantly higher in all measures of variation (ANOVA, p ranging from 0.013 for mean number of alleles to < 0.0001 for expected heterozygosity). Again, microsatellite diversity in desert samples exceeds that of coastal ones, a pattern similar to that found for the Tehachapi Transect (Table 16). There is, however, a strong correlation between sample size and mean number of alleles for this set of samples (Z-test, r = 0.951, p < 0.0001), although this does not affect the other diversity measures.

Table 24. Measures of diversity in 18 microsatellite loci for nine samples (five of the "coastal" and four of the "desert" morphological and mitochondrial groups; see Figs. 55 and 73) of the San Gorgonio Pass Transect. Samples are identified by their mtDNA subclade and locality number(s) (see Appendix).

Sample (clade, locality number)	Mean N	Mean # alleles	Gene diversity	H_o	H_e	F_{is}
1B - Lone Pine Canyon (CA-324, 325)	4.8	4.18	0.733	0.616	0.670	0.094
1B - Aguanga (CA-278)	3.8	4.19	0.679	0.676	0.669	-0.013
1B - Lamb Canyon (CA-222)	15.7	6.88	0.653	0.609	0.690	0.021
1B - Banning (CA-232)	12.6	6.11	0.672	0.589	0.688	0.049
1B - Cabezon (CA-230)	5.4	4.69	0.705	0.589	0.622	0.060
2A - Cactus Flat (CA-332-333)	15.8	8.78	0.775	0.751	0.789	0.050
2A - Berdoo Canyon (CA-291)	19.2	9.28	0.799	0.788	0.779	-0.012
2A - Orocopia Mts. (CA-300)	38.4	12.11	0.737	0.781	0.813	0.040
2A - Big Maria Mts. (CA-313-314)	7.9	6.44	0.815	0.696	0.784	0.118[1]

[1] significantly different from 0 at p < 0.05, based on bootstrapping over loci with 1000 repetitions

We examined further the transition in microsatellite loci across the San Gorgonio Pass Transect by asking if there is evidence of genetic admixture within and among any of our samples, with specific reference to individuals from the contact localities between "coastal" and "desert" samples (Fig. 74). As with the Tehachapi Transect, we used the model-based method described by Pritchard et al. (2000) and implemented in the program STRUCTURE (Pritchard and Wen, 2003) to calculate probabilities of membership in either the "coastal" or "desert" groups for each individual in the transect, including both non-contact and contact population samples. Again, because of high consistency (r > 0.974 for all comparisons) among different runs where k, the population parameter, was allowed to vary from two (the number of different mtDNA clades) to nine (the number of individual "parental" populations), we report only data from the k = 2 analysis. We then determined the likelihood of specific hybrid class individuals (F1, F2, and first generation backcross to both "coastal" and "desert" parental types) using the NewHybrid program (Anderson and Thompson, 2002). This latter analysis gives an indication of on-going hybridization as opposed solely to the retention of an earlier episode of hybridization in the genotypic arrays.

All individuals belonging to each of the five "coastal" and four "desert" samples (Fig. 74) were assigned to their respective groups with posterior probabilities greater than 0.949 by the STRUCURE analysis (Table 25). The average assignment probability within the "coastal" samples is 0.996 (0.0007 standard error, range 0.969 to 0.999); that of the "desert" samples is 0.994 (0.00096 standard error, range 0.949-0.999). Moreover, most individuals from the group of contact samples were likewise assigned as a member of either the "coastal" or "desert" group with equally high probabilities, typically with such assignments to the source population geographically closest (Fig. 75). For example, samples from the western side of the general contact area (from the vicinity of Whitewater [samples CA-247 and CA-248] and along the margins of the San Jacinto and Santa Rosa Mts. [CA-256, CA-261, and CA-281]) all have probabilities of belonging to the "coastal" group greater than 0.976. Similarly, all individuals from localities in the vicinity of Desert Hot Springs, on the desert slope immediately east of San Gorgonio Pass (localities CA-252, CA-263, and CA-265-266), with the exception of a single individual, are all assigned to the "desert" group at a probability > 0.963. The single exception (MVZ 206814, from locality CA-263) has a probability of assignment to the "desert" group of 0.885 (Table 25).

The separation of "coastal" and "desert" microsatellite groups in the Whitewater, San Jacinto-Santa Rosa Mts., and Desert Hot Springs contact areas (Fig. 75) is congruent with morphological and mtDNA assignments. For example, all adult males from samples near Desert Hot Springs that are classified by microsatellites as "desert" also have the "desert" phallic type and haplotypes of the "desert" mtDNA subclade 2A. Similarly, all males at the Whitewater and San Jacinto-Santa Rosa Mts. localities are "coastal" in their microsatellite assignments and have the "coastal" glans and mtDNA subclade 1B haplotypes. Consequently, there is little evidence of either sympatry or genetic admixture between the "coastal" and "desert" types of woodrats immediately east of San Gorgonio Pass or along the eastern slopes of the San Jacinto and Santa Rosa Mts., although the closest localities of each group are less than two miles apart (between Whitewater [CA-247-248] and Desert Hot Springs [CA-252]; Figs. 74 and 75).

Table 25. Assignment probabilities for 291 individual woodrats, based on 18 microsatellite loci, along the San Gorgonio Pass Transect. The number assigned to "coastal" or "desert" groups (probabilities > 0.95) or arbitrarily classified as "intermediate" (probabilities 0.89 to 0.11 to either "coastal" or "desert" are given. Both "coastal" and "desert" samples are those identified in Fig. 74; other samples are those from within the contact area, also identified in Fig. 74.

Sample	Probability to "coastal" > 0.95	Intermediate	Probability to "desert" > 0.95
"coastal" samples	47	0	0
"desert" samples	0	0	88
north and west of Palm Springs			
Whitewater (CA-247-248)	6	0	0
Desert Hot Springs (CA-252, CA-263, CA-265-266)	0	1*	48
San Jacinto and Santa Rosa Mts. (CA-256, CA-261, CA-281)	13	0	0
Morongo Valley			
west end (CA-338)	0	6**	1
mid valley (CA-340)	0	0	6
east end (CA-341)	5	9***	56
Pioneertown (CA-342)	1	1****	2

*	probability to "desert" of 0.885 (MVZ 206814, from locality CA-263)
**	probabilities to "desert" range from 0.318 to 0.892.
***	probabilities to "desert" range from 0.410 to 0.864.
****	probability to "desert" of 0.810 (MVZ 199814)

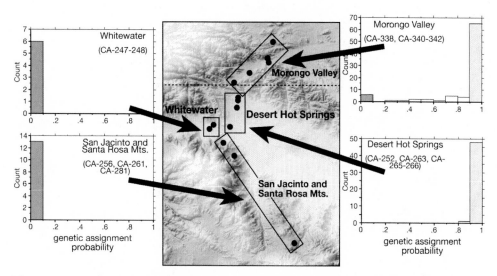

Figure 75. Map of contact region along the margins of San Gorgonio Pass, Riverside Co., California. Individual population samples are grouped into four regions, with histograms of probability assignments based on 18 microsatellite loci to the "desert" group illustrated for each. Only in the pooled Morongo Valley sample is there a clearly defined set of genetically intermediate individuals.

While there is no evidence of admixture in samples immediately east of San Gorgonio Pass in the transition between coastal and desert habitats, samples from Morongo Valley (Fig. 75) to the northeast between the San Bernardino and Little San Bernardino mountains include both "coastal" and "desert" individuals at the same localities, as defined by their morphology (Fig. 66) and mtDNA (Fig. 73). There is also evidence of genetic admixture at these localities, as 16 of 87 specimens exhibit microsatellite probability assignments between 0.89 and 0.11 to either "parental" group (Table 25). For example, six of seven individuals taken at the western edge of Morongo Valley (CA-338), nine of 70 individuals from the east end of Morongo Valley (CA-341), and one of four individuals from near Pioneertown (CA-342) have intermediate assignment probabilities. Individuals belonging to both the "coastal" mtDNA subclade 1B and "desert" subclade 2A co-occur at two of these localities (CA-338 and CA-341), with their houses completely intermixed at the latter (Fig. 67). There is also morphological evidence for past and possibly current hybridization at these localities, since adult males exhibit mixed mtDNA genotypes and glans morphologies (one of five individuals with "coastal" glans has a "desert" subclade 2A haplotype while 10 of 35 with the "desert" type of glans have "coastal" subclade 1B haplotypes).

Given the evidence of hybridization, we used the NewHybrid program (Anderson and Thompson, 2002) to determine the likelihood of F1, F2, and/or backcross individuals among the 87 specimens from Morongo Valley. This analysis identified no F1 or F2 individuals (no specimen has an assignment probability to these hybrid categories greater than 0.074 and 0.292, respectively). However, nine individuals were assigned as backcrosses to the "coastal" group (average probability = 0.811, range 0.497 to 0.968) and 21 were assigned as backcrosses to the "desert" group (average probability = 0.663, range 0.107-0.996). All 16 of the "intermediate" individuals identified in the STRUCTURE analysis are included in this group of 30 backcross hybrids. Clearly, therefore, specimens of both "coastal" and "desert" morphologies co-occur at local sites within Morongo Valley where evidence of hybridization is present. However, hybridization appears sporadic at the present time since all putative hybrids are of backcross origin and no F1 individuals were found in the available sample, even where the two species co-occur with intermixed houses in the eastern end of Morongo Valley (locality CA-338).

Importantly, hybridization is also asymmetrical, as the distribution of maternal genomes (as evidenced by mtDNA clade membership) is skewed in favor of "coastal" subclade 1B. For both backcross hybrid categories, most individuals have subclade 1B haplotypes (seven of nine of the "coastal" backcross individuals; 13 of 21 of the "desert" backcross individuals). These proportions are not significantly different (X^2 = 0.0370, p = 0.543), despite what might be expected given the opposite directions of backcrossing. Importantly, this bias towards subclade 1B maternal genomes among hybrid class individuals is the opposite of that of the pool of "pure" individuals (those with assignment probabilities of > 0.90 from the STRUCTURE analysis) at these same localities (five individuals of subclade 1B and 62 of subclade 2A), a highly significant skew (X^2 = 37.966, p < 0.0001). Hence, the bout(s) of hybridization that produced the class of hybrids must have been biased with female subclade 1B individuals preferentially mating with males of subclade 2A. Asymmetry in mating, with "coastal" females preferentially mating with "desert" males in Morongo Valley is opposite of the pattern observed at the Tehachapi Transect contact locality in Kelso Valley, although here it is not clear if it is coastal males or desert females that have the advantage, as all individuals share the same maternal genome.

Taxonomic considerations.—The San Gorgonio Pass Transect includes the holotypes of two named members of the *Neotoma lepida* group: *gilva* (type locality of Banning, Riverside Co. [locality CA-229]; ANSP 1665) and *bella* (type locality of Palm Springs, Riverside Co. [locality CA-267]; MCZ 5308). The holotype of *gilva* has a posterior probability of 1.0 of membership in the "coastal" morphological group and in the more restricted Coastal-e pooled locality that contains other specimens from the vicinity of Banning. This is true even if the holotype is removed from its pre-defined reference group and treated as an "unknown." This specimen also exhibits those qualitative characteristics of members of the "coastal" morphological group, namely a well-developed anteromedian flexus on M1 and relatively small and non-swollen tympanic bullae (BUL = 6.78 mm; BUW = 7.11 mm). The lacrimal bone is missing on both sides of the skull, so the position of this character relative to the fronto-maxillary suture cannot be determined.

The holotype of *bella* (Fig. 76), however, is one of the "intermediate" individuals from the Palm Springs sample, treated in all analyses as an "unknown" locality. The posterior probability of this specimen to the "desert" group is 0.574 and to the geographically adjacent Desert-e pooled sample is 0.623. Qualitatively, however, this specimen appears to be a rather typical member of the "desert" morphological group, as it combines a shallow to non-existent anteromedian flexus on M1 (Fig. 25), a lacrimal positioned so that the fronto-maxillary suture intersects the posterior one-third of the bone (Fig. 28), large and swollen tympanic bullae (BUL = 7.23 mm; BUW = 7.72 mm), and a large vomer exposed in the septum of the incisive foramen. Perceptively, Grinnell and Swarth (1913: 344) regarded the type of *bella* as "...an obvious intergrade between *desertorum* and *intermedia*..." although they concluded that it was "...nearest *desertorum*" (p. 345). Goldman (1910: 78) synonymized *bella* under his *N. desertorum*. Based on our results, both sets of previous authors were correct in their separate assessments of the holotype of *bella*.

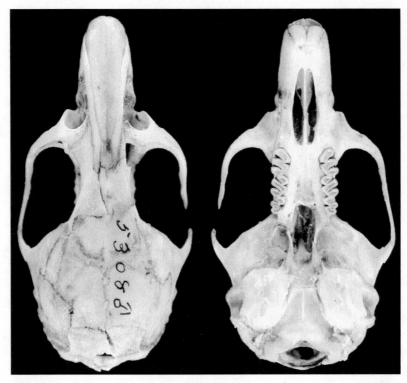

Figure 76. Dorsal (left) and ventral (right) views of the skull of the holotype of *Neotoma bella* (MCZ 5803, adult male). Note position of lacrimal relative to the fronto-maxillary suture in the dorsal view and both the smooth anterior-lingual border to the anteroloph of M1 in the ventral view and enlarged, swollen tympanic bullae in the ventral view, all characteristics of the "desert" morphological group.

San Diego Transect

This transect includes samples from coastal southern California in San Diego Co. and northwestern Baja California east through the Imperial Valley around the southern margins of the Salton Sea to the western side of the Lower Colorado River in Imperial Co. and northeastern Baja California. This set of localities again includes representatives of the "coastal" and "desert" global morphological groups we have described above as well as the coastal mtDNA subclade 1B and desert subclade 2A. In these two respects, the San Diego Transect is identical to the San Gorgonio Pass Transect immediately to the north, also described in detail above.

<u>Localities and sample sizes</u>.—Localities are reasonably densely packed on the western and eastern sides of the transect, but few samples are available from the large area of the Imperial Valley and sand dunes where woodrat habitat is effectively absent. We organized our samples by current subspecies designations (Grinnell, 1933; Hall, 1981) into three geographic groups for analysis (Fig. 77). A West sample encompasses localities along the coast and the foothills east of San Diego and Tijuana, each allocated to the subspecies *N. l. intermedia* (including its type locality at Dulzura, San Diego Co. [CA-157]). A Central sample includes localities from the drier mountainous region of eastern San Diego Co. south into the Sierra Juarez in Baja California, all assigned to *N. l. gilva*. Finally, an East sample includes localities from the Chocolate Mts. east of the Salton Sea to the lower Colorado River. This sample combines specimens allocated to both *N. l. lepida* and *N. l. grinnelli* (including its type locality near Picacho, Imperial Co. [CA-210]). Those specimens of this group from Riverside Co. are also included within the Desert-c group in the San Gorgonio Pass Transect. Unlike both the Tehachapi and San Gorgonio Pass Transects, in part because of the wide gap in the distribution of desert woodrats through the Imperial Valley, there are no geographically intermediate localities in this transect where both individuals of the "coastal" or "desert" morphological groups are either in apparent contact or nearby one another. Hence, we designate no locality samples to an "unknown" category.

We list individual localities by number, as in the Appendix, by pooled locality, along with sample size for craniodental morphometric (n_m), colorimetric (n_c), glans penis (n_g), and molecular samples (n_{DNA}), with respective museum catalog numbers for all specimens.

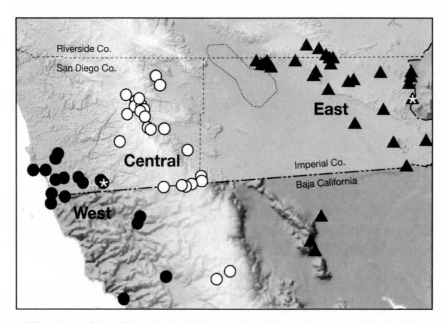

Figure 77. Map of localities included in the San Diego Transect, divided into three geographic groups. Circles identify localities from which all individuals examined possessed the "coastal" qualitative morphological characters; triangles represent those localities where all individuals are of the "desert" morphological type. The West group is coincident with the subspecies *intermedia*; the Central group with *gilva*; and the Desert group includes specimens allocated to both *lepida* and *grinnelli*. Localities marked by an asterisk denote the type localities of *intermedia* (Dulzura, San Diego Co., the circle in the West group) and *grinnelli* (Picacho, Imperial Co., the triangle in the East group).

West (total $n_m = 95$, $n_c = 65$, $n_g = 4$, $n_{DNA} = 6$)

CALIFORNIA:– S SAN DIEGO CO.: (1) CA-148: n_m=5, n_c=5, n_g = 2, n_{DNA}=4; MVZ 3771-3774, 3778, 197379-197382; (2) CA-149: n_m=1, n_c=8; MVZ 3109-3114, 3125; (3) CA-150: n_m=1; SDNHM 8031; (4) CA-151: n_m=3; SDNHM 2416, 16010-16011; (5) CA-152: n_m=5, n_c=7; MVZ 2857-2863; (6) CA-153: n_m=20, n_c=13; MVZ 3086-3090, 3092-3095, 3097-3099, 3101-3108; (7) CA-154: n_m=2; SDNHM 10783, 10785; (8) CA-154: n_m=1; SDNHM 22835); (9) CA-156: n_m=2; USNM 60697, 61000; (10) CA-157: n_m=21, n_c=13; ANSP 8343 [holotype of *N. l. intermedia*]; LACM 2054-2055, 75333; MVZ 3325, 7187-7188; UCLA 1167, 1184-1185, 1221, 3307, 3309-3310, 3323-2234; USNM 45098, 91567-91570; (11) not found: n_m=2; SDNHM 19588-19590.

MEXICO:– BAJA CALIFORNIA: (12) BCN-1: n_m=5, n_c=6; MVZ 39593-39595, 39597-39599; (13) BCN-2: n_m=1, n_c=2; USNM 81885-81886; (14) BCN-5: n_m=4; MVZ 39600-39603; (15) BCN-4: n_m=3; USNM 138280-138282; (16) BCN-12: n_m=3, n_c=6; USNM 137225, 137227, 137230, 137264-137266; (17) BCN-14: n_m=13, n_c=5, n_{DNA}=2; MVZ 148228-148232, 148238-148241, 184243-148245, 148250; (18) BCN-23: n_m=3; SDNHM 11822-11824.

Central (total n_m = 72, n_c = 55, n_g = 20, n_{DNA} = 18)
CALIFORNIA:– SAN DIEGO CO.: (1) CA-163: n_m=1; MVZ 150164; (2) CA-164: n_m=1; LACM 89273; (3) CA-165: n_m=1, n_c=1; MVZ 2785; (4) CA-166: n_m=1, n_c=1; SDNHM 23882; (5) CA-166a; n_g = 2; CSULB 10237-10238; (6) CA-167: n_m=1, n_c=1; MVZ 3775; (7) CA-168: n_m=1, n_c=2; MVZ 3776; (8) CA-169: n_m=1, n_g =1; MVZ 147685; (9) CA-170: n_m=1, n_c=4, n_g = 1, n_{DNA} = 4; MVZ 195241-195244; (10) CA-171: n_m=6; LACM 89275-89276, 89279-89282; (11) CA-172: n_m=2; SDNHM 2171, 2183; (12) CA-173: n_m=2, n_c=4, n_g =2; MVZ 147687, 147692; (13) CA-174: n_m=15, n_c=12; MVZ 7562-7563, 7590, 16624-16625; SDNHM 1237, 1533, 1871, 2276, 2337, 2514, 2519, 22674-22675; (14) CA-177: n_m=1, n_c=1, n_g =1; MVZ 95020; (15) CA-178: n_m=1; SDNHM 162; (16) CA-178a; n_g = 2; SDNHM 21168-21169; (17) CA-179: n_m=1, n_c=3; MVZ 7556-7557, 7574; SDNHM 2265; (18) CA-181: n_m=1, n_c=1; MVZ 122492; (19) CA-182: n_m=3; USNM 349448-349449, 349872; (20) CA-184: n_m=2; LACM 75823, 75873); (21) CA-185: n_m=14, n_c=14, n_g = 4, n_{DNA} = 14; MVZ 198335-198348; (22) CA-185a; n_g = 2; LACM 46672-46673; (23) CA-186: n_m=2, n_c=3; MVZ 7190-7191, 18940 (24) CA-187: n_m=1, n_c=1; MVZ 122491. IMPERIAL CO.: (25) CA-188a; n_g =2; LACM 49804-49805; (26) CA-189: n_g =3; MVZ 149351, 149354.

MEXICO:– BAJA CALIFORNIA: (27) BCN-3: n_m=5, n_c=6; MVZ 39589-39592, 39615-39616; USNM 60991; (28) BCN-8: n_m=4; SDNHM 12079-12080, 12095, 12121; (29) BCN-8a: n_g = 1; UNT 607; (30) BCN-9: n_m=3, n_c=1; MVZ 38165; SDNHM 4617, 5841; (31) BCN-10: n_m=1; MVZ 112883.

East (total n_m = 74, n_c = 34, n_g = 14, n_{DNA} = 26)
CALIFORNIA:– IMPERIAL CO.: (1) CA-190: n_m=10; LACM 75334, 75336, 75338-75345; (2) CA-195: n_m=1; LACM 991731; (3) CA-197: $_m$=1, n_c=1; MVZ 84768; (4) CA-198: n_m=2, n_c=2; MVZ 84766-84767; (5) CA-199: n_m=3; LACM 91647-91649; (6) CA-200: n_m=2; LACM 91654-91656; (7) CA-201: n_m=3; LACM 91651-91653; (8) CA-202: n_m=1; LACM 91650; (9) CA-204: n_m=4, n_c=3; MVZ 65885-65888; (10) CA-204a; n_g = 3; CSULB 10542-10544; (10) CA-205: n_m=12, n_c=5, n_g = 8, n_{DNA} = 26; MVZ 195259-195293, 215616-215640; (11) CA-206: n_m=5, n_c=8; MVZ 10446, 10448-10452, 10455-10456; (12) CA-207: n_m=2, n_c=2; MVZ 95023-95024; (13) CA-208: n_m=1, n_c=1; MVZ 10429; (14) CA-

209: $n_m=7$, $n_g = 1$; LACM 63700-63701, 63703, 63707, 63711, 75552, 75555; (15) CA-210: $n_m=5$, $n_c=6$; MVZ 10430, 10434-10435, 10437, 10438 [holotype of *N. l. grinnelli*], 10439, 10717; (16) CA-212: $n_m=1$, $n_c=1$, $n_g=1$; MVZ 95025. RIVERSIDE CO.: (17) CA-307: $n_m=3$; LACM 75485, 75487, 75491; (18) CA-308: $n_m=3$; LACM 75500, 75507, 75509; (19) CA-309: $n_m=3$; LACM 75521, 75523, 75526.

MEXICO:– BAJA CALIFORNIA: (20) BCN-100: $n_m=1$; USNM 136696; (21) BCN-101: $n_m=3$, $n_c=3$, $n_g = 1$; MVZ 111919-111921; (22) BCN-102: $n_m=1$, $n_c=2$; USNM 136648, 136996.

Morphometric differentiation.—We provide descriptive statistics for external and craniodental variables for each of the three sample groups in Table 26. There are significant differences ($p < 0.05$) for 24 of the 25 variable (excluding Mastoid Breadth [MB]) in pairwise comparisons between geographically adjacent samples, based on one-way ANOVA and using Fisher's PLSD for pairwise tests. The West and Central samples, which share the general gross morphological characteristics of the "coastal" morphological group differ in 17 of the 25 total variables, while the East sample (the sole representative of the "desert" morphological group) differs by 19 variables from each of the "coastal" samples. Most univariate character differences are grossly clinal across the transect, with character dimensions decreasing generally from west to east (Fig. 78, upper panels). However, within each pooled sample (West, Central, and East) this clinal pattern is true only for the two "coastal" samples, when taken together, not in the East sample. For example, in separate regression analyses of these two morphological groups, Condyloincisive Length (CIL; Fig. 78, upper right) exhibits a clinal pattern only from the coast to eastern San Diego Co. (samples West and Central: $r = -0.482$, $F_{(1,156)} = 47.283$, $p < 0.0001$; slope $= -1.814$), while there is no such relationship ($r = 0.007$, $F_{(1,68)} = 0.003$, $p = 0.956$; slope $= 0.031$) within the East sample, which is distributed over a somewhat broader range of longitude ($2.4°$ versus $1.6°$). Although there is clinal variation in most characters across the "coastal" group of samples (those belonging to our West and Central samples) are similar to one another in bullar dimensions, and they are markedly different relative to the East sample (Table 26; Fig. 78, bottom). This pattern of variation across the San Diego Transect mirrors that of the San Gorgonio Pass Transect; that is, gradually decreasing size from western to interior samples of the "coastal" morphological group with little to no differentiation among included "desert" samples, even though the latter span a similar range in longitude.

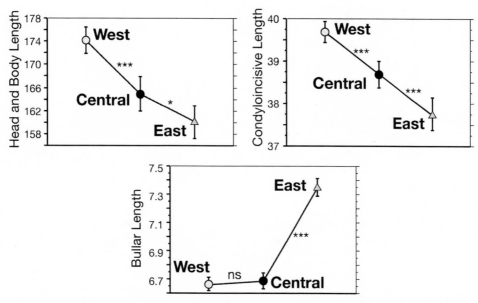

Figure 78. Plots of mean and 95% confidence limits for each sample of the San Diego Transect, arranged from west to east. Significance levels in pairwise comparisons between geographically adjacent samples are indicated: ns = non-significant, * p < 0.05, *** p < 0.001 (see text). Upper panels provide overall size shifts of the total body and skull; bottom panel illustrates changes in a bullar dimension.

Table 26. External and craniodental measurements of adult (age classes 1-5) specimens of *Neotoma lepida* along the San Diego Transect in southern California and northern Mexico (see Fig. 77). Means, standard error, sample size, and range are provided for each sample. Current subspecies allocations are given for each designated geographic group.

Variable	West (*intermedia*)	p[1]	Central (*gilva*)	p[1]	East (*grinnelli*)
external					
TOL	326.9±1.77		314.1±2.45		295.6±1.97
	82	***	65	***	62
	295-366		278-362		266-337
TAL	152.8±1.14		149.2±1.48		135.5±1.29
	82	*	65	***	62
	127-176		127-179		112-166
craniodental					
CIL	39.68±0.13		38.69±0.16		37.76±0.19
	97	***	71	***	70
	36.4-42.5		35.8-41.4		34.2-41.6
ZB	21.56±0.09		21.08±0.10		20.54±0.11
	97	***	71	***	70
	19.1-23.7		19.3-22.9		18.4-23.0
IOC	5.50±0.03		5.46±0.03		5.03±0.03
	97	ns	71	***	70
	5.0-6.1		4.9-6.1		4.6-5.5
RL	16.24±0.06		15.83±0.08		15.38±0.09
	97	***	71	***	70
	14.9-17.7		14.5-17.1		13.9-17.5

Variable	West (*intermedia*)	p[1]	Central (*gilva*)	p[1]	East (*grinnelli*)
HF	34.0±0.15		32.5±0.18		30.5±0.17
	90	***	71	***	67
	30-38		28-36		27-34
E	29.0±0.30		29.7±0.27		29.8±0.28
	68	ns	64	ns	64
	21-35		25-33		24-35
AW	7.65±0.03		7.47±0.03		7.00±0.03
	97	***	71	***	70
	7.1-8.2		6.9-8.4		6.5-7.7
OCW	9.66±0.05		9.49±0.04		8.90±0.04
	97	**	71	***	70
	8.9-10.4		8.9-10.4		8.0-9.6
MB	17.24±0.05		17.02±0.07		16.95±0.08
	97	ns	71	ns	70
	15.6-18.6		15.5-18.0		15.2-18.2
BOL	5.90±0.04		5.75±0.04		5.45±0.04
	97	**	71	***	70
	5.1-7.0		4.8-6.6		4.7-6.6

Table 26 (continued)

Variable	West (intermedia)	p¹	Central (gilva)	p¹	East (grinnelli)
NL	15.87±0.07	**	15.45±0.08	**	15.01±0.10
	14.1-17.6		14.1-16.9		13.5-17.0
	97		71		70
RW	6.63±0.03	ns	6.54±0.03	***	6.33±0.03
	5.6-7.4		6.0-7.2		5.7-7.0
	97		71		70
OL	14.36±0.05	***	13.95±0.05	**	13.64±0.07
	13.0-15.3		13.0-14.8		12.5-15.3
	97		71		70
DL	11.47±0.06	**	11.05±0.08	ns	10.99±0.09
	10.4-12.9		9.6-12.5		9.5-12.9
	97		71		70
MTRL	8.18±0.04	ns	8.10±0.04	***	7.87±0.04
	7.33-8.83		7.2-8.8		7.3-8.5
	97		71		70
IFL	8.76±0.04	*	8.59±0.06	*	8.41±0.06
	7.7-9.7		7.5-9.6		7.1-9.7
	97		71		70
PBL	18.25±0.07	***	17.69±0.08	ns	17.46±0.10
	16.9-19.7		16.0-19.5		15.7-19.5
	97		71		70

Variable	West (intermedia)	p¹	Central (gilva)	p¹	East (grinnelli)
MFL	7.98±0.05	**	7.73±0.05	*	7.56±0.05
	7.0-9.1		6.9-8.5		6.5-8.7
	97		71		70
MFW	2.61±0.02	ns	2.59±0.02	***	2.35±0.02
	2.1-3.2		2.0-3.1		1.8-2.8
	97		71		70
ZPW	4.23±0.02	**	4.09±0.03	ns	4.12±0.03
	3.8-4.9		3.3-4.6		3.6-7.7
	97		71		70
CD	15.85±0.05	**	15.6±0.06	ns	15.53±0.06
	14.7-17.0		14.4-16.6		14.1-17.0
	97		71		70
BUL	6.67±0.02	ns	6.69±0.03	***	7.35±0.03
	6.2-7.2		6.2-7.3		6.5-7.9
	97		71		70
BUW	7.04±0.02	ns	7.02±0.03	***	7.62±0.03
	6.6-7.9		6.4-7.5		7.0-8.3
	97		71		70

1 – ns = non-significant, * = p < 0.05, ** = p < 0.01, *** p < 0.001 (ANOVA, Fisher's PLSD posterior test of pairwise comparisons)

Multivariate analyses demonstrate the same degree of concordance shown by univariate character variation among samples of this transect. We performed both principal components and canonical variates analyses and provide factor loadings and standardized coefficients resulting from both in Table 27. The distribution of individuals of the three sample groups in a scatterplot of PC-1 versus PC-2 scores, which combine to represent 66.5% of the total pool of variation, as well as the vector plot of character loadings, is the same as presented for other transects that include both "coastal" and "desert" morphological groups (compare Fig. 79 with Figs. 40 and 63). The PC analysis once again contrasts largely uniformly high character loadings for all variables except the two bullar dimensions on the 1st axis against high loadings for the bullar characters on the 2nd, with PC-1 thus a general "size" axis and PC-2 scores representing differentiation largely in bullar size.

Significant size differences are present across the transect (ANOVA: West vs Central PC-1 scores, Fisher's PLSD, $p < 0.0001$; Central vs East, $p < 0.0001$), supporting the univariate analyses, which indicate clinal size differences among the three grouped samples (Fig. 74). Scores for PC-2, however, do not distinguish the West and Central samples (Fisher's PLSD, $p = 0.0631$), but both "coastal" samples are sharply different from the "desert" sample (Fisher's PLSD, $p < 0.0001$ in both comparisons). Again, the differences in PC-2 scores mirror the sharp contrast between the two "coastal" samples (West and Central) and the "desert" samples (East) in the bullar dimensions, BUL and BUW (Fig. 73, bottom).

Given the results of both univariate and PC analyses, it is thus not surprising that in a canonical analysis, whether a priori groups are the three West, Central, and East or the two-group "coastal" and "desert," discrimination among groups is complete with 100% correct assignments of all individuals based on their posterior probability scores. Nevertheless, one specimen (SDNHM 162, from the imprecise locality "near Borrego Spring" on the west side of the Salton Sea [locality CA-178]) is intermediate between "coastal" and "desert" samples in its CAN score and thus posterior probability of group assignment (Fig. 80). And, individuals with either Baja or western tip types of the glans penis co-occur at one locality near Ocotillo Wells (CA-178a; SDNHM 21167 and 21168). Thus, the ranges of both morphological groups may abut along the lower slopes of the Santa Rosa, Vallecito, and Jacumba mountains in eastern San Diego and western Imperial counties. It is possible that occasional hybrid individuals may result in these contact areas, as happens at other points on contact in areas to the north.

Table 27. Principal component factor loadings and standardized coefficients from the Canonical Variates Analysis for log-transformed cranial variables of the "coastal" and "desert" morphological groups of the San Diego Transect.

Variable	PC-1	PC-2	CAN-1*
log CIL	0.970	0.083	1.305
log ZB	0.922	0.019	0.207
log IOC	0.610	-0.496	0.254
log RL	0.892	0.058	0.013
log NL	0.854	0.060	0.173
log RW	0.771	-0.026	0.119
log OL	0.845	-0.024	-0.044
log DL	0.834	0.296	-0.658
log MTRL	0.370	-0.348	0.019
log IFL	0.781	0.180	0.078
log PBL	0.887	0.147	-0.176
log AW	0.638	-0.517	0.252
log OCW	0.747	-0.408	0.370
log MB	0.819	0.303	-0.242
log BOL	0.775	-0.118	-0.083
log MFL	0.737	0.092	-0.154
log MFW	0.442	-0.338	0.346
log ZPW	0.559	0.267	-0.044
log CD	0.785	0.249	-0.130
log BUL	-0.115	0.848	-0.746
log BUW	0.074	0.874	-0.709
eigenvalue	11.143	2.807	1.074
% contribution	53.06	13.37	100.00

* standardized canonical coefficients were determined from a two group analysis, with the West and Central samples included together as a "coastal" reference group for comparison to the single "desert" reference group.

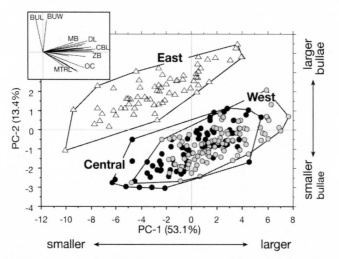

Figure 79. Scatterplot of individual scores on the first two principal components axes. Circles identify individuals of the West (gray-filled) and Central (black) samples of the San Diego Transect, those with a "coastal" morphology; triangles are specimens of the East sample, which is of the "desert" morphology. The inset box illustrates character vectors along both axes, which contrasts the highly positive character vectors for all variables exclusive of those of the bulla (BUL and BUW) on the 1st axis with the strongly positive bullar dimensions on the 2nd.

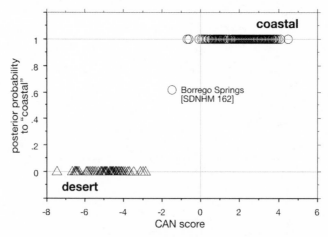

Figure 80. Plot of the posterior probability of membership to the "desert" morphological group from the Colorado Desert of Imperial Co. for each specimen examined in the San Diego Transect relative to the score of that individual on the first CAN axis.

Color variation.—We used the same three geographic groups to analyze colorimetric variation across the San Diego Transect, again limiting analysis to the X-coefficients of each of the four topographic regions of the study skin. As with other transects, all four variables are highly intercorrelated, with pairwise r-values ranging from 0.437 (Tail-X vs Chest-X; Z-value = 6.407, p < 0.0001) to 0.704 (Dorsal-X vs Lateral-X; Z-value = 11.978, p < 0.0001). There is also a rather uniform pattern of color change from the coast east to the Colorado River, as the mean value of each trichromatic X-coefficient is lower (=darker) in the "West" sample and progressively higher (paler) through the "Central" to the "East" sample (Table 28). This pattern of univariate color change from the coast to the interior deserts mirrors that observed for the other east-west transects presented above.

Because of the high intercorrelations between X-coefficients, we represent colorimetric variation along the San Diego Transect by a principal components analysis, presenting factor loadings for each of the four variables in Table 29. As with other colorimetric analyses, all four variables load highly and nearly equally on the first PC axis. On the second axis, however, Tail-X and Chest-X contrast strongly with one another, reflecting the relatively weak correlation between an individual's color score at these two topographic positions of the skin, while the other two variables remain relatively unimportant. Hence, the first PC axis, which explains more than 69% of the total pool of variation, expresses an overall trend of color variation across the entire body while the second axis, accounting for only 14% of the variation, suggests that specimens with dark tails may have paler sides, or the reverse.

Geographically, color, as summarized by the PC analysis, appears to vary clinally from west to east across the transect (Table 29). For example, the regression of PC-1 scores against the longitudinal position of each sample is highly significant (r = -0.777, $F_{(1,188)}$ = 285.942, p < 0.0001). However, as was true for the craniodental analysis above, the cline is really limited just to the pooled "coastal" samples (West and Central) and does not include the "desert" sample (East). Individual regressions for each of these two groups, which span nearly the same west-east distance (see Fig. 78 and presentation above for the craniodental PCA results) indicate a significant relationship between longitude and PC-1 scores for the "coastal" pooled sample (r = -0.607, $F_{(1,145)}$ = 84.477, p < 0.0001) but not for the collective samples of the "desert" group (r = 0.030, $F_{(1,41)}$ = 0.036, p = 0.851). No such clinal pattern is observed for PC-2 scores, whether all samples of the transect are included in an analysis or separate analyses are performed on the "coastal" and "desert" pooled groups.

Table 28. Descriptive statistics for the colorimetric X-measurement for the four regions of the woodrat study skins. Means ± one standard error, sample sizes, and ranges are given for each of three pooled geographic samples along the San Diego Transect (Fig. 77).

Variable	West	Central	East
Dorsal-X	7.95±0.16	10.73±0.30	14.36±0.39
	83	64	43
	4.7-11.8	6.5-16.1	9.0-20.9
Tail-X	5.15±0.15	8.44±0.33	9.29±0.35
	83	64	43
	2.5-9.4	3.4-16.2	5.5-14.2
Lateral-X	18.03±0.28	22.94±0.47	29.61±0.71
	83	64	43
	11.5-24.3	12.1-31.4	17.9-37.9
Chest-X	35.20±0.52	40.22±0.73	46.51±0.82
	83	64	43
	23.7-45.0	26.4-57.7	36.2-54.8

Table 29. Principal component eigenvalues and factor loadings of colorimetric variables from all samples of the San Diego Transect.

Variable	PC-1	PC-2	PC-3
Dorsal-X	0.875	0.029	-0.357
Tail-X	0.767	0.588	0.259
Lateral-X	0.882	-0.119	-0.197
Chest-X	0.805	-0.461	0.358
eigenvalue	2.779	0.573	0.361
% contribution	69.47	14.32	9.03

Two points are worth emphasizing regarding colorimetric change across the San Diego Transect. First, overall pelage color across the entire San Diego Transect gradually becomes paler from the coast to the mountains in eastern San Diego Co., after which there is a sharp and significant increase in paleness in comparisons between those interior mountains (the Central sample) and those of the desert ranges in Imperial Co. (the East sample; ANOVA, Fisher's PLSD pairwise comparison, p = 0.0017). Second, the strong clinal pattern of inter-locality variation in colorimetric variables in the "coastal" morphological samples contrasts strikingly with the lack of inter-locality differences among the collective "desert" samples.

Morphological – molecular concordance.—Our data for mtDNA haplotype variation is limited for the San Diego Transect and inadequate for strong conclusions. Nevertheless, we have multiple samples that span the geographic range of the "coastal" morphological group, all members of which belong to the coastal subclade 1B, but only one sample of the "desert" morphological group (Tumco Mines [locality CA-205]), specimens of which belong to the desert subclade 2A. Given the shift from subclade 1B to 2A along the eastern margins of San Gorgonio Pass and on opposite sides of the Coachella Valley to the north, we suspect that the entire geographic distribution of our East sample of the San Diego Transect will belong to the desert subclade 2A while the entirety of samples in San Diego Co. (our West and Central groups) will belong to the coastal subclade 1B. Thus, we are confident that future studies in this region will document strong concordance between the groups identified by morphological (external, craniodental, and colorimetric) criteria and the mtDNA clades to which these groups belong.

We have only four samples along this transect for which microsatellite data are available. Three of these represent the "coastal" morphological and mtDNA group (CA-148, n = 4, from our West pooled sample, and CA-170, n = 4, and CA-185, n = 14, of the Central sample). One sample is from the "desert" group and thus belongs to the East pooled sample (CA-205, n = 26). The few samples, and particularly the small sample sizes for two localities, preclude detailed analyses. However, all of the 18 loci at least are in Hardy-Weinberg equilibrium for the Jacuma (CA-185) and Tumco Mine (CA-205) samples. Moreover, the pattern of differentiation expressed between the four localities mirrors their respective placements in the "coastal" and "desert" groups. Average Fst values between the three "coastal" samples are 0.0295 while the average difference between these and the single "desert" sample is an order of magnitude larger, at 0.282 (range, 0.261 to 0.302). Consequently, the association of morphological assignment, mtDNA haplotype, and microsatellite characterization across the San Diego Transect is completely concordant among these three data sets as well as with the transitions observed above for the San Gorgonio Pass Transect to the geographically immediate north.

TRANSITIONS WITHIN THE "COASTAL" MORPHOLOGICAL GROUP

Coastal California

This transect includes samples along coastal California, from Alameda Co. in the north to the Mexican border in San Diego Co. All samples belong to the "coastal" morphological group defined by qualitative craniodental characters, and all specimens that were sequenced are members of either the mtDNA coastal subclade 1C (central coast ranges) or 1B (south coast). The sampled range spans the complete distribution of two currently recognized subspecies (*californica* and *petricola*) and part of two others (*gilva* and *intermedia*). Because the type localities of three of these (*californica*, *intermedia*, and *petricola*) are within the sample areas, the holotypes and/or topotypes of each are included in the analyses below. The colorimetric samples also include the lectotype of *N. intermedia sola*, currently listed as a synonym of *gilva* (Hall, 1981).

We grouped localities for the analysis of both craniodental morphometrics and colorimetric variables into nine pooled samples comprised of geographically adjacent samples (Fig. 81). From north to south are two pooled samples of *californica*, one composed of localities in the Diablo Range (Diablo sample) and a second with those localities in the Gavilan Range (Gavilan sample), and a group encompassing all known specimens of *petricola* (Santa Lucia sample). Those localities from the southern coastal ranges and the Mt. Pinos area are combined into west to east units that largely conform to the Coastal-w and Coastal-e groups used in the Tehachapi Transect (Central Coast and Tejon samples, respectively). These are successively followed to the south by a "south coast" sample that includes specimens from Ventura and western Los Angeles counties along with, immediately to the east, Coastal-w and Coastal-c samples, which are the same groups used in the San Gorgonio Pass Transect. The southernmost sample is that from western San Diego Co., the West sample used in the San Diego Transect. We included the same group structure used in other transect analyses to provide continuity and comparison among our separate analyses of geographic trends. For the northern geographic groups, localities were assembled either on explicit geographic grounds (i.e., because of a common distribution in the Diablo Range) or taxonomic reasons (samples of *californica* from the eastern side of the Salinas Valley and those of *petricola* from the western side).

Localities and sample sizes.—In the analyses below, we include craniodental measurements from 486 adult specimens and colorimetric variables from 352 individuals. A total of 123 of these were sequenced for the mtDNA *cyt-b* gene, with sequences from eight of the nine geographic samples (only the South Coast sample from Ventura and Los Angeles counties lacks sequenced specimens). As in previous analyses, we list specimens of each geographic sample for which we examined the craniodental (n_m), colorimetric (n_c), and glans penis (n_g) morphology, and mtDNA sequences (n_{DNA}), as well as the specific localities taken from specimen labels and museum catalog numbers (numbered as in the Appendix).

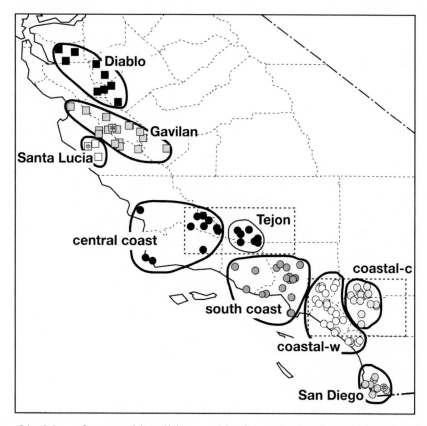

Figure 81. Map of grouped localities used in the analysis of craniodental and color characteristics of woodrats distributed along coastal California. The Central Coast and Tejon samples are largely the same as two of those employed in the Tehachapi Transect (dashed box), and the Coastal-w and Coastal-c samples are those used in the San Gorgonio Pass Transect (dashed box). The Gavilan sample includes the type locality of *californica*, the Santa Lucia sample includes the type locality of *petricola*, and the San Diego sample includes the type locality of *intermedia* (symbols with asterisks, respectively).

Diablo (total n_m = 49, n_c = 46, n_g = 7, n_{DNA} = 5)

CALIFORNIA:– ALAMEDA CO.: (1) CA-1: n_m = 14, n_c = 11, n_g = 1; MVZ 134194-134197, 134200-1324208, 134210; (2) CA-2: n_m = 10, n_c = 10, n_g =4; MVZ 102629-102638; (3) CA-3: n_m = 4, n_c = 8; MVZ 94796-94803, 94810. MERCED CO.: (4) CA-7: n_m = 4, n_c = 4, n_g = 2, n_{DNA} = 4; MVZ 195982-195985; (5) CA-8: n_m = 9, n_c = 5; LACM 3216, MVZ 14087-14092, 57236-57237. SANTA CLARA CO.: (6) CA-15: n_m = 4, n_c = 4; USNM 150455, 150871, 150873-150874. STANISLAUS CO.: (7) CA-5: n_m = 1 n_c = 1, n_{DNA} = 1; MVZ 197371; (8) CA-6: n_m = 3, n_c = 3; MVZ 101216-191218.

Gavilan (total n_m = 54, n_c = 50, n_g = 3, n_{DNA} = 28)

CALIFORNIA:– FRESNO CO.: (1) CA-22: n_m = 6; USNM 149770-149775. MONTEREY CO.: (2) CA-23: n_m = 1, n_c = 1; MVZ 198681; (3) CA-24: n_m = 1, n_c = 2; MVZ 108682-108683; (4) CA-25: n_m = 2, n_c = 3; MVZ 108684-108686; (5) CA-34: n_m = 9 n_c = 7, n_g = 2, n_{DNA} = 9; MVZ 195214-195222; (6) CA-35: n_m = 1, n_c = 1; MVZ 108680; (7) CA-36: n_m = 1, n_c = 1; MVZ 65111. SAN BENITO CO.: (8) CA-10: n_m = 1, n_c = 1; USNM 150875; (9) CA-11: n_m = 1, n_c = 1; MVZ 122321; (10) CA-12: n_m = 5, n_c = 4, n_g = 1, n_{DNA} = 5; MVZ 196061-196065; (11) CA-13: n_m = 2, n_c = 2, n_{DNA} = 2; MVZ 196072-196073; (12) CA-15: n_m = 1, n_c = 1; MVZ 28206; (13) CA-16: n_m = 4, n_c = 4; USNM 67144, 150884-150886; (14) CA-17: $_m$ = 2, n_c = 4; MVZ 73003-73006; (15) CA-18: n_m = 1, n_c = 3; MVZ 73007-73009; (16) CA-19: n_m = 1, n_c = 1; MVZ 73010; (17) CA-20: n_m = 12, n_c = 11, n_{DNA} = 12; MVZ 195223-195234; (18) CA-21: n_m = 3, n_c = 3; USNM 150878-150880.

Santa Lucia (total n_m = 15, n_c = 29, n_g = 1, n_{DNA} = 11)

CALIFORNIA:– MONTEREY CO.: (1) CA-27: n_c = 1; USNM 118138; (2) CA-28: n_c = 3; USNM 118123, 118133-118134; (3) CA-29: n_m = 3, n_c = 3; MVZ 30202 [holotype of *petricola* von Bloeker], 30203-30204; (4) CA-30: $_m$ = 11, n_c = 11, n_g = 1, n_{DNA} = 11; MVZ 186294-186298, 195326-195331; (5) CA-31: n_c = 5; USNM 118283-118284, 118285, 118289, 118382; (6) CA-32: n_c = 4; USNM 118293, 118385-118386; (7) CA-33: n_m = 1, n_c = 1; USNM 118384.

Central Coast (total n_m = 36, n_c = 30, n_g = 10, n_{DNA} = 29)

CALIFORNIA:– SAN LUIS OBISPO CO.: (1) CA-37: n_m = 3, n_c = 4; USNM 43469/31596, 43484-43485/31723-31725; (2) CA-38: n_m = 3, n_c = 3, n_{DNA} = 3; MVZ 196759-196761; (3) CA-39: n_m = 1, MVZ 128812; (4) CA-40: n_m = 6, n_c = 6, n_g = 4, n_{DNA} = 6; MVZ 195975-195980; (5) CA-41: n_m = 8, n_c = 8, n_g = 3, n_{DNA} = 8; MVZ 195967-195974; (6) CA-42: n_m = 5, n_c = 2, n_{DNA} = 5; MVZ 196754-196758; (7) CA-43: n_m = 1, n_c = 1, n_{DNA} = 1; MVZ 195966; (8) CA-44:

$n_m = 5$, $n_c = 5$, $n_g = 2$, $n_{DNA} = 5$; MVZ 195961-195965; (9) CA-45: $n_m = 1$, $n_c = 1$, $n_g = 1$, $n_{DNA} = 1$; MVZ 195981. SANTA BARBARA CO.: (10) CA-48: $n_m = 1$; LACM 48984; (11) CA-48a: $n_m = 1$; USNM 130135; (12) CA-48b: $n_m = 2$; LACM 20772-20773.

Tejon (total $n_m = 45$, $n_c = 38$, $n_g = 19$, $n_{DNA} = 39$)

CALIFORNIA:– KERN CO.: (1) CA-56: $n_c = 1$; USNM 31516 – skin, lectotype of *N. desertorum sola*; (2) CA-57: $n_m = 3$, $n_c = 3$, $n_g = 1$, $n_{DNA} = 3$; MVZ 198581-198583; (3) CA-58: $n_m = 1$; MVZ 28207; (4) CA-59: $n_m = 1$, $n_c = 1$; SDNHM 5988; (5) CA-59: $n_m = 4$ $n_c = 1$, $n_g = 2$, $n_{DNA} = 4$; MVZ 196097-196100; (6) CA-60: $n_m = 32$, $n_c = 32$, $n_g = 16$, $n_{DNA} = 32$; MVZ 196771-196779; 196809-196821, 200730-200739. VENTURA CO.: (7) CA-49: $n_m = 3$; MVZ 5331, 5376, 5378; (8) CA-50: $n_m = 1$; MVZ 5333.

South Coast (total $n_m = 58$, $n_c = 12$, $n_g = 2$)

CALIFORNIA:– LOS ANGELES CO.: (1) CA-103: $n_m = 2$, $n_c = 1$; MVZ 5335, LACM 48966; (2) CA-105: $n_m = 5$; LACM 10194, 21143, 22964, 29972, 48940; (3) CA-106: $n_m = 1$; LACM 48938; (4) CA-107: $n_m = 1$; LACM 48968; (5) CA-109: $n_m = 6$, $n_c = 5$; MVZ 9483-9484, USNM 5954, LACM 43650-43651, 44986, 44989-44990; (6) CA-110: $n_m = 3$, $n_c = 6$; MVZ 5550, 5580-5582, 6984-6985; (7) CA-111: $n_m = 2$; MVZ 5381, 6984; (8) CA-112: $n_m = 1$; LACM 91457; (9) CA-112: $n_m = 5$; LACM 20562, 48948, 48950-48952; (10) CA-113: $n_m = 1$; LACM 48953; (11) CA-113a; $n_g = 1$; CSULB 7617; (12) CA-114: $n_m = 3$; LACM 48957, 48959, 48961; (13) CA-115: $n_m = 1$; LACM 48967; (14) CA-116: $n_m = 13$; LACM 8430-8431, 8472, 10343-10344, 10346, 10349, 48932, 48944, 87760, 87774, 96146-96147; (15) CA-117: $n_m = 7$; LACM 48941-48943, 87469, 87471, 91055, 91441). VENTURA CO.: (15) CA-51: $n_m = 4$; UCLA 2402, 2456, 2461, 2466; (16) CA-52: $n_m = 2$; LACM 48971-48972; (17) CA-53: $n_m = 1$; LACM 3436; (18) CA-53a; $n_g = 1$; CSULB 3110.

Coastal-w (total $n_m = 90$, $n_c = 16$, $n_g = 6$, $n_{DNA} = 4$)

CALIFORNIA:– LOS ANGELES CO.: (1) CA-118: $n_m = 3$; LACM 88273-88274, 88276; (2) CA-119: $n_m = 3$; LACM 20635, 20637, 21234; (3) CA-120: $n_c = 1$; MVZ 9059; (4) CA-121: $n_m = 2$; LACM 29962, 44974; (5) CA-122: $n_m = 1$; LACM 20616; (6) CA-123: $n_m = 1$, $n_c = 1$; MVZ 25557; (7) CA-124: $n_m = 2$; LACM 49628, 96062; (8) CA-125: $n_m = 1$, $n_c = 1$; MVZ 65593. ORANGE CO.: (9) CA-128: $n_m = 1$; LACM 29940; (10) CA-129: $n_m = 1$; LACM 31729; (11) CA-130: $n_m = 1$; LACM 29938; (12) CA-131: $n_m = 1$; LACM 29939; (13) CA-132: $n_m = 2$; LACM 29960, 44969; (14) CA-133: $n_m = 7$, $n_c = 8$; MVZ 2359-2366; (15) CA-133a; $n_g = 1$; CSULB 2945; (16) CA-134: $n_m = 4$; LACM 29933,

29935, 29937, 44972; (17) CA-134a;-b; n_g = 2; CSULB 2580, 3132; (18) CA-135: n_m = 1; LACM 29951; (19) CA-136: n_m = 12, n_c = 5; MVZ 2342-2346; LACM 29950, 29952, 29954-29957, 29959; (20) CA-137: n_m = 2; LACM 29947-29948; (21) CA-138: n_m = 2; LACM 29941-29942; (22) CA-139: n_m = 24; LACM 44061-44064, 44066=44068, 44073-44074, 44076-44077, 44079, 44082-44086, 44088-44089, 44091, 44117, 44126-44127; (23) CA-140: n_m = 1; LACM 44970; (24) CA-141: n_m = 2; USNM 149849, 149851; (25) CA-142: n_m = 4, n_g = 2, n_{DNA} = 4; MVZ 197375-197378. SAN DIEGO CO.: (26) CA-143: n_m = 2; SDNHM 20609-20610; (27) CA-144: n_m = 5; MVZ 150157, 150159-150162; (28) CA-145: n_m = 1; SDNHM 20606; (29) CA-146: n_m = 3; SDNHM 20607-20608, 20613; (30) CA-147: n_m = 1; SDNHM 20612; (31) CA-147a, n_g = 1; SDNHM 19588.

Coastal-c (total n_m = 55, n_c = 72, n_g = 2, n_{DNA} = 2)
CALIFORNIA:– RIVERSIDE CO.: (1) CA-214: n_m = 2, n_c = 6; MVZ 2434-2437, 3410-3411; (2) CA-215: n_m = 5, n_c = 7; USNM 93982-93984, 93986, 93989, 93994-93995; (3) CA-216: n_m = 1, n_c = 3; MVZ 2534-2536; (4) CA-217: n_m = 1; SDNHM 6644; (5) CA-218: n_m = 1; USNM 70039; (6) CA-219: n_m = 2, n_c = 1, n_g =1; MVZ 121585-121586); (7) CA-220: n_m = 2, n_c = 2; MVZ 90673, 90720; (8) CA-221: n_m = 5, n_c = 5; MVZ 88525-88529. SAN BERNARDINO CO.: (9) CA-315: n_m = 6, n_c = 6; MVZ 24499-24504; (10) CA-316: n_m = 10, n_c = 19; MVZ 2663, 2666-2669, 2673-2679, 2681-2687; (11) CA-317: n_m = 9, n_c = 9; USNM 127985-127988, 127990-127994; (12) CA-318: n_m = 1; SDNHM 16015; (13) CA-319: n_m = 6, n_c = 9; MVZ 2590-2598; (14) CA-320: n_m = 1, n_c = 1; MVZ 77229; (15) CA-321: n_m = 1, n_c = 2; MVZ 77227-77228; (16) CA-322: n_m = 2, n_c = 2, n_g = 1, n_{DNA} = 2; MVZ 196052-196053.

San Diego (total n_m = 76, n_c = 58, n_g = 2, n_{DNA} = 3)
CALIFORNIA:– SAN DIEGO CO.: (1) CA-148: n_m = 6, n_c = 5, n_g = 2, n_{DNA} = 3; MVZ 3771-3774, 3778, 197380-197383; (2) CA-149: n_m = 5, n_c = 8; MVZ 3109-3114, 3125-3126; (3) CA-150: n_m = 1; SDNHM 8031); (4) CA-151: n_m = 3; SDNHM 2416, 16010-16011; (5) CA-152: n_m = 5, n_c = 7; MVZ 2857-2863; (6) CA-153: n_m = 20, n_c = 23; MVZ 3084, 3086-3090, 3092-3108; (7) CA-154: n_m = 2; SDNHM 10783, 10785); (8) CA-155: n_m = 1; SDNHM 22835; (9) CA-156: n_m = 2; USNM 60697, 61000; (10) CA-157: n_m = 22, n_c = 13; ANSP 8343 [holotype of *intermedia*], MVZ 7187-7188, 54095, UCLA 1167, 1184-1185, 1221, 2054-2056, 3307, 3309-3310, 3323-3325, USNM 45096, 45098, 91567-91570; (11) CA-161: n_m = 2, n_c = 2; MVZ 3039-3040; (12) CA-162: n_m = 2; SDNHM 148, USNM 54849; (13) CA-163: n_m = 1; MVZ 150164; (14) CA-165: n_m = 1; MVZ 2785; (15) CA-187a: n_m = 3; SDNHM 19588-15990.

Craniodental and colorimetric variation.—Each craniodental variable exhibits significant variation among the nine sample areas, with p < 0.001 in all cases (ANOVA, $F_{(8,473)} > 3.298$ for each variable). Of external variables, HF and E also vary significantly (ANOVA, $F_{(4,336 \text{ to } 4,448)} > 3.336$, p < 0.001), while TOL and TAL do not (ANOVA, $F_{(8,419)} < 1.463$, p > 0.169). We provide a summary of the descriptive statistics (mean, standard error, and range) for each of the nine geographic groups in Table 30.

In general, cranial length increases slightly from north to south (the correlation of CIL vs. latitude: r = 0.548, Z-value = 12.188, p < 0.001), although the Diablo sample at the northern end is not significantly different from those in San Diego (ANOVA, $F_{(1,123)} = 1.340$, p = 0.2492). Most of the trend of increasing size is from the Gavilan and Santa Lucia samples south to the Coastal-w sample, with little change in size from there further to the south (Fig. 82). CIL is only significantly different in three pairs of geographically adjacent samples across the transect (arrows in Fig. 82): (1) Diablo vs. Gavilan in the north (ANOVA, Fisher's PLSD, p = 0.0022), (2) Tejon vs. South Coast (p = 0.0101), and (3) Coastal-w vs. San Diego (p = 0.0138). Trends in other craniodental variables largely mirror the pattern exhibited by CIL.

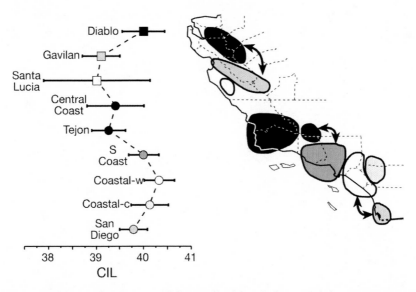

Figure 82. Mean and 95% confidence limits of the cranial measurement CIL (Condyloincisive Length) among geographically pooled samples of the Coastal California Transect, arranged from north to south as in Fig. 81. Arrows indicate those geographically adjacent samples that are significantly different (by ANOVA, Fisher's PLSD test).

Table 30. External and cranial measurements of adult (age classes 1-5) specimens of the nine geographic samples of *Neotoma lepida* along the Coastal California Transect in California (from Alameda Co. south through San Diego Co.; Fig. 81). Means, standard error, sample size, and range are provided for each sample.

Variable	Diablo	Gavilan	Santa Lucia	Central Coast	Tejon	South Coast	Coastal-w	Coastal-c	San Diego
external									
TOL	323.9±2.4	328.5±2.4	323.3±6.4	329.4±3.5	321.5±2.0	323.5±2.4	321.6±2.7	326.9±1.9	328.8±2.6
	45	40	10	28	36	55	66	50	68
	291-367	302-360	297-352	295-363	291-345	285-375	302-360	297-364	295-440
TAL	152.9±1.7	155.0±1.3	155.9±4.1	156.2±2.1	151.2±1.3	149.8±1.3	149.7±1.6	151.7±1.4	154.3±1.3
	45	40	10	28	37	55	66	50	68
	128-175	134-170	141-175	134-174	133-162	128-174	120-174	125-176	134-179
HF	33.6±0.2	33.8±0.2	33.6±0.5	33.8±0.2	33.3±0.2	33.6±0.2	33.6±0.2	34.6±0.2	33.7±0.1
	48	44	11	32	42	56	68	53	72
	31-37	30-37	31-36	31-36	31-35	30-38	27-37	30-38	30-36
E	28.9±0.3	31.9±0.5	29.5±0.7	32.8±0.3	30.9±0.3	29.2±0.3	29.8±0.3	29.4±0.4	29.0±0.3
	38	43	10	27	36	43	52	23	61
	20-32	26-41	25-32	29-36	25-34	25-35	26-35	26-34	21-35
craniodental									
CIL	39.99±0.	39.10±0.2	39.01±0.1	39.39±0.3	39.25±0.2	39.98±0.2	40.31±0.2	40.11±0.2	39.77±0.2
	49	44	11	32	41	58	91	54	73
	36.8-42.7	36.9-42.8	36.4-41.2	36.9-43.0	36.9-41.8	36.5-43.0	37.2-43.7	37.7-43.9	36.4-42.5
ZB	21.92±0.1	21.4±0.1	21.5±0.3	21.71±0.2	21.91±0.1	21.94±0.1	22.1±0.1	21.9±0.1	21.92±0.1
	49	44	11	32	42	58	91	54	73
	20.2-23.7	19.4-23.2	19.7-22.9	20.2-23.7	20.6-22.9	19.9-23.4	20.3-24.5	20.6-24.3	19.1-23.7

Table 30 (continued)

Variable	Diablo	Gavilan	Santa Lucia	Central Coast	Tejon	South Coast	Coastal-w	Coastal-c	San Diego
IOC	5.51±0.02	5.57±0.03	5.59±0.05	5.35±0.03	5.56±0.04	5.61±0.03	5.55±0.03	5.50±0.12	5.54±0.03
	49	44	11	32	42	58	91	54	73
	5.2-5.9	5.22-5.93	5.4-5.9	5.0-5.7	5.0-6.1	5.2-6.0	5.1-6.2	5.0-6.2	5.0-6.1
RL	16.38±0.1	16.01±0.1	16.17±0.3	16.32±0.1	16.23±0.1	16.37±0.1	16.5±0.1	16.64±0.1	16.35±0.1
	49	44	11	32	42	58	91	54	74
	14.8-17.8	14.2-18.3	14.9-17.3	15.0-18.2	15.2-17.4	15.3-17.7	15.1-18.9	15.2-18.6	14.9-17.7
NL	15.69±0.1	15.46±0.1	15.55±0.19	15.78±0.15	15.44±0.1	16.00±0.09	16.37±0.1	16.31±0.1	15.91±0.1
	49	44	11	32	42	58	91	54	74
	14.4-17.5	13.8-17.9	14.4-16.2	14.3-17.7	14.5-16.4	14.5-17.1	14.6-18.6	14.5-18.4	14.1-17.6
RW	6.57±0.03	6.35±0.04	6.30±0.19	6.30±0.04	6.53±0.04	6.57±0.03	6.84±0.04	6.76±0.05	6.63±0.04
	49	44	11	32	42	58	91	54	74
	6.0-7.0	5.9-7.1	5.8-6.6	5.8-6.7	5.8-7.2	6.0-7.0	5.9-7.5	6.0-7.7	6.0-7.4
OL	14.28±0.1	14.21±0.1	14.07±0.17	14.28±0.10	14.07±0.1	14.59±0.06	14.85±0.1	14.53±0.1	14.41±0.1
	49	44	11	32	42	58	91	54	73
	12.8-15.4	13.6-15.2	13.1-14.8	13.2-15.3	13.2-14.7	13.6-15.7	13.6-15.9	13.5-15.9	13.4-15.3
DL	11.64±0.1	11.33±0.1	11.19±0.22	11.29±0.13	11.27±0.1	11.51±0.06	11.79±0.1	11.49±0.1	11.50±0.1
	49	44	11	32	42	58	91	54	74
	10.2-13.1	10.2-13.2	10.2-12.2	10.0-12.9	10.2-12.2	10.4-12.6	10.3-13.5	10.1-13.3	10.4-12.7
MTRL	8.15±0.03	8.07±0.04	8.37±0.10	8.32±0.04	8.22±0.04	8.22±0.04	8.07±0.04	8.35±0.04	8.18±0.04
	49	44	11	32	42	58	91	54	74
	7.4-8.8	7.4-8.8	8.0-9.0	7.8-8.7	7.5-8.8	7.5-8.7	7.1-8.9	7.8-9.0	7.3-8.8
IFL	8.55±0.07	8.36±0.07	8.83±0.17	8.71±0.09	8.73±0.06	8.75±0.05	8.95±0.06	8.89±0.6	8.73±0.05
	49	44	11	32	42	58	91	54	74
	7.5-9.3	7.1-9.4	8.2-9.9	7.7-9.8	7.9-9.4	7.7-9.7	7.9-10.2	8.0-9.7	7.7-9.7

Table 30 (continued)

Variable	Diablo	Gavilan	Santa Lucia	Central Coast	Tejon	South Coast	Coastal-w	Coastal-c	San Diego
PBL	18.51±0.1	18.01±0.1	18.02±0.26	17.99±0.15	17.85±0.1	18.27±0.09	18.44±0.1	18.16±0.1	18.34±0.1
	16.6-20.1	16.9-19.9	16.8-19.3	16.6-19.8	16.2-19.2	15.4-19.8	16.8-20.2	15.7-20.3	17.1-19.7
	49	44	11	32	42	58	91	54	74
AW	7.60±0.03	7.61±0.03	7.53±0.06	7.60±0.05	7.51±0.03	7.70±0.03	7.73±0.03	7.62±0.03	7.62±0.03
	7.1-8.0	7.2-8.1	7.2-7.8	6.9-8.1	7.1-7.9	7.3-8.3	7.0-8.6	7.2-8.3	7.2-8.2
	49	44	11	32	42	58	91	54	74
OCW	9.92±0.04	9.70±0.04	9.64±0.11	9.66±0.06	9.50±0.04	9.64±0.03	9.68±0.04	9.52±0.04	9.71±0.03
	9.3-10.7	9.3-10.4	8.9-10.1	9.0-10.5	9.0-10.1	8.9-10.2	9.1-10.6	8.9-10.7	9.0-10.4
	49	43	11	32	42	58	91	54	73
MB	17.4±0.06	17.1±0.06	17.14±0.17	17.35±0.10	17.4±0.06	17.47±0.05	17.6±0.06	17.3±0.07	17.2±0.06
	16.5-18.3	16.2-17.9	16.3-17.6	16.5-18.4	16.5-18.2	16.0-18.2	16.2-18.9	16.3-18.9	15.6-18.6
	49	43	11	32	42	58	91	54	73
BOL	6.03±0.05	5.83±0.05	5.95±0.09	5.92±0.07	5.98±0.05	5.96±0.05	6.01±0.04	5.93±0.06	5.92±0.04
	5.26-6.80	5.1-6.5	5.5-6.3	5.3-6.8	5.3-6.8	5.0-6.6	5.2-7.0	5.1-7.0	5.1-7.0
	49	43	11	32	42	58	91	54	73
MFL	7.51±0.07	7.43±0.06	7.61±0.09	7.70±0.09	7.77±0.05	7.73±0.06	8.04±0.05	7.95±0.05	8.04±0.06
	6.59-8.71	6.6-8.0	7.2-8.0	6.8-8.4	6.8-8.4	6.9-8.6	6.6-9.4	7.2-8.7	7.0-9.1
	49	43	11	32	42	58	91	54	73
MFW	2.75±0.03	2.65±0.04	2.79±0.04	2.73±0.03	2.59±0.03	2.64±0.03	2.75±0.02	2.72±0.3	2.62±0.03
	2.3-3.3	2.1-3.2	2.6-3.1	2.4-3.0	2.2-3.0	2.2-3.0	2.2-3.6	2.3-3.2	2.1-3.1
	49	44	11	32	42	58	91	54	73
ZPW	4.17±0.03	4.05±0.04	4.12±0.09	4.06±0.04	4.05±0.03	4.12±0.03	4.17±0.03	4.20±0.03	4.21±0.3
	3.7-4.9	3.7-4.5	3.5-4.5	3.7-4.5	3.7-4.4	3.6-4.8	3.6-5.1	3.8-4.8	3.3-4.9
	49	44	11	32	42	58	91	54	74

Table 30 (continued)

Variable	Diablo	Gavilan	Santa Lucia	Central Coast	Tejon	South Coast	Coastal-w	Coastal-c	San Diego
CD	15.7±0.05	15.6±0.06	15.52±0.15	15.65±0.08	15.5±0.06	15.49±0.05	15.9±0.05	16.2±0.08	15.9±0.06
	49	43	11	32	42	58	91	54	73
	14.8-16.3	14.9-16.6	14.8-16.1	14.8-16.4	14.9-16.4	14.7-16.2	14.9-17.3	15.2-18.0	14.7-17.0
BUL	6.68±0.3	6.73±0.04	6.65±0.06	6.86±0.05	6.68±0.02	6.61±0.3	6.74±0.03	6.84±0.03	6.65±0.03
	49	43	11	32	42	58	91	54	73
	6.3-7.0	6.0-7.5	6.4-6.9	6.3-7.3	6.1-7.1	6.0-7.1	5.8-7.4	6.4-7.5	6.2-7.3
BUW	6.96±0.03	7.05±0.03	6.79±0.04	7.06±0.04	6.92±0.03	6.98±0.04	7.20±0.03	7.23±0.03	7.10±0.03
	49	43	11	32	42	58	91	54	73
	6.5-7.5	6.6-7.4	6.6-7.0	6.6-7.5	6.4-7.3	6.3-7.7	6.5-7.9	6.6-7.7	6.6-7.6

We investigated the general trends in craniodental characters along the Coastal California Transect further by principal components analysis. Only the first three axes have eigenvalues greater than 1.0. The first axis explains 48.7% of the variation but subsequent axes explain a maximum of only 6.4% (Table 31). PC-1 is a general size axis, as all variables except MTRL load highly and positively, with correlation coefficients of individual variables versus PC-1 scores always highly significant ($p < 0.0001$ in all comparisons, based on Z-values); that for MTRL vs. PC-1 is not significant (factor loading = 0.061, Z-value = 1.295, p = 0.1952). The best univariate indicator of overall size is CIL, which has both the highest factor loading (0.958, Table 31) and is most highly correlated with PC-1 scores (Z-value = 40.650, $p < 0.0001$). As a consequence, the pattern of PC-1 scores grouped into the pooled samples of the transect very closely mirrors that illustrated for cranial length (Fig. 82).

Our nine pooled samples of the Coastal California Transect broadly overlap in PC space (Fig. 83), without the kind of separation that is apparent in all PC analyses that include both "coastal" and "desert" morphological types (compare Fig. 83 to Figs. 40, 53, and 63). Despite this extensive overlap, however, there is an overall significant difference among all nine pooled samples (ANOVA comparison of all nine geographic samples for PC-1 scores, for example: $F_{(8,443)} = 6.073$, $p < 0.0001$) as well as between some geographically adjacent ones. For example, the Diablo sample is significantly larger in general size from the geographically adjacent Gavilan sample to the immediate south (ANOVA, Fisher's PLSD, p = 0.0157) and the Coastal-w sample in southern California is significantly different from those to the immediate northwest (South Coast, Fisher's PLSD, p = 0.0021) and south (San Diego, Fisher's PLSD, p = 0.0026). On the other hand, large geographic areas are similar in their position in PC space, including the two samples from opposite sides of the Salinas Valley (Santa Lucia vs. Gavilan, Fisher's PLSD, p = 0.9479). All samples from the middle of the transect (Gavilan, Santa Lucia, Central Coast, and Tejon) are similar in general size, although smaller than those to the immediate north or south; and those samples from south of the Transverse Ranges, with the exception of the slightly larger Coastal-w sample, are also uniform in general size.

Table 31. Principal component eigenvalues and factor loadings for log-transformed craniodental variables of adult specimens of the Coastal California Transect.

Variable	PC-1	PC-2	PC-3
log CIL	0.958	-0.018	-0.070
log ZB	0.872	-0.117	0.047
log IOC	0.238	-0.447	0.626
log RL	0.852	0.144	-0.123
log NL	0.830	0.109	-0.191
log RW	0.743	-0.061	0.080
log OL	0.780	-0.100	-0.094
log DL	0.873	-0.263	-0.191
log MTRL	0.061	0.777	0.344
log IFL	0.772	0.008	-0.239
log PBL	0.832	-0.035	-0.056
log AW	0.498	0.024	0.513
log OCW	0.498	-0.061	0.390
log MB	0.809	-0.090	0.194
log BOL	0.735	-0.070	0.151
log MFL	0.690	0.017	-0.242
log MFW	0.355	0.153	0.125
log ZPW	0.583	0.107	-0.062
log CD	0.762	-0.007	-0.015
log BUL	0.484	0.579	0.070
log BUW	0.651	0.122	-0.076
eigenvalue	10.236	1.337	1.246
% contribution	48.7	6.4	5.9

Overall the pattern of both univariate and multivariate comparisons among samples of the Coastal California Transect is one of slight geographic variation, mostly of increasing size from the middle parts of the transect both north and south. Where differences between adjacent samples do occur, these are slight. These patterns contrast sharply with the west-to-east transects that include both "coastal" and "desert" morphological groups, where sharp transitions in univariate and multivariate characters are consistently observed.

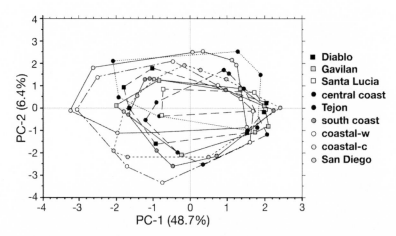

Figure 83. Ellipses encompassing all individual scores of each geographic sample identified by name and symbol (see Fig. 81) of the Coastal California Transect on the first two principal components axes. Components were extracted from the covariance matrix of log-transformed craniodental variables; factor loadings and eigenvalues are provided in Table 31.

The pattern of colorimetric variation across the Coastal California Transect is similar to that of craniodental variables, whether examined by univariate or multivariate methods. All X-coefficient values of the four topographic regions of the skin exhibit statistically significant differences among the nine geographic samples (ANOVA, where $F_{(8,343)}$ is always > 15.146 and p is always < 0.0001). However, most of the differences observed result from the inclusion of the very pale individuals from the vicinity of the Carrizo Plain in western Kern and eastern San Luis Obispo counties (the majority of the Central Coast sample in this transect; Fig. 81). This pooled sample was shown in the Tehachapi Transect to be substantially paler than all other "coastal" morphological samples and equivalent to the palest of the "desert" groups further east along that transect. Secondarily, because of the general increase in paleness from west to east as was evident in both the Tehachapi and San Gorgonio Pass transects, the Tejon, Coastal-w, and Coastal-c samples are also paler than those on the coast. If these more interior samples are excluded from the Coastal California Transect, then all remaining samples are homogeneous (by ANOVA, using Fisher's PLSD test for multiple comparisons, the lowest p-value obtained for any pair of samples is 0.1404 [Santa Lucia versus South Coast]). These differences are readily apparent by a simple inspection of the mean and range of X-coefficients for each topographic region of the skin (Table 32).

Table 32. Colorimetric X-coefficients for the four topographic regions of the study skin for geographic samples of the Coastal California Transect. Mean, standard error, sample size, and range are given for each sample.

Sample	Dorsal-X	Tail-X	Lateral-X	Chest-X
Diablo	7.954±0.180	7.968±0.257	20.041±0.336	38.404±0.700
	47	47	47	47
	5.61-10.51	4.89-11.71	14.45-26.59	28.54-50.18
Gavilan	8.215±0.187	8.166±0.271	20.396±0.397	39.112±0.714
	53	53	53	53
	5.48-11.43	4.62-14.10	14.01-27.51	24.89-52.84
Santa Lucia	8.46±0.278	9.277±0.408	19.818±0.645	40.228±1.114
	29	29	29	29
	6.00-11.47	4.54-14.16	10.68-25.66	25.97-45.02
Central Coast	10.842±0.388	11.232±0.454	27.554±1.056	47.350±1.430
	30	30	30	30
	5.63-14.31	5.28-17.07	12.79-37.130	31.84-62.11
Tejon	8.873±0.225	8.370±0.288	23.39±0.565	46.665±1.381
	41	41	41	41
	6.12-11.97	5.34-14.02	15.09-30.39	9.46-60.68
South Coast	7.592±0.391	5.612±0.446	20.844±0.515	37.716±1.158
	12	12	12	12
	5.95-10.11	3.50-8.50	17.83-23.79	32.70-43.79
Coastal-w	8.975±0.388	6.096±0.537	20.29±0.567	36.539±1.161
	16	16	16	16
	6.25-10.94	3.06-11.63	15.65-24.31	27.37-46.57
Coastal-c	9.353±0.161	6.444±0.279	20.446±0.329	37.924±0.655
	66	66	66	66
	6.53-12.56	3.22-15.66	13.43-27.79	25.04-50.29
San Diego	7.988±0.178	5.014±0.169	17.86±0.357	35.587±0.587
	58	58	58	58
	5.06-11.84	2.78-9.41	11.52-24.34	26.19-43.34

The pattern of differences among samples is similar across all four topographic regions of the skin, as the X-coefficient variables are all highly intercorrelated (p < 0.0001 in all comparisons. Correlation coefficients range from 0.263 [Dorsal-X versus Chest-X] to 0.462 [Tail-X versus Lateral-X]). Because of this, we used a principal components analysis to summarize colorimetric variation in a multivariate context. Of the four axes extracted, PC-1 is the only one where the eigenvalue is greater than 1.0, and it alone explains 53.9% of the variation (Table 33). Each X-coefficient loads positively and strongly, suggesting that PC-1 expresses primarily the degree of darkness or paleness around the entire body. PC-2 contrasts the Dorsal-X coefficient with Chest-X and Tail-X, PC-3 contrasts Tail-X with Chest-X, and PC-4 is primarily variation in Lateral-X.

Table 33. Principal component factor loadings for colorimetric variables of adult specimens of the Coastal California Transect.

Variable	PC-1	PC-2	PC-3	PC-4
Dorsal-X	0.669	0.685	-0.108	-0.269
Tail-X	0.733	0.268	0.590	-0.207
Lateral-X	0.821	0.075	-0.004	0.566
Chest-X	0.706	0.458	-0.507	-0.188
eigenvalue	2.157	0.759	0.617	0.471
% contribution	53.9	18.9	15.4	11.8

Scores of each PC colorimetric axis are significantly heterogeneous among the pooled samples, with p-values ranging from < 0.0001 (PC-1, PC-2, and PC-3) to 0.0126 (PC-4). Fig. 84 illustrates the geographic pattern of variation in PC-1. As with the univariate analyses, the three northern samples (Diablo, Gavilan, and Santa Lucia) are darker than those to the immediate south (Central Coast; Fig. 84). The very pale Central Coast sample contrasts with all others, as does the more intermediate color of the Tejon sample. The three northern samples are similar in degree of darkness to those from southern California, but the San Diego sample is substantially darker than all others, significantly so in all cases (ANOVA, Fisher's PLSD test, p < 0.0001 in all pairwise comparisons). The Central Coast sample includes four individuals from San Luis Obispo (locality CA-37), near the coast, as well as those from the more interior Carrizo Plain. The four coastal specimens are darker (mean PC-1 = 1.0835, range 0.818 to 1.378) than samples to the north (Santa Lucia or Gavilan, combined mean PC-1 = 0.073 and 0.214, respectively) or south (South Coast, mean PC-1 = 0.545), and do not overlap in their PC-1 scores with the Carrizo specimens (mean = -2.3277, range -4.077 to -1.157). Excluding the Carrizo Plain and Tejon individuals, color is overall uniformly dark along coastal California from Alameda Co. to San Diego Co.

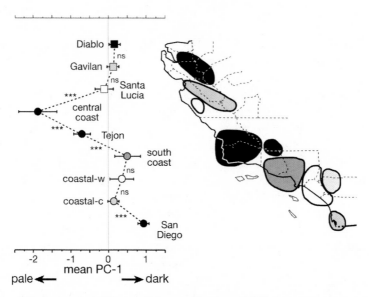

Figure 84. Means and 95% confidence limits of PC-1 scores for colorimetric variables among the nine geographic samples of the Coastal California Transect. The degree of statistical significance (based on ANOVA, Fisher's PLSD posterior tests) between geographically adjacent samples is indicated: ns = non-significant, *** = p < 0.001.

Morphological – molecular concordance.—There is a shift along coastal California in the distribution of mtDNA haplotypes, with individuals belonging to the coastal subclade 1C distributed from Tejon Pass north through the coastal ranges and those belonging to subclade 1B occurring south of Tejon Pass (Figs. 6 and 47, above). The area of clade overlap, in the vicinity of Tejon Pass between Mt. Pinos and the Tehachapi Mts., is included in our comparisons among coastal samples of the Coastal California Transect. We asked if mtDNA clade boundaries coincided with any morphological shift by pooling samples into their respective clade groups, assuming that specimens for which no sequence data are available had haplotypes of the same clade as those in the geographic area that had been sequenced. Comparisons between the mtDNA clades yielded significant differences for both the craniodental and colorimetric PC-1 scores (ANOVA, $F_{(1,404)} = 22.345$, $p < 0.001$ and $F_{(1,283)} = 24.560$, $p < 0.001$, respectively). However, the means are quite similar and their distributions are broadly overlapping in both cases: (1) craniodental 1B mean = -0.211 (range -1.931 to 2.341), 1C mean = 0.287 (range -2.021 to 2.186); and (2) colorimetric 1B mean = 0.491 (range -1.541 to 2.168), 1C mean = 0.113 (range -1.602 to 1.378). Thus, the difference in morphology, either craniodental or color, between mtDNA clades is slight, even if significant. As we described above, both craniodental and colorimetric variables either do not vary along the transect or exhibit only a clinal pattern.

A clinal pattern of differentiation from north to south along the Coastal California Transect is also apparent in the limited data we have for the 18 microsatellite loci. Pooled samples ranging in size from two to 38 are available for seven of the nine geographic groups identified in Fig. 81 (no samples are available for the Diablo or South Coast groups). Estimates of pairwise Fst among all samples generated by the GDA software (Lewis and Zaykin, 2002) are strongly correlated with the geographic distance among them. A Mantel test for the matrix correlation between log(Fst) and log(geographic distance) is highly significant, with $r = 0.678$, $Z = -38.030$, $p < 0.003$ (IBD, v. 1.52; Bohonak, 2002). There is no apparent step in this cline across the geographic boundary between mtDNA clades, but samples are not available to examine this possibility in detail.

Taxonomic considerations. —The Coastal California Transect includes the entire distributional range of two subspecies, *californica* and *petricola*, in the northern part of the range of woodrats along the California coast and all of the range of the southern coast subspecies *intermedia*, except the localities in northwestern Baja California (compare Fig. 81 to Map 435 in Hall, 1981). Because available samples include either holotypes or topotypes of each of these subspecies, we are able to compare each in detail and thus to evaluate their validity as formally recognizable taxa with our larger and more extensive geographic samples and with more sophisticated methods of character analysis.

Both *intermedia* and *californica* were described in 1894, and their respective descriptions make no mention of the other taxon. Rather, both descriptions provide only comparisons to a species, *N. mexicana*, which does not occur within California and is not even a close relative (Matocq et al., 2007). These comparisons by themselves, therefore, provide no basis for evaluating *californica* with respect to *intermedia*. The type locality *intermedia* is Dulzura, San Diego Co. (locality CA-157), part of our pooled San Diego geographic sample. The type locality of *californica* is Bear Valley, San Benito Co. (locality CA-16), included within our Gavilan sample. Von Bloeker (1938), in his description of *petricola*, did explicitly compare his new taxon to both *californica* and *intermedia* and thus provided a set of characters for us to evaluate. The type locality of *petricola* is Abbott's Ranch, Arroyo Seco, in Monterey Co. (locality CA-29), which is included within our Santa Lucia sample. Because *petricola* and the geographically adjacent *californica* have restricted geographic ranges and thus their comparisons are unlikely to be influenced by geographic variation, we evaluate *petricola* with respect to *californica* first.

Von Bloeker (1938, p. 203) differentiated his *petricola* from *californica* on the basis of smaller size, relatively longer tail, shorter hind foot and ear, shorter but broader skull, and darker overall color. However, in pairwise comparisons between our Santa Lucia and Gavilan samples (which include the type localities of both subspecies), we failed to substantiate any of these differences, except for a slight difference in ear height. For example, our measure of body size, TOL, is not significantly different between these two samples (ANOVA, $F_{(1,48)} = 0.842$, p = 0.3633), nor is relative Tail Length (the ratio of TAL to TOL: $F_{(1,48)} = 2.817$, p = 0.0998), Hind Foot Length (HF: $F_{(1,53)} = 0.152$, p = 0.6983), skull length (CIL: $F_{(1,52)} = 0.040$, p = 0.8426), or skull breadth (using either ZB [$F_{(1,53)} = 0.1181$, p = 0.7324] or MB [$F_{(1,52)} = 0.026$, p = 0.8726] as measures of breadth). Of the mensural characters considered diagnostic by von Bloeker, only ear height exhibited statistical significance between our samples (E: $F_{(1,36)} = 6.481$, p = 0.0129), but given the problems in comparing this measurement across temporal samples where different criteria for the measurement were likely used, even this difference is questionable. Further, the Santa Lucia and Gavilan samples cannot be distinguished by overall color (PC-1 scores: $F_{(1,89)} = 2.932$, p = 0.0903), with *petricola* actually somewhat paler than *californica* (mean PC-1, which scales darkness [see Fig. 84], is -0.115 for *petricola* and 0.124 for *californica*). This inability to differentiate between samples of *petricola* and *californica* extends to the nuclear genome, as the Fst between samples from near King City on the east side of the Salinas Valley (localities CA-12, CA-13, CA-20, and CA-34) and that from near the type locality of *petricola* in Arroyo Seco on the west side of the Valley (locality CA-30) is only 0.0065. Given all available data, therefore, *petricola* cannot be distinguished from other samples of "coastal" morphological form of the desert woodrat, regardless of the subspecies to which these samples are currently allocated (Hall, 1981).

The evaluation of *californica* (including *petricola*) and *intermedia* is somewhat more complex, because of the failure of previous authors to denote diagnostic differences among them. The much broader geographic area covered by their respective ranges further complicates this evaluation. An appropriate comparison, however, is critical to current management concerns, as the California Department of Fish and Game (http://www.dfg.ca.gov/) lists the San Diego Woodrat (the subspecies *N. l. intermedia* as per Hall, 1981) as a "Species of

Special Concern" for conservation reasons, based on presumed habitat conversion throughout much of the taxon's southern California distribution.

Interestingly, both Goldman (1932) in his revision of the *lepida* group and Grinnell (1933) in his synopsis of California mammals list *californica* as a junior synonym of *intermedia*, which has date of publication priority (January versus May, 1894). Neither author, however, provided reasons for their respective decisions. Similarly, while Hall (1981) regarded *intermedia* and *californica* as valid subspecies of *N. lepida*, he provided no rationale for his decision. However, Hall's mapped ranges of both subspecies (Map 435, p. 759) do provide a geographic hypothesis that we can test. He drew the distribution of *intermedia* along the southern and central California coast as far north as San Luis Obispo, an area that includes our San Diego, South Coast, Coastal-w, Coastal-c, and Central Coast geographic samples. And, he mapped the range of *californica* to encompass the Diablo and Gavilan ranges east and southeast of the San Francisco Bay Area, a region encompassed by our Diablo and Gavilan samples.

To compare *californica* and *intermedia*, we limit our analyses to coastal populations by excluding the Tejon geographic sample (which is allocated to the subspecies *gilva*) from the Coastal California Transect, and examine the mean scores (with 95% confidence limits) of the first PC axis for both the craniodental and colorimetric variables (Fig. 85). As above, most geographically adjacent population samples are not significantly different from one another, although a few differences do exist. In the craniodental analysis, for example, there are two areas of transition between neighboring samples: a decrease in overall general size between the Diablo and Gavilan samples in the north and a further decrease between the South Coast and Coastal-w samples in the south. Both of these transitions, however, lie within the ranges of each subspecies rather than at the hypothesized boundary between them. For the color variables there are again only two geographic points of statistical transition between adjacent samples. However, one of these (the transition between Santa Lucia [and Gavilan, data not shown] and the Central Coast, limited to the four specimens from San Luis Obispo [locality CA-37] in this analysis) is coincident with Hall's (1981) hypothesized boundary between *californica* and *intermedia*. This difference is highly significantly different (ANOVA, Fisher's PLSD test, $p = 0.0008$). The second significant shift is between samples in the eastern Los Angeles basin (Coastal-w) and San Diego ($p < 0.0001$). Overall, therefore, the pattern of both craniodental and colorimetric variation along coastal California (Fig. 85) is complex, with multiple geographically limited clines that may be parallel or opposite one another. There is no unified, sharp transition across the hypothesized subspecies boundary at the southern end of the coastal ranges.

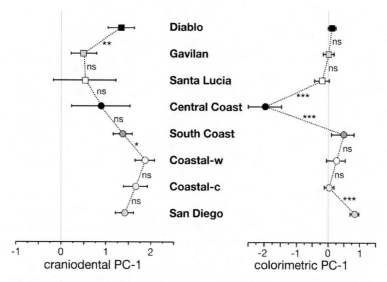

Figure 85. Summary of character variation among the coastal samples of central and southern California that span the collective ranges of the subspecies *californica* (Diablo and Gavilan samples), *petricola* von Bloeker (Santa Lucia sample), and *intermedia* (the five southern samples, from Central Coast to San Diego). The left panel shows means and 95% confidence limits for PC-1 scores based on the 21 craniodental variables; the right panel are means and 95% confidence limits for PC-1 scores based on the four X-coefficient colorimetric variables. Levels of significance between geographically adjacent samples are indicated (ns = non-significant, * = p < 0.05, *** = p < 0.001; ANOVA, Fisher's PLSD posterior tests for comparisons among multiple samples).

Peninsular and Insular Baja California

Woodrats of the *Neotoma lepida* group inhabit the entire length of Baja California, from the US border to the Cape region, across all ecoregions and at elevations, ranging from sea level to above 8,000 feet in the Sierra San Pedro Martír. They also occur (or did until recently; see Álvarez-Castañeda. and Cortés-Calva, 1999) on five islands on the Pacific side and nine on the Gulf side of the peninsula. It is not a surprise, therefore, that morphological variation is both substantial and reflected by a current taxonomy that recognizes 22 taxa either at the species or subspecies levels (e.g., Hall, 1981; Álvarez-Castañeda. and Cortés-Calva, 1999).

　　We began our analysis of morphometric diversity among samples of desert woodrats from Baja California, including both peninsular and insular populations, by asking: "to what degree are the "coastal" and "desert" morphologies recognized across the range of the *lepida* group within the US reflected in these samples?" To do this we initially performed a principal components analysis on the 21 log-transformed craniodental variables. We included all samples from the mainland and islands off both coasts of Baja California and those of "desert" morphology that formed the East pooled sample in the San Diego Transect (see Fig. 77). The latter included three localities from northeastern Baja California (localities BCN-100, BCN-101, and BCN-102) and those from southeastern California. The analysis included 1014 adult specimens, 70 of the "desert" group, 737 from the remainder of the peninsula, and 207 from the combined set of 11 insular taxa.

　　We illustrate a scatterplot of individual scores on the first two PC axes, which combine to explain 70.9% of the total pool of character variation, in Fig. 86. Individuals of the "desert" morphological group are separable from all others from the peninsula and islands, with the latter two groups overlapping broadly. The pattern of character vectors is also similar to that seen in comparisons that include both "coastal" and "desert" groups in other analyses, either globally within the USA (e.g., Fig. 23) or across each transect between the "coastal" and "desert" groups (Figs. 40, 53, 63, and 79). Extensive variation within the "non-desert" peninsular and insular samples is also apparent. Separation of individuals of the "desert" morphology from the remaining peninsular and all insular samples is emphasized in a canonical variates analysis using the same 21 craniodental variables (Fig. 87), where "desert," "peninsular," and "insular" samples are segregated into a priori defined groups. The "desert" sample is sharply segregated from the other two along the first axis (ANOVA, Fisher's PLSD, $p < 0.0001$), which explains 82% of the variation, while peninsular and insular samples overlap broadly on this axis (ANOVA, $p = 0.8252$) although weakly separated on the second canonical axis (ANOVA, $p < 0.01$).

　　Because of these analyses, we excluded all samples that exhibit the "desert" morphology (including the three from extreme northeastern Baja California [localities BCN-100, BCN-101, and BCN-102]) from further examination of craniodental variation throughout the peninsula and islands.

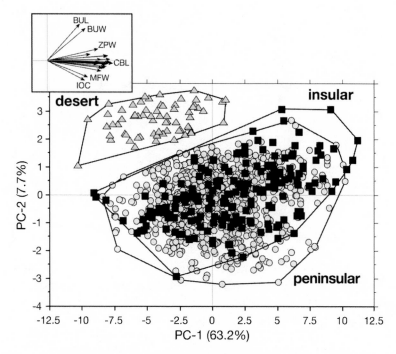

Figure 86. Scatterplot of PCA scores on the first two axes derived from 21 craniodental variables, with samples of the "desert" morphology from northeastern Baja California and southeastern California contrasted with other peninsular and insular samples from the remainder of Baja California. Inset on upper left is the character vector diagram illustrating how individual variables influence the distribution of specimens on both axes.

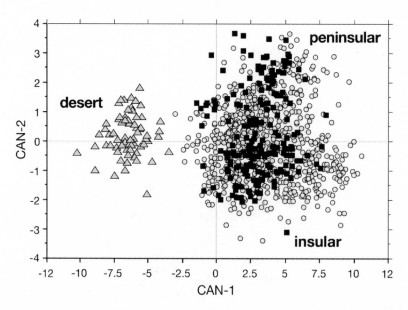

Figure 87. Scatterplot of the first two canonical variates axes comparing samples of the "desert" morphology from extreme northeastern Baja California and adjacent southeastern California (see San Diego Transect, above) and both peninsular (gray-filled circles) and insular (black squares) samples from the remainder of Baja California.

Localities and sample sizes.—Subsequent to excluding the few "desert" morphological samples from the northeastern corner of Baja California, we analyzed the remainder of our samples in an iterative manner. Both the many formally recognized taxa in this group of samples and the broad distribution of individual scores in the PC and CAN analyses underscore the substantial geographic variation present in this region. We began our analyses by segregating all peninsular localities into taxa based on mapped ranges (Hall, 1981; Alvarez-Castañeda and Cortés-Calva, 1999) and then determined if these samples constituted homogeneous groupings based on Fisher's PLSD posterior tests of PC-1 and PC-2 scores. This set of analyses resulted in 22 separate geographic groups distributed along the peninsula, each delineated by overall size and/or differential "shape" axes, and with each formal taxon but two (*aridicola* and *notia*) comprising two to three sample groups. Because all 11 insular populations are formally recognized taxa, we included each of these as a separate sample. The final groups of samples used in our analyses are indicated in the map, Fig. 88. We designate peninsular samples by letter, from Group A (the northwestern coast, subspecies *intermedia*) to Group V (the southeastern coast, a sample of *arenacea*), and insular samples by their taxon name. As in previous analyses, we list specimens of each geographic sample for which we examined the craniodental (n_m), colorimetric (n_c), and glans penis (n_g) morphology, and mtDNA sequences (n_{DNA}), as well as the specific localities and museum catalog numbers (numbered as in the Appendix).

Peninsular samples:

Group A [*intermedia*] (total $n_m = 48$, $n_c = 45$, $n_{DNA} = 3$)
      MEXICO:– BAJA CALIFORNIA: (1) BCN-1; $n_m = 5$, $n_c = 6$; MVZ 39593-39595, 39597-39599; (2) BCN-2; $n_m = 1$, $n_c = 2$; USNM 81885-81886; (3) BCN-4; $n_m = 3$, $n_c = 4$; USNM 138280-138282; (4) BCN-5; $n_m = 5$, $n_c = 5$; MVZ 39600-39603, 44190; (5) BCN-6; $n_{DNA} = 1$; CIB 8660; (6) BCN-12; $n_m = 3$, $n_c = 6$; USNM 137225, 137227, 137230, 137264-137266; (7) BCN-14; $n_m = 13$, $n_c = 5$, $n_{DNA}=2$; MVZ 148228-148232, 148238-148241, 184243-148245, 148250; (8) BCN-16; $n_m = 12$, $n_c = 12$; USNM 60688, 60690-60695, 60991, 60996-60999; (9) BCN-17; $n_m = 1$; SDNHM 23240; (10) BCN-34; $n_m = 1$, $n_c = 1$; MVZ 36129; (11) BCN-103g (not found); $n_m = 4$, $n_c = 4$; USNM 140682, 140684-140686.

Group B [*gilva*] (total $n_m = 24$, $n_c = 25$)
      MEXICO:– BAJA CALIFORNIA: (1) BCN-3; $n_m = 6$, $n_c = 7$; MVZ 39589-39592, 39615-39616, USNM 60991; (2) BCN-7; $n_m = 7$, $n_c = 8$; MVZ 39607-39614; (3) BCN-8; $n_m = 4$, $n_c = 4$; SDNHM 12079-12080, 12095, 12121; (4) BCN-9; $n_m = 3$, $n_c = 3$; MVZ 38165; SDNHM 4617, 5841; (5) BCN-10; $n_m = 1$; MVZ 112833; (6) BCN-11; $n_m = 1$, $n_c = 2$; MVZ 38174-38175; (7) BCN-33; $n_m = 1$; USNM 137275; (8) BCN-103b (not found); $n_m = 1$, $n_c = 1$; SDNHM 15849.

Group C [*intermedia*] (total $n_m = 29$, $n_g = 1$, $n_c = 36$)
MEXICO:– BAJA CALIFORNIA: (1) BCN-19; $n_m = 1$, $n_c = 3$; USNM 138313; (2) BCN-20; $n_m = 18$, $n_c = 18$, $n_g = 1$; LACM 13693; MVZ 36130, 38167-38172; SDNHM 6258, 11589, 11663, 11739-11744, 11748-11749; (3) BCN-22; $n_m = 1$, $n_c = 1$; MVZ 148227; (4) BCN-23; $n_m = 3$, $n_c = 3$; SDNHM 11822-11824; (5) BCN-24; $n_m = 4$, $n_c = 10$; USNM 138325, 138328, 139023-139024; (6) BCN-25; $n_m = 1$; MVZ 36130; (7) BCN-103c (not found); $n_m = 1$, $n_c = 1$; SDNHM 23233.

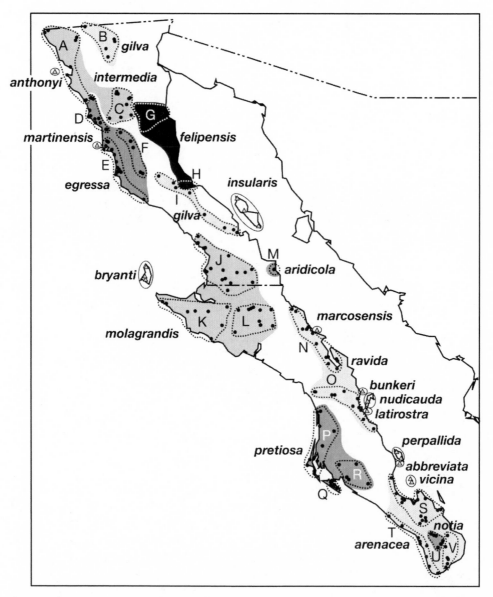

Figure 88. Map of Baja California depicting each recognized taxon (species or subspecies; see Hall, 1981; Álvarez-Castañeda. and Cortés-Calva, 1999) and grouped samples used in all morphometric analyses (letters designate peninsular samples; insular samples are identified by their respective trivial names).

Group D [*egressa*] (total $n_m = 41$, $n_c = 27$, $n_g = 1$, $n_{DNA} = 5$)
MEXICO:– BAJA CALIFORNIA: (1) BCN-15; $n_m = 1$; SDNHM 23222; (2) BCN-18; $n_{DNA} = 2$; CIB 7576-7577; (3) BCN-26; $n_m = 1$, $n_c = 1$; MVZ 97567; (4) BCN-27; $n_m = 3$, $n_c = 4$; MVZ 36125-36127; (5) BCN-27a; $n_m = 3$; CIB 7575-7577; (32) BCN-28; $n_m = 4$, $n_{DNA} = 3$; CIB 7578-7581; (6) BCN-29; $n_m = 13$, $n_c = 4$; MVZ 148223-148226, 148233-148237, 148246-148248; (7) BCN-30; $n_m = 9$, $n_c = 10$; MVZ 35843-35849, 36307-36208; (8) BCN-30a; $n_g = 1$; USNM 529405; (9) BCN-31; $n_m = 4$, $n_c = 5$; USNM 139370, 139647, 139650-139651; (10) BCN-32; $n_m = 1$; SDNHM 23225; (11) BCN-41; $n_m = 1$, $n_c = 2$; USNM 139663; (12) BCN-42; $n_m = 1$, $n_c = 1$; MVZ 35855.

Group E [*egressa*] (total $n_m = 52$, $n_c = 25$, $n_{DNA} = 4$)
MEXICO:– BAJA CALIFORNIA: (1) BCN-45; $n_m = 6$; CIB 4089-4094; (2) BCN-46; $n_m = 9$, $n_c = 2$; CIB 3857-3860, 3862-3865, MVZ 36311; (3) BCN-47; $n_m = 1$, $n_c = 5$; USNM 139642; (4) BCN-48; $n_m = 8$, $n_c = 13$; SDNHM 1184, USNM 139032-139034, 139036-139037, 139637, 564369; (5) BCN-50; $n_m = 5$; SDNHM 20088-20089, 20091-20092, 20095; (6) BCN-51; $n_m = 1$; CIB 7582; (7) BCN-52; $n_m = 3$, $n_c = 3$; MVZ 50138-50140; (8) BCN-55; $n_m = 1$; MVZ 111924; (9) BCN-56; $n_m = 3$, $n_{DNA} = 3$; CIB 2781-2782, 7583-7584; (10) BCN-57; $n_m = 2$, $n_{DNA} = 1$; CIB 2784, 2786; (11) BCN-58; $n_m = 1$, $n_c = 2$; MVZ 50142 [holotype of *N. lepida egressa*], 50143; (12) BCN-59; $n_m = 12$; CIB 4095-4107.

Group F [*egressa*] (total $n_m = 28$, $n_c = 13$)
MEXICO:– BAJA CALIFORNIA: (1) BCN-43; $n_m = 11$, $n_c = 10$; USNM 520400, 529397-529399, 529402, 529404-529409; (2) BCN-53; $n_m = 11$; SDNHM 18509-18511, 18515-18518, 18590-18592, 18605-18606; (3) BCN-54; $n_m = 1$; SDNHM 19775; (4) BCN-61; $n_m = 5$, $n_c = 3$; CIB 3395, 5009, MVZ 50144, USNM 139367, 139646.

Group G [*felipensis*] (total $n_m = 38$, $n_c = 62$, $n_{DNA} = 1$)
MEXICO:– BAJA CALIFORNIA: (1) BCN-25; $n_m = 1$, $n_c = 1$; USNM 138286;
(2) BCN-36; $n_m = 1$, $n_c = 1$; SDNHM 2548; (3) BCN-44; $n_m = 6$, $n_c = 8$; MVZ 37890, 37892-37893, 37896, 37901-37902; (4) BCN-37; $n_m = 18$, $n_c = 27$; MVZ 37904, 37906-37921, 37941, SDNHM 5118-5120, 5137-5138, 5154, 5167, 5192, 5195-5197, 22671, 22673; (5) BCN-38; $n_m = 12$, $n_c = 25$; MVZ 111922-111923, USNM 138287-138290, 138292-138301); (6) BCN-39; $n_{DNA} = 1$; CIB 3377.

Group H [*felipensis*] (total $n_m = 11$, $n_c = 12$)
MEXICO:– BAJA CALIFORNIA: (1) BCN-66; $n_m = 1$, $n_c = 2$; MVZ 50154-50155; (2) BCN-67; $n_m = 10$, $n_c = 10$; MVZ 50145-50153, 111925.

Group I [*gilva*] (total $n_m = 11$, $n_c = 7$, $n_g = 1$, $n_{DNA} = 5$)
MEXICO:– BAJA CALIFORNIA: (1) BCN-62; ($n_m = 2$, $n_c = 2$; USNM 139368-139369; (2) BCN-62a; $n_g = 1$, CSULB 6434; (3) BCN-63, $n_m = 1$; CIB 3378; (4) BCN-64; $n_{DNA} = 1$; CIB 2788; (4) BCN-65; $n_m = 1$, $n_{DNA} = 1$; CIB 3380-3381; (6) BCN-68; $n_m = 1$; CIB 2787; (7) BCN-69; $n_m = 3$, $n_c = 3$; USNM 139652-139654; (8) BCN-71; $n_m = 3$, $n_c = 2$; MVZ 159790-159791; (9) BCN-72; $n_{DNA} = 1$; CIB 3396; (10) BCN-73; $n_m = 3$, $n_{DNA} = 2$; MVZ 159790-159792; (11) BCN-75; $n_m = 1$, $n_{DNA} = 1$; CIB 7585; (12) BCN-76; $n_c = 1$; MVZ 159798; (13) BCN-77; $n_m = 1$, $n_c = 1$; SDNHM 18907, 19058.

Group J [*molagrandis*] (total $n_m = 62$, $n_c = 48$, $n_g = 10$, $n_{DNA} = 3$)
MEXICO:– BAJA CALIFORNIA: (1) BCN-74; $n_m = 1$, $n_c = 1$, $n_{DNA} = 1$; CIB 2790, 9246; (2) BCN-78a; $n_g = 1$; SDHNM 23235; (3) BCN-79; $n_m = 5$, $n_c = 6$; USNM 139657, 139659-139660, 139662; (4) BCN-80; $n_m = 2$; CIB 4109-4110; (5) BCN-81; $n_m = 1$; USNM 81076; (6) BCN-82; $n_m = 1$, $n_c = 1$; USNM 555330; (7) BCN-83; $n_m = 1$, $n_c = 1$, $n_g = 1$; MVZ 113810; (8) BCN-84; $n_m = 2$, $n_c = 2$; SDNHM 14065 [holotype of *N. lepida molagrandis*, 1945], 14066; (9) BCN-85; $n_m = 1$, $n_c = 3$; MVZ 38271-38273; (10) BCN-86; $n_m = 5$, $n_c = 5$; USNM 555301-555305; (11) BCN-87; $n_m = 14$, $n_c = 4$, $n_g = 7$; MVZ 111926-111939; (12) BCN-88; $n_m = 1$, $n_c = 1$; USNM 139665; (13) BCN-89; $n_m = 2$, $n_g = 1$; SDNHM 23232, 23234; (14) BCN-90; $n_m = 1$; CIB 9249; (15) BCN-93; $n_m = 1$, $n_c = 1$, $n_{DNA} = 1$; CIB 9248, SDNHM 14053; (16) BCN-94; $n_{DNA} = 1$; CIB 9250; (17) BCN-103a; $n_m = 1$; SDNHM 23237; (18) BCN-103c; $n_c = 2$; SDNHM 7027, 22678; (19) BCN-103e; $n_m = 1$; LACM 22482; (20) BCN-103f; $n_m = 1$; LACM 22480.   BAJA CALIFORNIA SUR: (20) BCS-8; $n_m = 20$, $n_c = 21$; USNM 523025, 532000-532006, 532010-532013, 532017-532019, 532023-532028.

Group K [*molagrandis*] (total $n_m = 33$, $n_c = 7$, $n_g = 1$, $n_{DNA} = 5$)
MEXICO:– BAJA CALIFORNIA SUR: (1) BCS-1; $n_m = 4$; LACM 22484-22486, 22551; (2) BCS-2; $n_m = 4$; LACM 22469-22470, 22479; MVZ 50137; (3) BCS-3; $n_m = 5$, $n_c = 7$, $n_g = 1$; MVZ 113805-113809, 113826, 115326-115327; (4) BCS-4; $n_m = 8$; LACM 22471-22475, 22547-22549; (5) BCS-5; $n_m = 1$; LACM 22478; (6) BCS-6; $n_m = 3$; CIB 3405, 3409-3410; (7) BCS-7; $n_{DNA} = 1$; CIB 9836; (8) BCS-9; $n_m = 2$; MVZ 35852-35853; (9) BCS-27; $n_{DNA} = 1$; CIB 9832; (10) BCS-28; $n_{DNA} = 1$; CIB 9838; (11) BCS-29; $n_m = 2$, $n_{DNA} = 2$; CIB 7587-7588; (12) BCS-120g; $n_m = 4$; LACM 22487-22490.

Group L [*molagrandis*] (total n_m = 29, n_c = 23, n_{DNA} = 8)
MEXICO:– BAJA CALIFORNIA SUR: (1) BCS-10; n_m = 11, n_c = 11; USNM 555268-555278; (2) BCS-11; n_m = 3; CIB 3414-2416; (3) BCS-12; n_m = 1, n_{DNA} = 2; CIB -8650-8651; (4) BCS-13; n_m = 1; CIB 8650; (5) BCS-17; n_m = 2, n_c = 2; USNM 139674-139675; (6) BCS-18; n_m = 2; LACM 22466-22467; (7) BCS-20; n_m = 7, n_c = 8; SDNHM 6851, 6892, USNM 138672-139671, 139673; (8) BCS-21; n_{DNA} = 2; CIB 9257-9258; (9) BCS-22; n_{DNA} = 3; CIB 9259-9261;   (10) BCS-23; n_m = 2, n_c = 2; USNM 139752-139753; (11) BCS-26; n_{DNA} = 1; CIB 9262.

Group M [*aridicola*] (total n_m = 6, n_c = 6, n_{DNA} = 3)
MEXICO:– BAJA CALIFORNIA: (1) BCN-91; n_m = 5, n_c = 5, n_{DNA} = 3; SDNHM 15506, 15567, 15569, 15580, 15595 [holotype of *N. lepida aridicola*]; (2) BCN-91a; n_m = 1, n_c = 1; SDNHM 15676.

Group N [*ravida*] (total n_m = 29, n_c = 24, n_g = 5, n_{DNA} = 15)
MEXICO:– BAJA CALIFORNIA SUR: (1) BCS-15; n_m = 1, n_{DNA} = 8; CIB 8652-8659; (2) BCS-16; n_m = 4 n_{DNA} = 3; CIB 45, 2796-2798, 2800, 2802; (3) BCS-19; n_m = 1; CIB 9256; (4) BCS-20; n_m = 1, n_c = 2; MVZ 38270; (5) BCS-22; n_m = 2; CIB 9259-9260; (6) BCS-24; n_m = 2, n_{DNA} = 1; CIB 7593-7594; (7) BCS-25; n_m = 4, n_c = 4; USNM 531987-531990; (8) BCS-31; n_{DNA} = 3; CIB 9263-9265; (9) BCS-31a; n_g = 4; UNT 610-613; (10) BCS-32; n_m = 2, n_c = 2; USNM 140687-140688; (11) BCS-33; n_m = 4, n_c = 4, n_g = 1; USNM 111940-111943; (12) BCS-34; n_m = 3, n_c = 2; LACM 58505, UCLA 20016-20017; (13) BCS-35; n_c = 4; USNM 555296-555299; (14) BCS-36; n_c = 1; MVZ 149564; (15) BCS-37; n_m = 1; USNM 555299; (16) BCS-38; n_m = 4, n_c = 4; SDNHM 14829, 14849-14851; (17) BCS-120a ; n_c = 1; USNM 139666.

Group O [*ravida*] (total n_m = 48, n_c = 51, n_{DNA} = 4)
MEXICO:– BAJA CALIFORNIA SUR: (1) BCS-39; n_{DNA} = 1; CIB 9840; (2) BCS-40; n_{DNA} = 2; CIB 9842, 9844; (3) BCS-41; n_m = 1, n_{DNA} = 1; CIB 7595; (4) BCS-42; n_m = 1, n_c = 1; USNM 140689; (5) BCS-43; n_m = 3, n_c = 3; USNM 555279-555281; (6) BCS-44; n_m = 19, n_c = 26; MVZ 50197, 50199, 50201-50206, SDNHM 14771-14772, 14786-14788, USNM 79064-79066, 79068-79072, 140690-140691, 140692 [holotype of *N. intermedia ravida*], 140693-140694; (7) BCS-45; n_m = 3, n_c = 2; USNM 529395-529396, 531942; (8) BCS-46; n_m = 1, n_c = 1; USNM 261712; (9) BCS-48; n_m = 2, CIB 851, 854; (10) BCS-51; n_m = 3, n_c = 3; USNM 531984-531986; (11) BCS-52; n_m = 3, n_c = 3; USNM 531980-531982; (12) BCS-53; n_m = 1, n_c = 1; SDNHM 19385; (13) BCS-59; n_m = 1; LACM 58506; (14) BCS-60; n_m = 6, n_c = 6; USNM 531992, 531994, 531996-531999; (15) BCS-75; n_c = 1; USNM 79062; (16) BCS-120b; n_m = 1, n_c = 1; UCLA 18121; (17) BCS-120d;

$n_m = 1$, $n_c = 1$; USNM 261701; (18) BCS-120c; $n_m = 1$, $n_c = 1$; USNM 261702; (19) BCS-120e; $n_m = 1$, $n_c = x$; USNM 261712; (20) BCS-120h; $n_c = 1$; SDNHM 19140.

Group P [*pretiosa*] (total $n_m = 102$, $n_c = 77$, $n_g = 5$, $n_{DNA} = 1$)
        MEXICO:– BAJA CALIFORNIA SUR: (1) BCS-55; $n_m = 1$, $n_c = 1$; USNM 529394; (2) BCS-56; $n_m = 20$, $n_c = 16$, $n_g = 2$; CIB 7596, MVZ 50156-50160, 111944-111948, SDNHM 14768-14769, USNM 140695-140701; (3) BCS-57; $n_m = 1$, $n_c = 1$; MVZ 35842; (4) BCS-58; $n_m = 6$, $n_c = 4$; USNM 529386-529391, 529394; (5) BCS-61; $n_m = 22$, $n_c = 16$; MVZ 50161-50164, 50166-50171, 50173-50179, USNM 140123 [holotype of *N. intermedia pretiosa*, 1909], 146121-146125, 146132-156134, 146136-146138; (6) BCS-63; $n_m = 1$; CIB 7597; (7) BCS-65; $n_m = 51$, $n_c = 39$, $n_g = 3$, $n_{DNA} = 1$; CIB 5147-5148, 5152, 5163-5165, 5167, 5353, 6088-6093, 7598, MVZ 50182-50190, 115328-115334, 115344, SDNHM 1187-1189, 19372-19382, USNM 146127, 146139-146141, 146143, 146790-146791, 146816; (8) BCS-66; $n_m = 1$; CIB 5168.

Group Q [*pretiosa*] (total $n_m = 32$, $n_c = 16$, $n_g = 1$, $n_{DNA} = 3$)
        MEXICO:– BAJA CALIFORNIA SUR: (1) BCS-67; $n_m = 7$, $n_{DNA} = 1$; CIB 5156-5161; (2) BCS-68; $n_m = 5$, $n_{DNA} = 2$; CIB 6094-6098; (163) BCS-69; $n_m = 18$, $n_c = 16$, $n_g = 1$; MVZ 50191-50194, 115335-115336, SDNHM 1185-1186, USNM 145145, 146131, 146144, 146146, 146789, 146811, 146813, 146815, 198413-198414; (3) BCS-120j; $n_m = 2$; USNM 198413-198414.

Group R [*pretiosa*] (total $n_m = 11$, $n_c = 4$, $n_{DNA} = 5$)
        MEXICO:– BAJA CALIFORNIA SUR: (1) BCS-62; $n_m = 1$; CIB 848; (2) BCS-71; $n_m = 2$, $n_{DNA} = 2$; CIB 7599-7600; (3) BCS-72; $n_m = 6$, $n_c = 2$; USNM 555284-555289; (4) BCS-70; $n_m = 1$, $n_c = 1$; MVZ 50180; (5) BCS-73; $n_{DNA} = 2$; CIB 7706-7707; (6) BCS-74; $n_{DNA} = 1$; CIB 8662; (7) BCS-76; $n_m = 1$, $n_c = 1$; USNM 555290.

Group S [*arenacea*] (total $n_m = 50$, $n_c = 33$, $n_g = 1$, $n_{DNA} = 13$)
        MEXICO:– BAJA CALIFORNIA SUR: (1) BCS-77; $n_m = 1$; CIB 5008; (2) BCS-82; $n_m = 2$; CIB 793-794; (3) BCS-83; $n_m = 1$, $n_c = 1$, $n_g = 1$; MVZ 111040; (4) BCS-84; $n_m = 9$, $n_{DNA} = 9$; CIB 791, 7708-7709, 8662, 8664-8667, 8687; (5) BCS-85; $n_m = 2$, $n_c = 2$; USNM 555294-555295; (6) BCS-86; $n_m = 1$; CIB 7603; (7) BCS-87; $n_m = 8$, $n_c = 7$; USNM 146797-146798, 146806, 146809-146810, 146821, 146827-146828; (8) BCS-88; $n_m = 2$, $n_c = 3$; MVZ 43142-43144; (9) BCS-89; $n_m = 4$, $n_c = 3$; USNM 531943-531945, 531947; (10) BCS-90; $n_{DNA} = 1$; CIB 10908; (11) BCS-91; $n_{DNA} = 1$; CIB 10907; (12) BCS-95; $n_m = 1$, $n_c = 1$; USNM 555293; (13) BCS-96; $n_m = 1$, $n_{DNA} = 1$; CIB 5341; (14) BCS-97; $n_m = 10$, $n_c = 8$. $n_{DNA} = 1$;

CIB 7572, MVZ 111964-111972; (15) BCS-98; n_m = 2, n_c = 2; USNM 555291-555292; (16) BCS-106; n_m = 6, n_c = 6; MVZ 43165, 43167-43168, 43170-43172.

Group T [*arenacea*] (total n_m = 56, n_c = 49, n_{DNA} = 3)
   MEXICO:– BAJA CALIFORNIA SUR: (1) BCS-94; n_m = 8, n_c = 7; USNM 529375-529378, 529380-529381, 529383, 529385; (2) BCS-99; n_m = 1, n_c = 2; MVZ 111951-111952; (3) BCS-101; n_m = 9, n_c = 8; USNM 531949-531953, 531955, 531957-531959; (4) BCS-102; n_m = 7, n_c = 8; MVZ 43125-43132; (5) BCS-103; n_m = 10, n_c = 6; USNM 531969-531972, 531974-531979; (6) BCS-104; n_{DNA} = 2; CIB 7573=7574; (7) BCS-105; n_m = 5, n_c = 5; USNM 529322-529324, 529327-529328; (8) BCS-108; n_{DNA} = 1; CIB 10909; (9) BCS-109; n_m = 7, n_c = 5; USNM 529329-529330, 529334-529335, 529338-529340; (10) BCS-113; n_m = 4, n_c = 5; USNM 529341-529343, 529346-529347; (11) BCS-115; n_m = 5, n_c = 3; USNM 4145, 71793, 146708-146710.

Group U [*notia*] (total n_m = 66, n_c = 61, n_g = 1, n_{DNA} = 4)
   MEXICO:– BAJA CALIFORNIA SUR: (1) BCS-116; n_m = 11, n_c = 10; USNM 525516-525517, 531960-531968; (2) BCS-117; n_m = 7, n_c = 7; USNM 529366-529367, 529369-529374, 529396; (3) BCS-118; n_m = 1; LACM 70195; (4) BCS-119; n_m = 3, n_c = 3; USNM 74250-74252; (5) BCS-120; n_m = 44, n_c = 41, n_g = 1, n_{DNA} = 4; CIB, 7589-7592, MVZ 43173-43175, 43177-43178, 43180-43182, 111958-111963, SDNHM 20252-20266, USNM 146793, 146794 [holotype of *N. intermedia notia*], 146795-146796, 146817-146820, 525520-525521, 525524-525525.

Group V [*arenacea*] (total n_m = 41, n_c = 33, n_g = 2)
   MEXICO:– BAJA CALIFORNIA SUR: (1) BCS-100; n_m = 3, n_c = 3, n_g = 2; MVZ 111953-111955; (2) BCS-107; n_m = 2, n_c = 2; MVZ 50195-50196); (3) BCS-110; n_m = 1, n_c = 2; USNM 74249, 146723; (4) BCS-111; n_m = 1, n_c = 1; MVZ 111957; (5) BCS-112; n_m = 27, n_c = 22; MVZ 43161-43164, SDNHM 14567-14572, USNM 71791, 146167, 146711, 146713-146715, 146717-146719, 146721-146722; (6) BCS-114; n_m = 7, n_c = 3; USNM 529349, 529352-529356.

Insular samples (arranged, from north to south, first on the Pacific side followed by the Gulf side):

   anthonyi (total n_m = 32, n_c = 42, n_{DNA} = 2)
   MEXICO:– BAJA CALIFORNIA: (1) BCN-13; n_m = 32, n_c = 42, n_{DNA} = 2; MVZ 38176-38179, SDNHM 5285, 5334-5337, USNM 137156-137157,

137159-137166,   137168-137169,   137171-137173,   137175-137176,   138199-137204, 137207, 137209-137213, 137216-217217, 137221-137222.

martinensis (total $n_m = 25$, $n_c = 29$, $n_{DNA} = 1$)
    MEXICO:– BAJA CALIFORNIA: (1) BCN-49; $n_m = 25$, $n_c = 29$, $n_{DNA} = 1$; MVZ 35986-30991, USNM 81062-81073, 81074 [holotype of *N. martinensis*], 81075, 139027-139035.

bryanti (total $n_m = 29$, $n_c = 36$, $n_g = 1$, $n_{DNA} = 1$)
    MEXICO:– BAJA CALIFORNIA: (1) BCN-97; $n_m = 7$, $n_c = 12$, $n_g = 1$; MVZ 36000-36004, 106738, 106740, USNM 530141-530144, 530146; (2) BCN-98; $n_m = 16$, $n_c = 19$; CIB 765, SDNHM 1179-1180, UCLA 19300, USNM 81078-81092, 186481 [holotype of *N. bryanti*]; (3) BCN-99; $n_m = 6$, $n_c = 5$, $n_{DNA} = 1$; CIB 764, SDNHM 19391-19392, USNM 81078, 530138-530140, 557708.

insularis (total $n_m = 13$, $n_c = 11$, $n_g = 6$, $n_{DNA} = 1$)
    MEXICO:– BAJA CALIFORNIA SUR: (1) BCN-95; $n_m = 6$, $n_c = 10$, $n_g = 6$, $n_{DNA} = 1$; LACM 20150-20151, SDNHM 19127-19128, 19198-19202, 19911, USNM 557708; (2) BCN-96; $n_m = 7$, $n_c = 1$; SDNHM 19127-19128, UCLA 19911, 20150-20151, USNM 198405 [holotype of *N. insularis*], 530147.

marcosensis (total $n_m = 33$, $n_c = 13$, $n_{DNA} = 10$)
    MEXICO:– BAJA CALIFORNIA SUR: (1) BCS-30; $n_m = 33$, $n_c = 13$, $n_{DNA} = 10$; CIB 808-820, 822, MVZ 59658-59659, SDNHM 19130-19192, UCLA 18086, 18088, 18090-18091, 20008-20009, 20010 [holotype of *N. lepida marcosensis*], 20011.

bunkeri (total $n_m = 8$, $n_c = 9$, $n_{DNA} = 1$)
    MEXICO:– BAJA CALIFORNIA SUR: (1) BCS-47; $n_m = 8$, $n_c = 9$, $n_{DNA} = 1$; UCLA 19720-19724, 19725 [holotype of *N. bunkeri*], 19726-19728.

nudicauda (total $n_m = 16$, $n_c = 9$, $n_{DNA} = 4$)
    MEXICO:– BAJA CALIFORNIA SUR: (1) BCS-49; $n_m = 16$, $n_c = 9$, $n_{DNA} = 4$; CIB 828, 2863-2865, 5347-5350, MVZ 59657, SDNHM 19133-19136, UCLA 18055-18018059, USNM 79073 [holotype of *N. nudicauda*].

latirostra (total $n_m = 21$, $n_c = 8$, $n_{DNA} = 6$)
    MEXICO:– BAJA CALIFORNIA SUR: (1) BCS-54; $n_m = 21$, $n_c = 8$, $n_{DNA} = 6$; CIB 795-807, SDNHM 19129, 19386-19390, 22818, UCLA 19718 [holotype of *N. lepida latirostra*].

perpallida (total n_m = 32, n_c = 20, n_{DNA} = 10)
    MEXICO:– BAJA CALIFORNIA SUR: (1) BCS-78; n_m = 32, n_c = 20, n_{DNA} = 10; CIB 831, 833, 835-837, 839-847, 5014-5015, MVZ 43145-43151, SDNHM 19137-19138, 19837-19839, UCLA 17987, 18001, 18003, 18005-18007, 19615, USNM 79061 [holotype of *N. intermedia perpallida*];

abbreviata (total n_m = 42, n_c = 21, n_g = 1, n_{DNA} = 6)
    MEXICO:– BAJA CALIFORNIA SUR: (1) BCS-79; n_m = 42, n_c = 21, n_g = 1, n_{DNA} = 6; CIB 766-787, MCZ 12260 [holotype of *N. abbreviata*], MVZ 43152-43156, 43158-43160, SDNHM 19125-19126, UCLA 18009-18010, 18014-18019, USNM 243417.

vicina (total n_m = 41, n_c = 31, n_g = 2, n_{DNA} = 8)
    MEXICO:– BAJA CALIFORNIA SUR: (1) BCS-80; n_m = 2, n_c = 2; SDNHM 19383-19384; (224) BCS-81; n_m = 39, n_c = 29, n_g = 2, n_{DNA} = 8; CIB 861-867, 2866, 3488, MVZ 43133-43140, 154152-154158, SDNHM 19141-19142, UCLA 17970, 19586, USNM 79060, 146799, 146802, 146803 [holotype of *N. intermedia vicina*], 146804, 146824-146826.

Habitat.—Woodrats of the *Neotoma lepida* group are found along the length of Baja California and in nearly all habitats across the eight phytogeographic regions (e.g., Wiggins, 1980; Grismer, 1994), from sea level in the very arid northeastern coastal plain of the Lower Colorado Valley (Fig. 89), the Vizcaino Desert of west-central Baja (Fig. 90), the Central Gulf Coast (Fig. 91), the Arid Tropical habitats of the Cape region (Fig. 92), and the oak woodland of the Sierra La Laguna (Fig. 93), as well as insular desert scrub, as on Isla San Marcos (Fig. 94).

Figure 89. Colorado Desert 1 km W San Felipe, Baja California (locality BCN-39), near type locality of *N. l. felipensis*.

Figure 90. Northern end of Vizcaino Desert at 5 km N & 6 km E El Rosario, Baja California (locality BCN-56), near type locality of *N. l. egressa*.

Figure 91. Central Gulf Coast phytogeographic region at El Barril, Baja California (locality BCN-91), type locality of *N. l. aridicola*.

Figure 92. Arid Tropical phytogeographic region near Cabo San Lucas, Baja California Sur (locality BCS-115).

Figure 93. Oak woodland of the Sierra La Laguna phytogeographic region, Baja California Sur (locality BCS-120), near type locality of *N. l. notia*.

Figure 94. Slopes of Isla San Marcos, Baja California Sur (locality BCS-30), type locality of *N. l. marcosensis*.

Morphometric variation.—We provide standard descriptive statistics for four external and 21 craniodental variables for each of the 22 peninsular groups and all 11 insular taxa in Table 34. Highly significant differences are present among peninsular and insular samples for each univariate variable (MANOVA, Wilks' λ approximate F = 5.2371, p < 0.0001, with each individual variable significant at p < 0.0001). As a result, we examined character trends among our samples using the first principal components axis to represent overall size and size-free canonical variates analysis (see methods) to examine cranial shape variation.

We provide factor coefficients for the first PC axis for each craniodental variable in Table 35, with the Pearson correlation coefficient of each eigenvector and individual log-transformed variables. As with our analyses of other geographic regions, PC-1 scores reflect a dominant influence of overall body size, as all character coefficients are positive and significantly related to each univariate character. For example, even the variable with the lowest loading on PC-1, Interorbital Constriction (IOC), is still significantly correlated with PC-1 scores (r = 0.543, Z-value = 22.594, p < 0.0001). This first PC-axis explains 64% of the total pool of variation, while the second axis explains less than 7%. The proportion of variance attributed to PC-1 scores is higher for this group of samples than for any other geographic area we have analyzed, either those including "coastal" and "desert" samples or those within either of these morphological groups (see separate transect analyses in the sections above and below). Clearly, overall size is a dominant component to cranial variation among all samples, both peninsular and insular, along the length of Baja California.

There is a general geographic trend in body size along the peninsula, as PC-1 scores are significantly and negatively correlated with the latitudinal position of each separate locality, including insular forms (r = -0.311, Z-value = -11.089, p < 0.0001). The correlation is slightly higher if specimens from all insular samples are excluded (r = -0.359, Z-value = -10.929, p < 0.0001). Overall, individuals are smallest in the north and become larger to the south. However, as is apparent by the relatively weak correlation coefficient, a latitudinal "effect" is limited. Indeed, individuals from the central gulf coast (Group M; *aridicola*) average smallest in body size while those from the mid-peninsular Isla Coronados (*bunkeri*) are the largest (Fig. 95).

Table 34. External and cranial measurements of adult (age classes 1-5) specimens from 22 samples of *Neotoma lepida* from mainland Baja California and adjacent islands (see Fig. 88). Samples are organized largely from north to south, with insular samples adjacent to their mainland counterparts. Means, standard error, sample size, and range are provided for each sample. Current subspecies of mainland samples are indicated in the map, Fig. 88.

Variable	Group A	Group B	*anthonyi*	Group C	Group D	Group E	Group F	*martinensis*	Group G
TOL	325.3±2.7	311.6±3.8	327.9±2.8	317.7±3.3	323.9±3.5	328.1±4.0	318.3±3.3	340.7±3.5	314.3±2.7
	39	22	27	28	33	49	27	24	35
	286-360	285-355	305-355	295-369	295-369	295-363	295-350	304-370	304-370
TAL	150.6±1.7	145.0±2.4	144.5±1.5	145.4±2.0	145.1±2.2	146.6±2.5	139.9±2.0	154.1±1.8	145.8±2.1
	39	22	27	28	33	49	27	24	35
	127-175	125-175	128-164	129-173	116-171	123-170	125-160	135-168	115-172
HF	35.3±0.3	33.3±0.3	37.0±0.1	33.7±0.3	33.7±0.3	33.9±0.2	33.9±0.2	39.3±0.3	32.7±0.2
	42	24	32	33	35	50	28	25	39
	31-35	31-36	35-39	31-37	29-37	33-37	32-36	36-42	29-35
E	29.9±0.4	27.7±0.5	28±0.0	27.5±0.5	29.7±0.7	32.2±0.9	30.4±0.5	32.0±2.1	29.3±0.7
	15	16	2	22	39	42	27	3	21
	28-33	25-31	28-28	23-31	23-37	27-38	26-34	29-36	25-35
CIL	39.6±0.19	38.29±0.33	41.27±0.26	39.38±0.22	40.67±0.32	41.15±0.33	39.60±0.28	42.27±0.36	38.82±0.25
	49	24	32	34	40	54	27	25	39
	36.2-42.3	35.4-41.7	38.7-43.8	37.4-41.8	37.0-44.1	38.1-44.4	37.3-43.9	39.2-45.5	36.1-42.4
ZB	21.66±0.11	20.72±0.18	22.28±0.18	21.51±0.17	22.36±0.19	21.97±0.13	21.67±0.20	22.64±0.23	21.09±0.17
	50	24	32	34	40	54	27	25	39
	19.6-23.4	19.1-22.5	20.1-24.4	20.2-23.6	20.2-24.2	20.8-23.5	18.7-24.4	21.0-24.3	18.5-23.3
IOC	5.41±0.04	5.43±0.05	5.54±0.03	5.46±0.05	5.36±0.05	5.51±0.06	5.35±0.04	5.46±0.03	5.31±0.04
	50	24	32	34	40	54	27	25	39
	5.0-5.9	5.1-6.1	5.2-5.7	4.9-6.1	4.9-5.9	4.9-6.0	5.1-5.8	5.1-5.7	4.9-5.9
RL	16.21±0.11	15.82±0.16	17.17±0.12	16.05±0.12	16.36±0.12	16.67±0.15	15.77±0.15	16.57±0.19	15.70±0.12
	50	24	32	34	40	54	27	25	39
	14.4-18.7	14.5-17.4	15.8-18.3	14.8-17.7	15.1-17.9	15.5-17.9	14.2-17.3	14.9-18.4	14.0-16.9

Table 34 (continued)

Variable	Group A	Group B	anthonyi	Group C	Group D	Group E	Group F	martinensis	Group G
NL	15.95±0.10	15.44±0.16	16.34±0.09	15.69±0.13	16.05±0.12	16.38±0.17	15.44±0.17	17.56±0.19	15.22±0.12
	50	24	32	34	40	54	27	25	39
	14.3-17.5	14.1-17.2	15.4-17.4	14.5-17.4	14.8-17.7	15.0-18.0	13.1-17.4	15.6-19.4	13.8-16.5
RW	6.58±0.04	6.49±0.06	7.00±0.05	6.58±0.05	6.88±0.06	6.84±0.05	6.71±0.04	7.09±0.06	6.60±0.05
	50	24	32	34	40	54	27	25	39
	6.0-7.1	5.9-7.1	6.6-7.6	6.1-7.0	6.2-7.5	6.4-7.3	6.4-7.3	6.6-7.9	6.0-7.3
OL	14.36±0.07	13.78±0.10	14.80±0.08	14.23±0.10	14.69±0.11	14.57±0.10	14.42±0.13	15.32±0.11	12.73±0.09
	50	24	32	34	40	54	27	25	39
	13.0-15.6	13.0-14.7	13.8-15.6	13.2-15.3	13.5-15.9	13.9-15.6	12.7-15.8	14.1-16.5	12.7-15.8
DL	11.53±0.09	10.99±0.13	12.18±0.13	11.26±0.09	12.60±0.14	11.75±0.16	11.21±0.21	12.50±0.16	10.95±0.12
	50	24	32	34	40	54	27	25	39
	10.4-13.1	9.9-12.0	10.9-13.5	10.4-12.2	9.9-12.8	10.5-13.1	10.1-15.4	11.2-13.9	9.4-12.5
MTRL	8.15±0.05	8.03±0.07	8.36±0.03	8.25±0.06	8.67±0.06	8.58±0.07	8.61±0.06	8.43±0.05	8.14±0.05
	50	24	32	34	40	54	27	25	39
	7.3-8.8	7.4-8.6	7.9-8.8	7.5-8.8	7.7-9.3	8.1-9.3	8.0-9.4	7.7-8.9	7.3-8.7
IFL	8.86±0.07	8.86±0.07	9.62±0.07	8.61±0.07	9.08±0.09	9.20±0.12	8.61±0.09	9.59±0.10	8.50±0.08
	50	24	32	34	40	54	27	25	39
	7.8-10.0	7.4-9.5	8.9-10.3	8.0-9.3	7.7-10.0	8.1-10.4	7.7-9.5	8.8-10.5	7.4-9.5
PBL	18.05±0.10	17.52±0.10	19.49±0.13	17.93±0.09	18.57±0.16	18.79±0.17	18.12±0.13	19.65±0.17	17.66±0.12
	50	24	32	34	40	54	27	25	39
	16.4-19.5	16.3-18.5	18.1-20.8	17.1-19.0	16.8-20.2	17.4-20.3	16.6-19.7	18.1-21.0	16.1-19.5
AW	7.58±0.04	7.43±0.04	8.09±0.04	7.60±0.05	7.78±0.05	7.97±0.05	7.79±0.07	8.19±0.05	7.36±0.04
	50	24	32	34	40	54	27	25	39
	6.8-8.2	6.9-8.4	6.8-8.2	6.9-8.2	7.1-8.4	7.5-8.4	7.0-8.4	7.8-8.6	6.6-7.8
OCW	9.58±0.04	9.39±0.08	10.06±0.03	9.41±0.06	9.67±0.07	9.79±0.06	9.72±0.08	9.94±0.06	9.24±0.04
	50	24	32	34	40	54	27	25	39
	8.9-10.2	8.8-10.3	9.6-10.4	8.8-10.0	8.9-10.5	9.3-10.4	8.7-10.5	9.2-10.3	8.7-9.9

Table 34 (continued)

Variable	Group A	Group B	anthonyi	Group C	Group D	Group E	Group F	martinensis	Group G
MB	17.17±0.08	16.87±0.13	17.25±0.07	17.24±0.11	17.68±0.11	17.51±0.11	17.44±0.10	18.18±0.13	16.95±0.11
	50	24	32	34	40	54	27	25	39
	15.8-18.6	15.5-18.1	16.4-18.3	16.0-18.2	16.5-18.8	16.6-18.6	16.7-18.3	17.0-19.1	15.6-18.5
BOL	5.87±0.05	5.72±0.07	6.02±0.05	5.85±0.06	6.11±0.08	6.10±0.08	5.82±0.08	6.39±0.07	5.72±0.06
	50	24	32	34	40	54	27	25	39
	5.2-6.9	4.8-6.2	5.5-6.8	5.3-6.5	5.2-7.2	5.3-7.0	5.1-6.9	5.9-7.1	5.0-6.5
MFL	7.91±0.06	7.63±0.09	7.33±0.07	7.98±0.08	8.20±0.10	8.35±0.10	8.01±0.09	8.22±0.09	7.74±0.08
	50	24	32	34	40	54	27	25	39
	7.0-9.0	6.9-8.4	6.7-8.2	7.0-8.7	7.1-9.2	7.6-9.0	7.3-9.0	7.6-9.0	6.8-8.9
MFW	2.64±0.03	2.53±0.03	2.52±0.02	2.74±0.04	2.77±0.04	2.76±0.05	2.72±0.05	2.59±0.04	2.58±0.04
	50	24	32	34	40	54	27	25	39
	2.2-3.2	2.3-3.0	2.3-2.8	2.2-3.4	2.3-3.3	2.3-3.1	2.3-3.2	2.3-2.9	2.1-3.1
ZPW	4.16±0.04	4.12±0.06	4.74±0.04	4.23±0.05	4.48±0.06	4.38±0.05	4.29±0.06	4.41±0.04	4.11±0.05
	50	24	32	34	40	54	27	25	39
	3.5-4.8	3.6-4.8	4.2-5.2	3.7-4.6	3.7-5.1	4.0-4.9	3.5-5.1	4.0-4.7	3.5-4.8
CD	15.89±007	15.51±0.10	15.89±0.07	15.77±0.07	16.32±0.10	16.34±0.08	16.04±0.10	16.87±0.13	15.69±0.07
	50	24	32	34	40	54	27	25	39
	14.9-17.3	14.4-16.2	14.9-17.3	15.0-16.5	15.2-17.2	15.8-17.2	15.4-17.6	15.5-17.8	14.8-16.6
BUL	6.67±0.04	6.58±0.04	6.88±0.03	6.77±0.04	7.01±0.04	7.06±0.05	6.96±0.05	6.70±0.05	6.77±0.03
	50	24	32	34	40	54	27	25	39
	5.9-7.2	6.2-6.9	6.4-7.1	6.4-7.2	6.6-7.8	6.7-7.5	6.5-7.3	5.9-7.1	6.3-7.1
BUW	7.13±0.04	7.02±0.06	7.43±0.04	7.14±0.06	7.44±0.04	7.44±0.05	7.39±0.05	6.98±0.05	7.19±0.04
	50	24	32	34	40	54	27	25	39
	6.5-8.0	6.5-7.5	7.0-8.3	6.5-7.6	6.9-8.0	6.9-8.0	6.7-8.0	6.4-7.4	6.7-7.7

Table 34 (continued)

Variable	Group H	Group I	*insularis*	*bryanti*	Group J	Group K	Group L	Group M	*marcosensis*
TOL	299.6±3.4	316.8±4.0	311.9±6.1	355.8±4.4	326.9±2.7	325.9±3.5	315.2±3.4	298.3±7.5	344.2±8.5
	9	17	11	27	67	30	30	6	22
	285-315	297-338	287-340	309-393	285-363	288-372	288-372	270-325	298-380
TAL	140.0±2.6	150.8±2.2	130.3±1.5	158.0±2.3	146.7±1.4	145.7±1.6	143.4±1.9	139.3±3.1	157.9±3.7
	9	19	11	27	60	30	19	6	22
	126-152	143-160	113-162	138-187	120-174	130-170	129.160	125-145	135-168
HF	33.0±0.3	33.3±0.3	35.0±0.7	38.9±0.3	35.2±0.3	33.9±0.3	33.5±0.4	32.2±05	38.2±0.3
	10	32	11	26	57	29	24	6	13
	32-34	31-36	30.5-37	36.5-42	29-40	31-38	31-37	31-34	37-40
E	32.3±0.5	27.7±0.5	32.3±0.5	32.4±0.5	33.8±0.3	34.3±0.4	33.2±0.4	26.8±6	33.8±0.6
	10	16	10	14	50	29	24	6	13
	30-34	31-36	27-38	30-36	29-38	28-37	32-37	24-28	31-38
CIL	38.11±0.47	38.42±0.34	39.94±0.55	44.1±042	40.60±0.24	41.23±0.26	39.17±0.29	37.21±0.66	42.16±0.47
	11	10	12	29	68	30	22	6	13
	35.1-39.9	36.7-40.0	37.0-42.3	39.1-48.4	36.5-44.4	38.6-44.0	36.7-41.9	35.0-39.4	39.8-45.2
ZB	20.45±0.26	20.96±0.26	22.78±0.31	23.64±0.24	21.96±0.12	22.36±0.18	21.20±0.19	20.0±0.31	22.52±0.34
	11	17	12	29	68	30	29	6	22
	19.0-21.6	20.0-22.4	20.8-24.6	21.0-26.0	20.3-23.7	20.6-24.0	19.3-22.5	18.9-21.1	20.7-24.5
IOC	5.38±0.07	5.36±0.06	5.26±0.04	5.58±0.04	5.49±0.04	5.44±0.05	5.48±0.04	5.28±0.06	5.65±0.07
	11	17	13	29	68	30	29	6	22
	5.1-5.7	5.0-5.7	5.0-5.6	5.2-6.0	4.8-6.1	5.1-6.2	5.1-5.8	5.1-5.5	5.3-6.0
RL	15.40±0.24	15.49±0.16	15.45±0.16	18.02±0.17	16.39±0.13	16.80±0.13	16.07±0.17	15.07±0.31	15.07±0.31
	11	17	13	29	68	30	29	6	22
	14.4-18.7	14.9-16.3	14.1-16.7	16.3-19.6	14.3-18.4	14.3-18.4	14.5-17.8	14.4-17.8	14.4-16.4
NL	14.88±0.25	15.20±0.21	14.96±0.17	17.39±0.18	16.12±0.12	16.44±0.13	15.81±0.15	15.06±0.41	17.18±0.27
	11	17	13	29	68	30	29	6	22
	13.2-16.1	14.0-16.3	14.0-16.0	15.2-19.1	14.3-18.6	15.3-17.8	14.6-17.3	14.1-16.5	15.8-19.4

Table 34 (continued)

Variable	Group H	Group I	insularis	bryanti	Group J	Group K	Group L	Group M	marcosensis
RW	6.38±0.09	6.53±0.10	7.30±0.11	7.30±0.11	6.77±0.04	6.73±0.06	6.55±0.05	6.27±0.13	7.05±0.11
	11	17	13	29	68	30	29	6	22
	6.0-6.9	6.0-7.2	6.4-7.9	6.4-7.9	6.4-7.9	6.2-7.6	6.1-7.1	5.9-6.7	6.6-7.6
OL	13.93±0.13	13.76±0.15	15.24±0.19	15.45±0.14	14.60±0.09	14.71±0.09	14.14±0.11	13.71±0.27	14.82±0.15
	11	17	13	29	68	30	29	6	22
	13.3-14.8	12.9-14.5	14.1-16.1	13.2-16.5	13.0-15.9	13.7-15.6	13.2-15.1	12.9-14.7	13.9-15.6
DL	10.85±0.24	10.56±0.13	10.85±0.24	12.60±0.18	11.49±0.09	11.77±0.11	11.11±0.14	10.23±0.39	11.29±0.23
	11	17	13	29	68	30	29	6	22
	9.5-11.7	9.9-11.4	9.7-11.9	10.9-14.3	10.1-13.1	10.6-12.8	10.0-12.5	9.1-11.8	10.3-13.1
MTRL	8.06±0.06	8.40±0.10	8.98±0.07	9.67±0.08	8.58±0.04	8.72±0.05	8.37±0.06	7.97±0.10	9.22±0.03
	11	17	13	29	68	30	29	6	22
	7.7-8.5	7.6-8.9	8.6-9.4	8.6-10.2	7.7-9.3	8.2-9.3	7.8-9.0	7.6-8.3	9.0-9.4
IFL	8.45±0.13	8.43±0.11	8.65±0.19	9.68±0.12	8.99±0.07	9.26±0.09	8.70±0.11	8.30±0.19	9.12±0.19
	11	17	13	29	68	30	29	6	22
	7.7-9.0	7.9-9.0	7.3-9.6	8.5-11.3	8.0-10.1	8.1-10.1	7.5-9.5	7.7-8.9	8.3-10.8
PBL	17.30±0.21	17.28±0.18	18.36±0.32	20.44±0.20	18.23±0.12	18.67±0.13	17.68±0.14	16.94±0.35	19.17±0.24
	11	17	13	29	68	30	29	6	22
	15.9-18.3	16.5-18.2	16.3-19.9	18.0-22.7	16.4-20.0	17.3-19.8	16.3-19.9	16.0-18.2	17.9-20.6
AW	7.28±0.07	7.37±0.07	8.34±0.07	8.49±0.05	7.70±0.04	7.93±0.06	7.52±0.05	7.24±0.06	7.99±0.08
	11	17	13	29	68	30	29	6	22
	6.8-7.5	7.2-7.9	8.1-9.0	8.1-9.0	7.1-8.5	7.3-8.5	7.1-8.0	7.0-7.4	7.6-8.3
OCW	9.29±0.10	9.34±0.06	9.54±0.07	10.26±0.05	9.71±0.04	9.74±0.06	9.45±0.06	9.18±0.12	10.26±0.05
	11	17	13	29	68	30	29	6	22
	8.8-9.7	9.2-9.9	9.2-9.9	9.8-10.8	9.0-10.4	9.1-10.4	8.8-10.2	8.9-9.6	9.9-10.5
MB	16.66±0.16	16.92±0.14	17.98±0.13	18.60±0.12	17.76±0.08	17.76±0.08	17.18±0.12	16.66±0.16	18.29±0.18
	11	17	13	29	68	30	29	6	22
	16.0-17.8	16.0-17.7	17.3-18.7	17.4-19.8	16.4-18.8	16.7-18.8	16.0-18.3	16.0-17.0	17.0-19.4

Table 34 (continued)

Variable	Group H	Group I	insularis	bryanti	Group J	Group K	Group L	Group M	marcosensis
BOL	5.39±0.05	5.74±0.10	6.07±0.12	6.23±0.09	5.97±0.05	5.98±0.07	5.78±0.08	5.52±0.10	6.25±0.12
	11	17	13	29	68	30	29	6	22
	5.1-6.9	5.3-6.2	5.6-6.6	5.2-7.0	5.2-6.8	5.4-6.9	5.3-6.2	5.2-5.9	5.5-6.8
MFL	7.74±0.08	7.68±0.13	7.83±0.10	8.86±0.12	8.20±0.07	8.40±0.09	7.95±0.09	7.49±0.22	8.78±0.13
	11	17	13	29	68	30	29	6	22
	6.9-8.3	7.3-8.4	7.3-8.4	7.8-10.4	7.2-9.7	7.4-9.8	7.2-8.7	6.9-8.5	7.8-9.5
MFW	2.74±0.07	2.67±0.05	2.38±0.02	2.89±0.05	2.86±0.03	2.83±0.04	2.70±0.05	2.55±0.06	3.08±0.07
	11	17	13	29	68	30	29	6	22
	2.4-3.1	2.3-3.0	2.3-2.5	2.4-3.4	2.3-3.5	2.4-3.2	2.1-3.3	2.3-2.7	2.6-3.5
ZPW	4.13±0.09	4.10±0.07	4.13±0.09	4.80±0.06	4.32±0.04	4.49±0.06	4.16±0.05	3.71±0.09	4.63±0.07
	11	17	13	29	68	30	29	6	22
	3.7-4.6	3.8-4.5	3.6-4.6	4.1-5.5	3.5-5.0	4.0-5.0	3.7-4.7	3.5-4.0	4.1-4.9
CD	15.48±0.12	15.63±0.11	16.18±0.16	17.44±0.17	16.24±0.08	16.36±0.11	15.98±0.08	15.40±0.14	16.25±0.15
	11	17	13	29	68	30	29	6	22
	14.8-16.6	15.0-16.2	15.6-17.5	15.8-19.2	15.0-17.4	15.4-17.8	15.1-16.8	15.0-17.4	15.4-17.2
BUL	6.79±0.08	6.96±0.05	7.28±0.06	7.36±0.06	7.08±0.03	7.25±0.05	6.93±0.05	6.42±0.08	7.44±0.04
	11	17	13	29	68	30	29	6	22
	6.4-7.1	6.8-7.2	6.9-7.6	6.7-8.0	6.3-7.5	6.7-7.6	6.5-7.4	6.2-6.7	7.2-7.7
BUW	7.30±0.06	7.26±0.06	7.61±0.08	7.79±0.08	7.51±0.04	7.60±0.05	7.40±0.06	6.95±0.08	7.77±0.10
	11	17	13	29	68	30	29	6	22
	6.9-7.7	6.9-7.7	7.1-8.1	7.1-8.7	6.8-8.1	7.2-8.5	6.6-7.7	6.8-7.2	7.3-8.3

Table 34 (continued)

Variable	Group N	Group O	*bunkeri*	*nudicauda*	*latirostra*	Group P	Group Q	Group R	*perpallida*
TOL	327.0±3.7	319.0±2.7	383.8±7.1	345.3±5.0	369.8±3.8	349.0±2.5	319.0±2.7	323.0±12.5	340.0±6.0
	28	43	6	17	16	92	20	10	32
	298-362	283-362	360-408	315-360	358-380	300-399	334-378	282-365	293-400
TAL	153.8±2.6	144.3±1.8	166.5±3.9	155.3±2.5	162.4±2.7	155.3±1.4	159.7±2.7	147.5±5.8	162.8±2.9
	28	43	6	17	16	92	20	10	32
	120-170	110-172	160-185	140-165	155-170	129-187	142-178	125-165	140-190
HF	33.0±0.2	33.5±0.2	41.5±0.5	38.4±0.4	36.6±0.5	37.9±0.2	37.6±0.6	35.6±0.8	36.2±0.4
	34	48	8	17	20	101	20	10	32
	31-35	31-37	40-43	36-40	35-39	35-42	33-41	32-40	33-40
E	31.4±0.6	31.2±0.4	30.5±0.3	32.3±0.7	32.6±1.0	34.4±0.3	34.5±0.6	32.0±0.7	31.0±0.5
	30	35	8	15	19	74	11	10	24
	26-35	25-34	20-32	30-36	27-35	28-38	33-37	30-36	28-34
CIL	39.17±0.32	39.39±0.21	44.65±0.60	41.82±0.47	43.54±0.44	43.20±0.21	43.40±0.43	39.53±0.80	40.63±0.61
	35	48	7	17	21	105	25	11	31
	36.5-42.8	36.6-43.2	43.2-45.8	39.5-43.5	42.0-45.6	38.2-48.1	40.4-47.6	36.6-42.4	35.3-45.8
ZB	20.86±0.16	21.14±0.15	24.12±0.60	22.89±0.19	23.83±0.19	23.56±0.12	23.38±0.30	21.43±0.47	21.99±0.34
	35	48	7	17	21	105	25	11	31
	19.4-22.8	19.2-23.7	20.8-25.4	21.9-23.8	23.1-24.7	21.1-26.3	21.5-26.5	20.0-23.3	19.3-24.7
IOC	5.34±0.05	5.60±0.04	5.89±0.04	5.75±0.04	5.72±0.08	5.66±0.03	5.61±0.06	5.55±0.05	5.58±0.05
	35	48	7	17	21	105	25	11	31
	4.8-5.7	5.1-6.2	5.7-6.0	5.6-5.9	5.4-6.2	5.2-6.2	5.2-6.0	5.4-5.9	5.2-5.9
RL	15.95±0.17	16.04±0.10	17.03±0.82	17.40±0.19	18.05±0.28	17.42±0.09	17.72±0.19	16.10±0.40	16.49±0.30
	35	48	7	17	21	105	25	11	31
	14.6-17.9	14.8-17.4	13.5-18.7	16.3-18.2	17.1-19.6	15.5-19.5	16.3-19.3	14.5-17.6	13.7-19.1
NL	15.67±0.13	15.78±0.11	17.19±0.78	16.58±0.25	18.00±0.27	17.20±0.10	17.60±0.19	15.73±0.41	15.98±0.31
	35	48	7	17	21	105	25	11	31
	14.5-17.6	14.5-17.6	13.9-19.3	15.4-17.6	17.2-19.5	14.9-19.1	16.2-19.2	14.3-17.5	14.0-18.8

Table 34 (continued)

Variable	Group N	Group O	bunkeri	nudicauda	latirostra	Group P	Group Q	Group R	perpallida
RW	6.38±0.06	6.49±0.04	7.77±0.25	7.09±0.06	7.67±0.08	7.23±0.04	7.16±0.08	6.88±0.13	6.62±0.10
	35	48	7	17	21	105	25	11	31
	5.9-7.1	5.7-7.0	6.9-8.7	6.8-7.4	7.3-7.9	6.2-7.9	6.5-7.7	6.5-7.5	6.0-7.7
OL	14.24±0.13	14.31±0.08	15.63±0.43	15.20±0.16	15.10±0.12	15.48±0.07	15.24±0.12	14.33±0.22	14.80±0.23
	35	48	7	17	21	105	25	11	31
	13.1-15.2	13.3-15.8	13.2-16.5	14.4-15.93	14.7-15.6	13.7-16.8	14.4-16.3	13.5-15.1	13.0-16.9
DL	11.15±0.15	11.21±0.10	11.78±0.69	12.03±0.23	12.63±0.18	12.21±0.09	12.47±0.22	12.47±0.27	11.58±0.26
	35	48	7	17	21	105	25	11	31
	9.8-12.7	10.0-12.6	9.0-13.6	10.8-13.0	12.0-13.5	10.5-14.1	11.1-14.8	9.9-11.8	9.1-14.2
MTRL	8.14±0.05	8.08±0.05	9.38±0.10	8.83±0.07	8.54±0.09	9.15±0.04	9.03±0.09	8.82±0.11	8.44±0.06
	35	48	7	10	21	105	25	11	31
	7.5-8.6	7.3-8.7	8.9-10.2	8.4-9.2	8.3-9.0	8.1-8.9	8.3-9.6	8.3-9.1	8.1-10.0
IFL	8.67±0.12	8.78±0.07	9.59±0.52	9.25±0.12	9.54±0.21	9.57±0.06	9.76±0.12	8.65±0.22	9.08±0.20
	35	48	7	17	21	105	25	11	31
	7.6-9.8	8.1-9.8	7.6-11.1	8.4-9.7	9.0-10.6	8.2-11.0	8.8-10.7	7.8-9.7	7.3-10.8
PBL	17.75±0.16	17.65±0.10	19.15±0.79	18.89±0.24	19.59±0.25	19.53±0.10	19.66±0.25	17.74±0.43	18.24±0.28
	35	48	7	17	21	105	25	11	31
	16.5-19.4	16.2-19.0	16.2-21.2	17.5-20.0	18.9-20.9	17.4-21.6	18.3-22.4	16.3-19.3	16.0-20.8
AW	7.27±0.05	7.37±0.03	8.44±0.13	8.29±0.05	8.22±0.06	8.28±0.04	8.10±0.08	7.80±0.11	7.66±0.06
	35	48	7	17	21	105	25	11	31
	6.9-8.0	6.8-7.8	7.8-8.8	8.1-8.6	7.9-8.4	7.4-9.0	7.7-8.6	7.2-8.3	7.2-8.2
OCW	9.39±0.07	9.53±0.05	10.69±0.07	10.03±0.08	10.70±0.09	10.21±0.04	10.10±0.05	9.90±0.12	9.79±0.08
	35	48	7	17	21	105	25	11	31
	8.8-10.1	8.5-10.5	10.4-10.9	9.4-10.3	10.4-11.1	9.4-11.3	9.5-10.5	9.4-10.2	9.2-10.3
MB	17.03±0.09	17.26±0.10	19.34±0.16	17.84±0.20	18.95±0.16	18.41±0.07	18.54±0.15	17.33±0.28	17.51±0.19
	35	48	7	17	21	105	25	11	31
	16.2-18.0	15.8-19.0	18.7-19.8	16.4-18.5	18.5-19.7	17.3-20.0	16.9-19.8	16.4-18.5	16.2-18.8

Table 34 (continued)

Variable	Group N	Group O	bunkeri	nudicauda	latirostra	Group P	Group Q	Group R	perpallida
BOL	5.87±0.10	5.91±0.05	6.78±0.06	6.44±0.11	7.07±0.13	6.53±0.05	6.45±0.08	6.06±0.23	6.13±0.09
	35	48	7	17	21	105	25	11	31
	5.0-7.1	5.1-6.7	6.5-6.9	5.7-6.8	6.5-7.7	5.5-8.6	5.9-7.2	5.2-7.0	5.7-6.9
MFL	7.94±0.09	7.96±0.08	9.52±0.10	8.93±0.12	9.09±0.16	8.81±0.06	8.86±0.10	7.73±0.17	8.37±0.16
	35	48	7	17	21	105	25	11	31
	7.4-8.8	6.9-9.2	9.2-9.9	8.3-9.6	8.4-9.8	7.4-10.1	7.9-9.7	7.0-9.8	7.0-9.8
MFW	2.79±0.04	2.82±0.03	3.08±0.16	2.68±0.05	3.17±0.06	3.20±0.03	3.24±0.07	2.99±0.09	2.87±0.08
	35	48	7	17	21	105	25	11	31
	2.3-3.2	2.4-3.3	2.3-3.3	2.4-3.0	2.9-3.4	2.5-4.0	2.5-3.6	2.7-3.3	2.1-3.6
ZPW	4.11±0.05	4.09±0.04	4.89±0.23	4.45±0.10	5.04±0.05	4.72±0.04	4.88±0.09	4.20±0.12	4.37±0.10
	35	48	7	17	21	105	25	11	31
	3.6-4.4	3.9-4.4	4.0-5.5	3.9-4.8	4.9-5.3	4.0-5.6	4.2-5.7	3.8-4.7	3.5-4.8
CD	15.56±0.09	15.91±0.07	17.13±0.34	16.86±0.13	17.37±0.09	16.98±0.08	16.89±0.13	15.92±0.18	16.28±0.18
	35	48	7	17	21	105	25	11	31
	14.9-16.4	14.9-16.9	15.7-18.1	16.0-17.4	17.1-17.8	15.5-18.7	16.1-18.0	15.1-16.7	15.2-18.0
BUL	6.68±0.05	6.71±0.03	7.17±0.16	7.25±0.05	7.30±0.06	7.32±0.03	7.21±0.06	7.02±0.12	6.86±0.07
	35	48	7	17	21	105	25	11	31
	6.3-7.2	6.2-7.2	6.5-7.9	7.0-7.5	7.1-7.5	6.6-7.9	6.7-7.6	6.6-7.6	6.4-7.5
BUW	7.12±0.05	7.19±0.05	7.86±0.22	7.61±0.08	7.89±0.07	7.68±0.03	7.84±0.06	7.48±0.09	7.32±0.07
	35	48	7	17	21	105	25	11	31
	6.7-7.6	6.7-8.0	6.8-8.4	7.2-8.1	7.7-8.2	7.0-7.7	7.3-8.2	7.0-7.8	6.8-8.0

Table 34 (continued)

Variable	abbreviata	vicina	Group S	Group T	Group U	Group V
TOL	284.3±2.7	325.3±3.9	346.1±4.6	343.2±3.4	329.8±2.6	359.6±4.6
	286-360	280-366	289-385	292-407	284-372	310-400
	37	35	43	50	66	36
TAL	123.6±1.4	150.7±2.1	164.4±2.4	157.8±1.9	156.0±1.5	171.1±2.7
	110-135	120-174	139-190	133-190	134-184	137-198
	37	35	43	50	66	36
HF	32.32±0.3	34.9±0.3	36.6±0.3	37.0±0.2	35.3±0.2	37.3±0.3
	31-35	32-39	32-40	33-39	31-39	32-41
	41	38	47	55	68	43
E	30.0±0.5	32.2±0.6	30.8±0.4	30.2±0.3	29.7±0.2	29.8±0.6
	27-32	25-34	25-33	22-35	26-33	26-37
	31	22	39	51	50	32
CIL	37.97±0.35	39.64±0.29	41.16±0.33	40.97±0.26	40.19±0.20	42.49±0.30
	34.7-40.0	36.9-43.2	37.1-45.0	36.9-45.9	36.0-43.21	38.2-46.2
	40	39	47	56	68	47
ZB	20.96±0.17	21.28±0.20	22.48±0.21	22.32±0.15	22.02±0.10	23.03±0.16
	19.1-21.8	19.1-23.35	20.1-24.5	20.3-24.7	19.2-23.7	21.0-25.8
	40	39	47	56	68	47
IOC	5.09±0.04	5.51±0.05	5.61±0.05	5.73±0.04	5.52±0.03	5.69±0.05
	4.9-5.3	5.1-6.2	5.1-6.5	5.2-6.6	5.0-5.9	5.0-6.4
	40	39	47	56	68	47
RL	14.81±0.14	16.01±0.14	16.62±0.15	16.73±0.12	16.27±0.11	17.32±0.15
	13.3-15.5	14.5-18.0	14.6-18.4	14.7-19.3	14.3-17.9	15.4-19.0
	40	39	47	56	68	47
NL	14.89±0.16	15.77±0.14	16.26±0.14	16.30±0.13	15.90±0.13	16.77±0.16
	13.5-16.0	14.0-17.6	14.2-17.7	14.6-18.8	13.6-18.0	14.4-19.1
	40	39	47	56	68	47

Table 34 (continued)

Variable	*abbreviata*	*vicina*	Group S	Group T	Group U	Group V
RW	6.41±0.06	6.53±3.06	6.79±0.05	6.83±0.04	6.71±0.04	7.12±0.06
	40	39	47	56	68	47
	5.9-7.0	5.8-7.2	6.3-7.4	6.2-7.7	6.0-7.5	6.4-8.0
OL	14.38±0.09	14.33±0.10	14.88±0.12	14.76±0.10	14.57±0.08	15.26±0.11
	40	39	47	56	68	47
	13.5-15.0	13.4-15.5	13.7-16.3	13.5-16.4	13.2-15.9	13.8-16.5
DL	10.82±0.16	11.13±0.12	11.77±0.15	11.67±0.12	11.45±0.10	12.28±0.13
	40	39	47	56	68	47
	9.4-11.7	9.9-12.4	10.0-13.3	10.3-14.0	9.6-13.2	10.7-13.8
MTRL	8.22±0.02	8.32±0.05	8.50±0.06	8.43±0.04	8.47±0.04	8.38±0.05
	40	39	47	56	68	47
	8.0-8.3	7.9-8.9	7.8-9.1	7.7-9.2	7.8-9.1	7.6-9.0
IFL	8.11±0.09	9.02±0.08	9.36±01107	9.22±0.09	9.17±0.08	9.66±0.09
	40	39	47	56	68	47
	7.2-8.7	8.3-8.9	7.9-10.3	7.9-10.8	7.8-10.3	8.3-11.0
PBL	17.78±0.17	17.73±0.15	18.72±0.16	18.36±0.13	18.16±0.10	19.04±0.15
	40	39	47	56	68	47
	16.3-18.8	16.3-19.7	16.7-20.2	16.2-21.1	15.9-19.6	17.2-21.7
AW	7.62±0.04	7.48±0.06	8.09±0.05	7.78±0.04	7.71±0.03	7.84±0.04
	40	39	47	56	68	47
	7.2-7.9	6.8-8.3	7.2-8.2	7.1-8.4	7.1-8.2	7.4-8.4
OCW	8.74±0.05	9.72±0.06	9.89±0.06	10.04±0.05	9.75±0.05	10.19±0.05
	40	39	47	56	68	47
	8.4-9.1	9.0-10.4	9.2-10.6	9.3-10.9	9.0-10.7	9.5-11.1
MB	16.14±0.11	17.08±0.09	17.58±0.12	17.59±0.08	17.179±0.07	17.94±0.11
	40	39	47	56	68	47
	15.1-16.9	15.9-18.0	16.2-19.2	16.2-19.2	15.8-18.3	16.3-19.2

Table 34 (continued)

Variable	*abbreviata*	*vicina*	Group S	Group T	Group U	Group V
BOL	5.33±0.06	5.91±0.07	6.11±0.10	6.12±0.06	5.96±0.04	6.46±0.08
	40	39	47	56	68	47
	4.8-5.7	5.3-6.5	5.1-8.2	5.1-7.1	5.2-6.8	5.4-7.6
MFL	8.06±0.09	8.08±0.09	8.37±0.09	8.41±0.08	8.09±0.06	8.70±0.09
	40	39	47	56	68	47
	7.3-8.7	7.3-9.5	7.2-9.6	7.3-9.9	6.9-9.3	7.4-9.9
MFW	2.45±0.04	2.88±0.04	3.10±0.05	3.09±0.03	2.94±0.03	3.09±0.05
	40	39	47	56	68	47
	2.2-2.7	2.4-3.4	2.5-4.0	2.6-3.7	2.6-3.4	2.5-3.7
ZPW	3.96±0.04	3.99±0.06	4.21±0.06	4.22±0.04	4.04±0.04	4.40±0.06
	40	39	47	56	68	47
	3.4-4.2	3.4-4.6	3.3-4.8	3.5-4.9	3.2-4.7	3.7-5.7
CD	15.24±0.06	15.81±0.08	16.05±0.08	16.18±0.07	15.94±0.06	16.50±0.10
	40	39	47	56	68	47
	14.8-15.8	14.7-16.6	15.3-17.0	15.1-17.3	14.9-17.1	15.4-18.2
BUL	6.15±0.04	6.55±0.04	6.72±0.04	6.63±0.04	6.46±0.03	6.59±0.04
	40	39	47	56	68	47
	5.9-6.5	6.1-7.1	6.2-7.2	6.0-7.4	6.0-7.1	6.1-7.1
BUW	6.88±0.05	7.13±0.05	7.15±0.05	7.11±0.05	6.92±0.03	7.09±0.04
	40	39	47	56	68	47
	6.6-7.3	6.6-7.6	6.6-7.9	6.4-7.8	6.4-7.6	6.5-7.7

The distribution of cranial size we describe here is in accordance with previous studies (Lawlor, 1982; Smith, 1992; see review in Lawlor et al., 2002). These studies, based on external body dimensions, documented the tendency for insular populations to be larger than their mainland counterparts. This observation, however, is not generalizable in the comparison of all insular to mainland samples. The three samples that average largest in cranial size are insular (*bunkeri* from Isla Coronados, *latirostra* from Isla Danzante, and *bryanti* from Isla Cedros). But, these insular samples are only marginally larger than the mainland Groups P and Q (*pretiosa*), with only *bunkeri* significantly so. (Note that Group Q is from Isla Margarita, but is not statistically larger than Group P, which itself includes both mainland samples and those from Isla Magdalena.) Other insular samples are scattered across the range of sizes, with one insular sample (*abbreviata* from Isla San Francisco) the same size as the smallest peninsular sample, Group M (Fig. 89).

We provide character eigenvectors contributing to size-free discrimination among all peninsular and insular samples for the first two canonical axes in Table 35, which combine to explain 53.7% of the total pool of variation. A scatterplot of scores for these two canonical axes is illustrated in Fig. 96. For simplicity, peninsular samples are grouped while each insular population is individually identified. The Isla Ángel de la Guarda sample (*insularis*) is well separated on CAN-1, significantly so in comparison to all other samples, either insular or peninsular (ANOVA, Fisher's PLSD $p < 0.0001$). Samples from Isla Todos Santos (*anthonyi*), Isla San Martín (*martinensis*), and Isla San Francisco (*abbreviata*) all differ marginally from the pooled peninsular samples on CAN-1 (ANOVA, Fisher's PLSD $p < 0.05$ in both cases) but highly so on CAN-2 ($p < 0.0001$). All other insular samples either broadly or completely overlap with the pooled peninsular sample.

Table 35. Coefficients of principal components (PC) analysis and size-free canonical discriminant (CAN) analysis of 21 craniodental log-transformed variables for 22 peninsular and 11 insular samples of the desert woodrat complex from Baja California.

Variable	PC-1	r[1]	CAN-1	CAN-2
logCIL	0.2655	0.9762	0.0408	0.0106
logZB	0.2529	0.9300	-0.0522	0.0093
logIOC	0.1477	0.5430	-0.0872	0.9362
logRL	0.2494	0.9172	0.0567	1.3495
logNL	0.2435	0.8954	0.0013	0.7393
logRW	0.2307	0.8483	-0.3018	-0.2587
logOL	0.2347	0.8631	-0.0841	-0.4182
logDL	0.2356	0.8664	0.1969	-0.2082
logMTRL	0.1591	0.5851	-0.5746	-0.0459
logIFL	0.2312	0.8499	0.0023	0.8436
logPBL	0.2505	0.9210	0.4147	-2.8129
logAW	0.2000	0.7352	-0.1672	-0.4045
logOCW	0.2157	0.7930	0.2767	0.4851
logMB	0.2457	0.9035	-0.5030	-0.1550
logBOL	0.2252	0.8281	0.1513	0.2924
logMFL	0.2164	0.7956	-0.4997	-0.4015
logMFW	0.1662	0.6110	-0.3227	0.8101
logZPW	0.2065	0.7591	-0.2542	-0.2149
logCD	0.2228	0.8194	1.1042	0.0085
logBUL	0.1515	0.5571	-0.6161	-0.0645
logBUW	0.1721	0.6328	-0.0866	-0.2908
eigenvalue	13.5198		2.0589	1.6223
% contribution	64.38		28.52	22.47

[1] Pearson correlation coefficients of PC-1 scores on original variables; all significant at $p < 0.0001$.

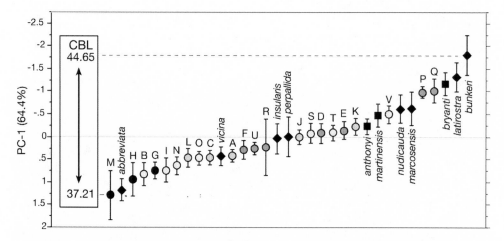

Figure 95. Mean and 95% confidence limits of cranial sizes (as represented by the first principal components axis based on 21 craniodental variables; see text) among peninsular and insular samples of desert woodrats from Baja California. Samples are organized from smallest (Group M, on the left) to largest (*bunkeri* Burt, on the right). Letters and circles identify peninsular samples, with the fill keyed to the map, Fig. 88; insular samples are identified by their respective trivial name, with squares indicating islands on the Pacific coast and triangles identifying those islands in the Gulf of California. The range in mean cranial length, as reflected by the measurement Condyloincisive Length (CIL; see Methods), is indicated in the box to the left.

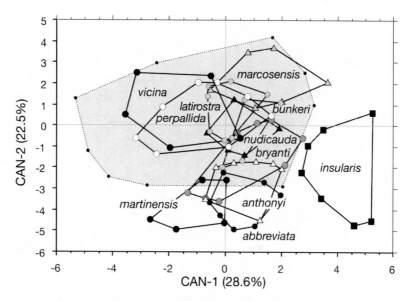

Figure 96. Ellipses encompassing scores on the first two axes of the size-free canonical analysis for Baja California samples. Ellipses enclose individual scores for each insular taxon and for the pooled peninsular samples (black circles enclosing gray ellipse).

To illustrate trends in size-free cranial variation among all samples, we map the statistically significant transitions between geographically adjacent sample groups for the scores of both canonical axes in Fig. 97. These maps provide the most effective means to visually represent the degrees of difference that are apparent in cranial structure, independent of overall body size, along the length of Baja California, including the islands on both the Pacific and Gulf coasts where woodrat populations are (or were) found.

We examine the relationship of each insular sample with respect to adjacent mainland samples in greater detail below. However, several overall patterns in size-free cranial characters in comparisons of insular samples, both to samples on the adjacent mainland and among themselves, deserve mention at this point (Fig. 97). In the summary below, we use "weakly", "moderately", and "strongly" to denote statistical significance equivalent to $p < 0.05$, 0.01, and 0.001, respectively, based on Fisher's PLSD comparisons from an overall ANOVA. Most importantly, the sample from Isla Ángel de la Guarda (*insularis*) is the most sharply differentiated of all 33 peninsular and insular groups. It is not only highly significantly separable from groups on the immediately adjacent eastern coast of

Baja California on both CAN-1 and CAN-2 axes (sample Groups G, H, I, J, and N), but this sample is highly divergent from all others anywhere along the entire length of the peninsula (Fisher's PLSD, p < 0.0001 in all pairwise comparisons). Similarly, the samples from both Isla Todos Santos (*anthonyi*) and San Martín (*martinensis*) off the northwestern Pacific coast are also strongly defined on both CAN-1 and CAN-2 axes with respect to their adjacent mainland samples. The sample from Isla San Marcos (*marcosensis*) differs strongly from geographically adjacent samples in CAN-1 scores, but not on the second axis. However, unlike *insularis*, these three samples/taxa do overlap with other sample groups elsewhere along the peninsular (data not provided). The sample from Isla Cedros (*bryanti*) is not differentiated from those samples of *molagrandis* (especially Groups K and L) from the adjacent Vizcaino Desert on the peninsula on the first axis, but is weakly separate on the second, although it is well-defined relative to sample Group J on both axes. Finally, the six remaining insular samples are only marginally separable, if at all, from other adjacent insular populations. The three Loreto Bay island taxa (*bunkeri* from Coronados, *nudicauda* from Carmen, and *latirostra* from Danzante) do not differ from one another on either axis, and only *bunkeri* and *nudicauda* are weakly to moderately different with respect to the adjacent mainland Group O sample; *latirostra*, from Danzante, cannot be distinguished from this sample in size-free craniodental measurements. Of the three southeastern mid-rift taxa, *perpallida*, from Isla San José, is likewise inseparable from the mainland Group R or S samples. This insular form does not differ from *abbreviata* from the adjacent Isla San Francisco on CAN-1 although the two are moderately differentiated on the second axis. Finally, *vicina*, from Isla Espíritu Santo is sharply separable from *abbreviata* to the immediate north on both axes but cannot be distinguished from the adjacent mainland Group S sample.

Equally sharp transitions occur among peninsular samples, but these are limited in number, with only four general geographic area shifts readily defined by CAN-1 scores; no significant shifts occur in CAN-2 scores (Fig. 97). The most sharply defined transition separates cape region samples from those along the remainder of the peninsula. Geographic Groups S, T, U, and V (which represent the currently recognized subspecies *arenacea* and *notia*) are homogeneous in size-free cranial characters but highly differentiated (p < 0.001 in all comparisons) with respect to sample Groups P, Q, and R (*pretiosa*) to the immediate north. Of the remaining peninsular samples, those from north of La Paz to the border with California, weak to strong transitions do occur between adjacent grouped samples in three regions. For example, samples A, B, and C (*intermedia* and *gilva*) differ strongly (p < 0.01 or < 0.001) in all pairwise combinations from Groups D, E, and F (*egressa*) to their immediate south along the northwest coast and montane regions. Similarly, samples from the mid-peninsular Vizcaino Desert (Groups J and K [*molagrandis*]) are weakly to moderately separable from those in the mountains to the east (Group L) or along the Gulf coast (Group M [*aridicola*] or Group N [*ravida*]). Finally, the Pacific coastal Groups P, Q, and R (*pretiosa* from the mainland and both Magdalena and Margarita islands) are separable from Group O (*ravida*) to the immediate east in the central mountains and Gulf coast.

Comparisons between the distributions of cranial sizes (Fig. 95) and the transitions in size-free cranial shape (Fig. 97) underscore the fact that some of the formally described taxa are separable only in overall size (e.g., *notia* [Group U] versus adjacent *arenacea* [Groups S, T, and V]; *molagrandis* [Groups J and K] and both *felipensis* [Group H] and *gilva* [Group I]; *bunkeri, nudicauda,* and *latirostra* and adjacent *ravida* [Groups N and O]; and *vicina* versus *arenacea* [Group S]), while others differ sharply in cranial shape. It remains to be determined if size itself has a substantial genetic component in woodrats and thus that differences in size reflect underlying adaptive divergence, or whether size is largely ecophenotypic, responding locally to habitat quality, as is true for pocket gophers (Patton and Brylski, 1987; Smith and Patton, 1988; Patton and Smith, 1990).

Cranial size and shape differentiation among insular samples, and the relationship of insular to mainland populations.—The 11 taxa restricted to islands on either the Pacific or Gulf sides of Baja California have received only cursory attention, either since their initial descriptions 70 to 120 years ago or subsequent to Goldman's 1932 revision of the *Neotoma lepida* group. Included within this group of taxa are four that have been recognized as separate species since their respective descriptions (*N. bryanti* [Isla Cedros]; *N. anthonyi* [Isla Todos Santos]; *N. martinensis* [Isla San Martín]; and *N. bunkeri* [Isla Coronados]). Two additional taxa were originally described as distinct species (*insularis* [Isla Ángel de la Guarda] and *abbreviata* [San Francisco]), but were later relegated to subspecies of *N. lepida* by Burt (1932, p. 182). All remaining insular taxa were described as subspecies of *N. lepida* (or *N. intermedia,* before Goldman [1932] formally included this taxon within his concept of *N. lepida*). As is apparent from the brief descriptions in the paragraphs above, there appears to be little relationship between either cranial size or size-free shape and the current taxonomic status of many of these forms. We examine the concordance and discordance between cranial size and shape among all insular samples, both among themselves and in relation to adjacent mainland samples immediately below.

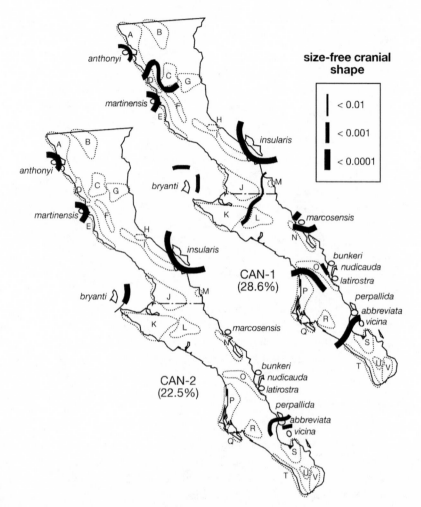

Figure 97. Geographic differentiation among peninsular and insular samples of the *Neotoma lepida* group in Baja California based on a size-free canonical variates analysis. Separate maps detail differences in CAN-1 and CAN-2 axes, which combine to explain 51.1% of the total pool of variation. Line thickness, as per the inset box in the upper right, indicates the degree of statistical significance between adjacent samples (based on ANOVA using Fisher's PLSD pairwise test).

We compare the 11 insular taxa for both size (PC-1 scores) and size-free cranial shape (CAN-1, using PC axes as variables) in Fig. 98. As described above (Fig. 95), individuals from Isla Coronados (*bunkeri*) are largest, although they are statistically equivalent in size to those from Isla Cedros (*bryanti*) and Isla Danzante (*latirostra*). Despite overlap in overall size with these two other insular samples, *bunkeri* stands as the giant within the desert woodrat complex in Baja California (Smith 1992). However, when size is removed, *bunkeri* is indistinguishable from *nudicauda* from Isla Carmen, a short distance to the south, or even from *marcosensis* from Isla San Marcos further to the north along the Gulf coast of the peninsula (Fig. 96). On the Pacific coast, *anthonyi* (Isla Todos Santos) and *martinensis* Goldman (Isla San Martín) overlap completely in cranial size but differ substantially in shape, with members of each taxon significantly different in both overall size and shape from *bryanti* (Isla Cedros) further to the south. Finally, the central and southern mid-rift insular taxa are different from one another, if at all, only in cranial size (e.g., *nudicauda* [Isla Carmen] versus either *bunkeri* [Coronados] or *latirostra* [Danzante] and *abbreviata* [San Francisco] versus *perpallida* [San José] or *vicina* [Espíritu Santo]). All of these taxa are rather uniform in cranial shape. The exception to this set of six taxa is *vicina*, which differs sharply from all others in size-free shape but not in overall size.

Smith (1992), in addition to documenting that the average insular individual is larger than those on the adjacent mainland, also asked whether size was related to biotic characteristics or other features of the islands (such as the presence of potential mammalian competitors and absence of mammalian predators, island area, time since isolation, and distance of island from mainland). She rejected (p. 268) the hypothesis that body size of the insular populations had altered randomly due to drift or local adaptation, but she found a weak relationship between size and the absence of predators. None of the insular physical attributes of the islands were useful in predicting body size of the rats inhabiting them.

We performed the same type of multiple regression analysis, but expanded the island characteristics to include depth of water channel separating an island from the mainland, maximum elevation, and plant species diversity (data in Murphy et al., 2002; Rebman et al., 2002). We used our expanded dataset and PC-1 scores as an index of multivariate size. Our results are completely consistent with those of Smith (1992); none of the physical or biotic variables, individually or collectively, were significantly related to body size. Nor does any relationship exist between any of these variables and the size difference between each insular sample and that from the closest neighboring sample, either an adjacent island or that nearest on the mainland. We could also find no relationship between mean size-free cranial shape and these variables, again using either mean CAN-1 scores or the difference in these between insular and mainland samples. In short, if Smith's (1992) rejection of random differentiation as an explanation for the pattern of body sizes among all insular populations of these woodrats is correct, we remain completely ignorant of the process, or processes, that have underscored either size differentials or variation in size-free cranial shape. Her hypothesis that the absence of mammalian predators may have driven a selective increase in size for the physiological advantage of increased energy intake from microbial fermentation remains the only available, but as yet untested, explanation for that relationship.

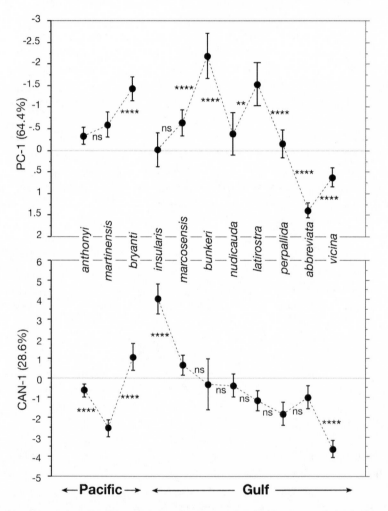

Figure 98. Means and 95% confidence limits for PC-1 scores (overall size; above) and CAN-1 scores (size-free shape; bottom) for the 11 insular taxa on both sides of Baja California (Pacific islands on left and Gulf of California islands on right). Significance levels between geographically adjacent samples are indicated: ns = non-significant, ** $p < 0.01$, **** $p < 0.0001$).

We illustrate cranial size and shape relationships between each insular taxon and those on the geographically adjacent mainland in figures 99 through 104, arranged from north to south along the length of Baja California. As is apparent from Fig. 99 (PC-1 scores, left), both *anthonyi* (Todos Santos) and *martinensis* (San Martín) overlap extensively with both coastal samples of *egressa* (Groups D and E) in size, but are larger than the coastal sample of *intermedia* (Group A). The relationship suggested by size-free cranial dimensions is, however, quite different (Fig. 99, CAN-1 scores, right). The two insular taxa differ between themselves and each is markedly different from all adjacent mainland samples. Allen (1898, p. 151) in his original description of *anthonyi* noted that his new taxon "...is too distinct, both in coloration and cranial details, to require comparison with any of its congeners." This certainly appears true in cranial shape characters, which we examine here. Goldman (1905, p. 28) compared his *martinensis* to Allen's *anthonyi*, describing it as closely resembling that taxon "...but tail more scantily haired, and cranial characters very different." Among the latter, he noted in particular the long nasals and small bullae of his *martinensis*, both characters that indeed exhibit highly significant differences in comparison between these two taxa (see Table 34 for means; NL, ANOVA, Fisher's PLSD, p < 0.0001; BUL, p = 0.0064; BUW, p < 0.0001). Although it is curious that neither Allen nor Goldman compared their respective insular taxa to mainland forms that had been described by the time of their respective studies, it seems clear that both *anthonyi* and *martinensis* are well separated in their cranial characteristics, both with respect to each other and collectively to all adjacent mainland samples. Commonalities in color and color pattern will be described below.

The insular sample from Cedros (*bryanti*), considered a species distinct from mainland representatives of the *N. lepida* group since its initial description in 1887, is substantially larger in overall cranial size compared to all peninsular samples along the central west coast of Baja California, including the southernmost sample of *egressa* (Group E) and the two Vizcaino Desert samples of *molagrandis* (Groups J and K; Fig. 100, PC plot on left). However, these taxa are almost indistinguishable in size-free cranial shape. Only the mainland Group J sample is weakly different (p = 0.009) from the insular *bryanti* on CAN-1 (Fig. 100, CAN plot on right); other mainland samples express the same overall cranial shape on this axis, although there are more substantial differences on the second CAN axis (Fig. 97). Merriam (1887) compared *bryanti* only to *N. floridana*, a species from the southeastern United States that is not closely related to the *N. lepida* group (Goldman, 1932; Edwards and Bradley, 2001). Goldman (1910), however, clearly placed *bryanti* within his *intermedia* group (= *lepida* group of Goldman, 1932), noting its large size and broader frontal area between the lacrimal bones than typical of other members of this group. The larger size is clearly apparent in Figs.

95, 98, and 100, but our set of cranial variables are inadequate to test the difference in frontal breadth mentioned by Goldman. The two characters in our dataset that could be considered surrogates for frontal breadth (Rostral Width [RW] and Interorbital Constriction [IOC]) give conflicting results: *bryanti* does not differ from any of the adjacent mainland samples in Interorbital Constriction (IOC, $p >$ 0.3703 in all comparisons) but does have a significantly broader rostrum (RW, $p <$ 0.0001 in all comparisons). In summary, *bryanti* differs from all adjacent mainland samples in size, but cannot be distinguished from the nearest-neighbor Group K from the Vizcaino Desert in size-free cranial shape.

The nearly opposite pattern to that described above for *bryanti* from Isla Cedros is exhibited by *insularis* from Isla Ángel de la Guarda (Fig. 101). This taxon, described originally as a distinct species by Townsend in 1912 but allocated to a subspecies of *N. lepida* by Burt (1932), is significantly larger than all adjacent peninsular samples along the Gulf coast (*felipensis* [Groups G and H] and *gilva* [Group I] on the immediately adjacent coast and *aridicola* [Group M] to the south), although it is of the same size as the Vizcaino Desert sample of *molagrandis* on the Pacific versant (PC-1 plot, left side of Fig. 101). In size-free cranial shape, however, the sample of Townsend's *insularis* is statistically distinct from all mainland samples, regardless from which side of the peninsula they are from (CAN-1 plot, right side of Fig. 101). Indeed, Townsend (1912, p. 125) noted that the skull of *insularis* was "...relatively shorter and broader, with heavier rostrum, heavier dentition and larger auditory bullae..." than mainland samples of *gilva* to which he compared it (our sample I). The skull of *insularis* is certainly broader across the zygomatic arches with a broader rostrum, longer maxillary tooth row, and larger bullae than the other samples to which we compare it here (Table 34), especially those of Group I with which it differs at $p < 0.001$ in all comparisons. As we documented earlier, the sample of *insularis* is overall the most distinctive in cranial morphology of all 33 Baja California samples we compare, including other insular and all mainland ones.

Similar to the pattern we observed for skulls of *insularis*, those of *marcosensis* from Isla San Marcos off the central Gulf coast of Baja California are both larger and of a different shape in comparison to those on the peninsular mainland (Fig. 102). These comparisons include those with *aridicola* [Group M] to the north on the coast and *ravida* [Group N] from the immediately adjacent coast; *marcosensis* Burt differs in shape less so in comparison to the easternmost sample of the Pacific coast taxon, *molagrandis* (Group L). Burt (1932) described three taxa of insular woodrats, including *marcosensis*, either as distinct species or subspecies of *N. lepida*. He noted, in particular, that the skull of *marcosensis* was "large and angular; supraorbital ridges prominent; interpterygoid fossa relatively wide; audital bullae medium" (Burt, 1932, p. 180), although among peninsular taxa he compared it only to *felipensis* from the north Gulf coast and *arenacea* from the Cape region, not to those from the central coast. Burt also believed that the closest relative of *marcosensis* was *nudicauda* from Isla Carmen, from which it differed only in darker color, wider interpterygoid fossa, and shorter incisive foramina. Individuals of *marcosensis* do have a significantly wider mesopterygoid fossa (p < 0.0001 in all comparisons; Table 34) than peninsular specimens, including those on the adjacent coast as well as specimens of *felipensis* to the north, although not those of *arenacea* (p > 0.7299). In comparison to *nudicauda*, *marcosensis* also has a broader mesopterygoid fossa (ANOVA, $F_{(1,38)}$ = 35.10, p < 0.0001) and shorter incisive foramen, although not significantly so in the latter case (p = 0.5993).

The trio of islands off the south-central Gulf coast (Coronados, Carmen, and Danzante) is home to three taxa currently regarded as distinct species (*N. bunkeri*) or subspecies of *N. lepida* (*nudicauda* and *latirostra*). Animals of each of these taxa differ in size from their neighbors on adjacent island, but *bunkeri* from the northernmost island and *latirostra* from the southernmost are similar (Fig. 103, PC-1 scores, left). All three are larger then samples on the immediately adjacent Gulf coast of the peninsula (Group N and especially Group O of the subspecies *ravida*). However, in size-free cranial shape, all three insular taxa are similar (Fig. 103, CAN-1 scores, right). Moreover, the insular taxa are also similar to, or only marginally different from, peninsular samples. Both *bunkeri* and *nudicauda* are both significantly different in size-free shape compared to the members of Group O on the adjacent mainland (*ravida*; p < 0.003 in both cases, ANOVA, Fisher's PLSD for pairwise comparisons), although *latirostra* is not (p = 0.0593). None of the insular taxa differ from the two samples of *pretiosa* (Groups P and R) from the Pacific coast. Burt (1932) described both *bunkeri* and *latirostra*, noting that the latter differed from *nudicauda* Goldman in larger size and heavier rostrum. However, he had but a single specimen of *latirostra* for comparison. Based on the larger sample available to us, *latirostra* is indeed larger than *nudicauda* (in total length and condyloincisive length, p < 0.0001 in both cases) and the skull does have a heavier rostrum, using rostral width as a proxy (p < 0.0001; see Table 34). Curiously, Burt made no comparison of his *latirostra* to any peninsular taxon. Rather, he considered *bunkeri* to be a representative of the *Neotoma fuscipes* group and thus made no comparisons of his new species to any of the taxa of Goldman's (1932) *N. lepida* group. As is evident by the placement of *bunkeri* in the PC and CAN plots presented in Fig. 103, however, this taxon overlaps extensively with other insular taxa in its immediate vicinity in both size and especially in size-free cranial shape, and it is only marginally different from mainland samples of the *N. lepida* group in shape. It also exhibits no phylogenetic relationship to either *N. fuscipes* or *N. macrotis*, members of Goldman's (1910) *N. fuscipes*-group, based on our molecular analyses.

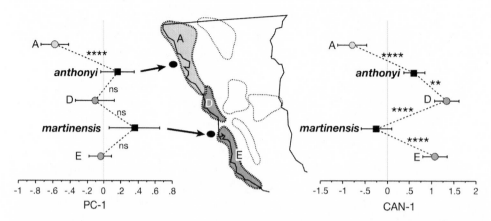

Figure 99.   Grouped samples of the insular *anthonyi* (Todos Santos) and *martinensis* (San Martín) and three from the northwestern Pacific coast of Baja California (Group A, *intermedia*; Groups D and E, *egressa*). The mean, range, and significance levels among adjacent samples are illustrated by diagrams of the mean and 95% confidence limits for overall cranial size (PC-1 scores, left) and size-free cranial shape (CAN-1 scores, right).

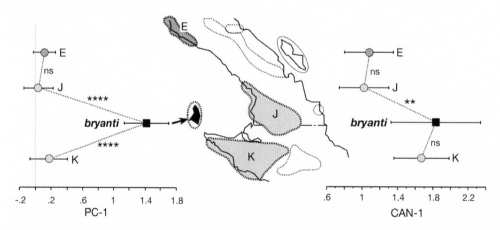

Figure 100.   Grouped samples of the insular *bryanti* (Cedros) and three on the Pacific side, west-central Baja California (Group E, *egressa*; Groups J and K, *molagrandis*). The mean, range, and significance levels among adjacent samples are illustrated by diagrams of the mean and 95% confidence limits for overall cranial size (PC-1 scores, left) and size-free cranial shape (CAN-1 scores, right).

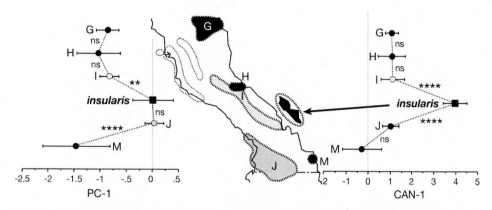

Figure 101. Grouped samples of the insular *insularis* (Ángel de la Guarda) and five on the adjacent Gulf side of the northcentral Baja California (Groups G and H, *felipensis*; Group I, southern *gilva*; Group J, *molagrandis*; and Group M, *aridicola*). The mean, range, and significance levels among adjacent samples are illustrated by diagrams of the mean and 95% confidence limits for overall cranial size (PC-1 scores, left) and size-free cranial shape (CAN-1 scores, right).

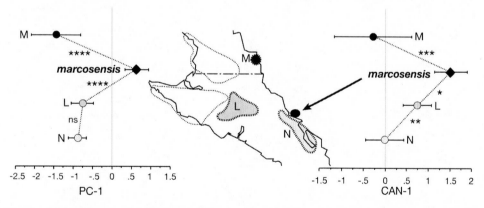

Figure 102. Grouped samples of the insular *marcosensis* (San Marcos) and three from the Gulf coast of the central Baja California (Group M, *aridicola*; Group L, *molagrandis*; and Group N, *ravida*). The mean, range, and significance levels among adjacent samples are illustrated by diagrams of the mean and 95% confidence limits for overall cranial size (PC-1 scores, left) and size-free cranial shape (CAN-1 scores, right).

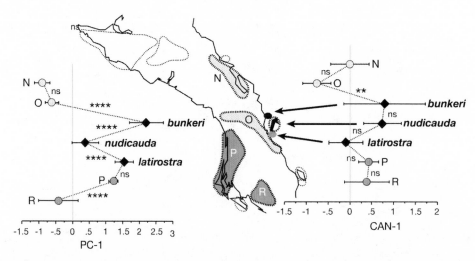

Figure 103.   Grouped samples of the insular *bunkeri* (Coronados), *nudicauda* (Carmen), and *latirostra* (Danzante) and four from east-central Baja California (Groups N and O, ravida; Groups P and R, pretiosa).   The mean, range, and significance levels among adjacent samples are illustrated by diagrams of the mean and 95% confidence limits for overall cranial size (PC-1 scores, left) and size-free cranial shape (CAN-1 scores, right).

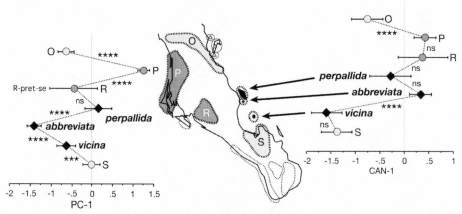

Figure 104. Grouped samples of the insular *perpallida* (San José), *abbreviata* (San Francisco), and *vicina* (Espíritu Santo and Partida) and peninsular southern Baja California (Group O, ravida; Groups P and R, pretiosa; and Group S, arenacea). The mean, range, and significance levels among adjacent samples are illustrated by diagrams of the mean and 95% confidence limits for overall cranial size (PC-1 scores, left) and size-free cranial shape (CAN-1 scores, right).

The final set of insular-mainland comparisons are those of the islands in the vicinity of La Paz Bay (San José, San Francisco, Partida, and Espíritu Santo), each taxon of which is currently regarded as a valid subspecies of *N. lepida* except for the woodrats from Isla Partida, which have not been previously reported. Specimens of *latirostra* from Isla San Francisco are among the smallest found along the entire peninsula, either mainland or insular (Fig. 95). This taxon is significantly smaller in cranial size both in comparison to all adjacent samples on the peninsula (e.g., Group S, *arenacea*, from the La Paz area) or islands (*perpallida* from San José and *vicina* from Espíritu Santo and Partida; Fig. 104, PC-1 scores, left). On the other hand, *perpallida* and *vicina* overlap in size and are at most slightly different from adjacent samples on the Gulf side of the peninsula (Group O, *ravida*; Group S, *arenacea*; or Group R, southern *pretiosa*). In size-free cranial shape, however, neither *abbreviata* nor *perpallida* differ in any respect from one another or from peninsular samples on the Pacific side nor does *vicina* differ from the sample of *arenacea* (Group S) from the adjacent Gulf coast.

Goldman (1909, p. 140-141) recognized that his *abbreviata* was similar to *N. intermedia* (= *N. lepida*) in color and cranial characters, but initially described it as a distinct species based on its overall small size and short tail. The tail of this taxon is indeed shorter than other taxa in the southern peninsular or islands (p < 0.0001 in all comparisons; Table 34) and proportionately shorter relative to body length than the two other insular samples (*perpallida*; p < 0.0001; *vicina*; p = 0.00657) as well as most mainland samples (Group S, a*renacea*, p < 0.0001; Group O, *ravida*, p = 0.0002; Group P, *pretiosa*, p < 0.0001). Burt (1932, p. 182) regarded the difference in proportionality relative to the other insular taxa (about 4%) to be "insignificant when one considers the variation in the group." He concluded that *abbreviata* was best treated as a subspecies. Both *perpallida* and *vicina* were described as subspecies of *N. intermedia* (Goldman, 1909, p. 139-140) and considered nearly the same in cranial characters but different in color tones. However, in both cranial size and especially in size-free shape, *vicina* is the least distinctive of this trio of insular taxa. It is, for example, indistinguishable from the adjacent mainland sample of *arenacea* Allen (Group S) in size-free cranial features, including comparisons at all canonical axes (data not shown). Goldman (1909, p. 140) did recognize that *vicina* is "similar to *N. i. arenacea*, but smaller," a view completely concordant with the results presented in Fig. 104.

Color variation.—We provide descriptive statistics for the colorimetric X-coefficients for the dorsum, tail, lateral, and chest regions of the pelage for each of the 33 peninsular and insular samples of Baja California woodrats in Table 36. All four coefficients exhibit significant geographic differentiation (ANOVA, p < 0.0001). Darkest samples are those from the northwestern coast (Groups A and C [*intermedia*] and D, E, and F [*egressa*]), the insular samples of from Todos Santos (*anthonyi*), Isla Cedros (*bryanti*), and Coronados (*bunkeri*), Group O (*ravida*), and Group U (*notia*, from the Sierra La Laguna). Palest samples are those from the north-central Gulf coast (Groups G [*felipensis*] and M [*aridicola*]) and the Vizcaino Desert (Groups J, K, and L [*molagrandis*]). Color characteristics of the four topographic regions of the skin are positively correlated (p < 0.0001 in all cases; r ranges from 0.605 [Dorsal-X versus Lateral-X; Z-value = 21.301] to 0.270 [Dorsal-X versus Chest-X; Z-value = 8.421]), indicating that the general dorsal color is reflected over the entire body.

Table 36. Descriptive statistics of four colorimetric variables of the pelage of the 33 peninsular and insular samples and taxa of the *Neotoma lepida* group from Baja California. Mean ± standard error, sample size, and range are given for each sample.

Sample	Dorsal-X	Tail-X	Lateral-X	Chest-X
A-*intermedia*	8.03±0.24	5.87±0.20	18.77±0.41	37.55±0.91
	45	45	45	45
	4.84-12.94	3.53-9.72	13.29-25.49	23.71-53.47
B-*gilva*	11.14±0.66	8.40±0.53	22.96±0.92	41.53±1.06
	25	25	25	25
	5.33-19.01	3.93-12.94	12.77-31.97	32.92-53.38
anthonyi	7.24±0.17	4.89±0.14	17.69±0.65	32.24±0.79
	42	42	42	42
	4.52-9.73	3.68-7.42	10.13-25.45	19.13-42.49
C-*intermedia*	9.18±0.23	6.17±0.29	20.19±0.59	37.44±1.02
	42	42	42	42
	5.87-11.54	3.26-10.82	13.42-29.5	16.50-48.89
D-*egressa*	8.58±0.34	5.92±0.45	19.08±0.55	36.88±1.13
	27	27	27	27
	5.85-14.36	2.69-12.13	12.74-23.88	23.96-48.18
E-*egressa*	8.32±0.27	5.09±0.23	18.67±0.61	37.58±1.12
	26	26	26	26
	6.15-12.13	3.34-7.53	12.74-23.92	24.26-48.16
F-*egressa*	8.04±0.23	5.54±0.34	20.07±0.86	37.68±0.94
	13	13	13	13
	6.36-9.58	3.94-7.45	14.51-24.57	32.08-43.66
martinensis	9.43±0.21	4.17±0.17	19.49±0.45	38.86±0.85
	29	29	29	29
	7.21-12.24	2.84-6.6	15.04-24.71	26.45-47.05

Table 36 (continued)

Sample	Dorsal-X	Tail-X	Lateral-X	Chest-X
G-*felipensis*	16.76±0.47 52 10.98-25.03	11.92±0.41 52 5.29-18.69	28.58±0.58 52 20.52-43.25	39.50±0.77 52 28.79-52.25
H-*felipensis*	11.16±0.39 12 8.79-13.57	9.11±0.51 12 6.09-11.25	23.09±0.92 12 17.93-29.05	41.32±1.14 12 33.46-45.86
I-*gilva*	11.52±0.87 9 8.23-13.57	8.05±0.57 9 5.72-10.87	24.10±1.49 9 15.49-29.62	43.47±2.21 9 35.22-56.14
insularis	13.68±0.58 11 9.46-15.99	14.34±0.59 11 11.29-16.79	31.42±1.43 11 25.18-42.79	41.41±1.93 11 30.83-54.51
bryanti	8.01±0.26 36 5.21-12.62	9.84±0.61 36 5.05-19.49	18.44±0.52 36 11.78-23.22	36.28±0.87 36 26.73-46.74
J-*molagrandis*	13.25±0.58 33 8.25-19.30	9.03±0.50 33 4.40-14.26	24.75±0.86 33 15.39-34.91	44.19±0.97 33 34.64-58.50
K-*molagrandis*	11.85±0.71 7 9.41-14.33	10.20±0.89 7 5.96-12.48	25.34±0.75 7 22.99-28.30	40.91±1.69 7 34.71-47.13
L-*molagrandis*	10.83±0.52 40 5.67-20.71	8.66±0.62 40 2.97-15.61	22.67±0.74 40 13.73-35.50	40.65±1.31 40 23.35-56.08
M-*aridicola*	13.21±0.96 6 10.84-16.79	12.05±1.31 6 7.69-16.54	28.71±1.58 6 23.19-34.50	43.04±1.58 6 35.98-46.64
marcosensis	9.59±0.48 13 6.38-12.69	8.43±0.63 13 4.75-11.14	24.37±0.58 13 20.54-27.38	38.23±1.95 13 22.35-47.22
N-*ravida*	9.51±0.35 24 6.24-13.75	7.09±0.48 24 4.00-14.17	20.71±0.76 24 12.46-26.66	40.04±1.36 24 28.94-54.12
O-*ravida*	8.35±0.28 51 5.29-15.41	6.26-0.32 51 2.97-14.16	18.88±0.57 51 11.30-31.97	34.98±0.94 51 20.68-55.18
bunkeri	8.42±0.34 9 6.66-9.79	6.78±0.31 9 5.35-8.18	20.01±1.18 9 14.97-24.15	32.61±1.13 9 28.07-38.64
nudicauda	10.65±0.62 9 7.39-12.41	8.41±0.48 9 6.00-10.30	22.81±0.60 9 20.28-26.56	38.09±1.95 9 30.89-50.96

Table 36 (continued)

Sample	Dorsal-X	Tail-X	Lateral-X	Chest-X
latirostra	11.02±0.39	8.53±0.38	26.07±0.92	42.86±1.29
	8	8	8	8
	8.59-12.04	6.51-9.72	22.39-30.54	37.52-47.33
P-*pretiosa*	10.74±0.19	7.71±0.21	22.34±0.40	40.94±0.58
	99	99	99	99
	7.26-16.44	3.57-14.94	10.58-32.78	24.46-52.83
Q-*pretiosa*	10.81±0.47	8.36±0.39	24.17±0.95	42.00±1.25
	16	16	16	16
	8.13-15.18	6.53-12.20	15.62-31.34	30.03-49.92
R-*pretiosa*	10.29±1.16	10.63±1.40	30.88±2.34	39.56±4.45
	4	4	4	4
	7.92-12.39	8.99-14.80	18.12-27.87	31.83-51.64
perpallida	12.23±0.37	10.31±0.46	27.38±0.84	43.63-1.19
	20	20	20	20
	8.45-15.94	6.24-14.16	20.41-34.04	33.88-54.64
abbreviata	12.78±0.25	11.66±0.35	20.49±0.65	32.01±1.33
	21	21	21	31
	9.72-14.32	9.12-14.07	14.80-28.97	22.83-43.50
vicina	10.27±0.29	7.10±0.43	22.77±0.57	40.33±0.92
	31	31	31	31
	7.18-13.8	4.54-16.09	14.07-29.1	29.99-52.94
S-*arenacea*	9.63±0.35	8.24±0.34	21.52±0.50	38.84±1.12
	33	33	33	33
	5.84-14.71	5.16-13.54	16.50-27.44	28.39-51.99
T-*arenacea*	10.75±0.29	8.43±0.34	23.39±0.46	42.42±0.99
	38	38	38	38
	7.63-14.95	5.22-13.51	17.07-28.38	29.10-52.57
U-*notia*	8.77±0.20	6.89±0.29	19.14±0.37	37.36±0.80
	63	63	63	63
	5.93-13.19	3.41-14.82	12.62-26.72	23.17-51.42
V-*arenacea*	10.61±0.25	7.52±0.42	22.73±0.54	37.90±1.10
	33	33	33	33
	6.67-13.69	3.80-14.85	16.55-28.96	20.87-52.74

We summarized overall color variation among all peninsular and insular samples by a principal components analysis using the four topographic X-coefficients (Table 37). Both the eigenvalues and factor loadings are similar to those found in other geographic transect regions we have analyzed. All four coefficients load positively and significantly ($p < 0.0001$ for each variable) on the first PC axis, which explains 57.9% of the total variation. The position of the individual scores on the second PC axis is most strongly influenced by Chest-X, while Dorsal-X and Tail-X coefficients contrast with this variable and Lateral-X has no statistical influence (PC-2 scores versus Lateral-X, Z-value = 1.322, p = 0.1862).

Table 37. Principal component eigenvalues and factor loadings of colorimetric variables from all samples of the Baja California Transect.

Variable	PC-1	PC-2
Dorsal-X	0.832	-0.266
Tail-X	0.797	-0.348
Lateral-X	0.843	0.044
Chest-X	0.569	0.797
eigenvalue	2.318	0.829
% contribution	57.945	20.714

Woodrats become paler from north to south and from the Pacific to Gulf coasts (latitude: r = 0.115, F = 11.078, p = 0.0009; longitude: r = -0.190, F = 30.775, p < 0.0001). These relationships are weak, however, and there are notable exceptions. To illustrate more effectively the overall geographic trends in color, we plotted the level of statistical significance in PC scores for the first and second axes between adjacent samples along the length of the peninsula in Fig. 105, based on an ANOVA and Fisher's PLSD posterior tests for each pairwise comparison. Along the peninsula, relatively sharp transitions in PC-1 color scores separate those samples from the Pacific coast that are notably dark overall (*intermedia* [Groups A and C] and *egressa* [Groups D, E, and F]) from paler specimens on the eastern side of the peninsula (*gilva*, Group B, and especially *felipensis*, Group H) or those to the south in the Vizcaino Desert (*molagrandis*, Groups J, K, and L). A second and equivalently sharp transition occurs between the mid-peninsular samples of *molagrandis*, *aridicola* (Group M), and *ravida* (Groups N and O) in comparison to those of *pretiosa* (Groups P, Q, and R) and *arenacea* (Groups S, T, and V) further to the south. Finally, the montane *notia* (Group U) from the Sierra La Laguna is notably darker overall when compared to all adjacent samples of *arenacea* in the lowlands surrounding this "sky island." PC-2 scores effectively separate only the pale coastal sample of *felipensis* from around its type locality near San Felipe on the northeastern Gulf coast (Group G) from all other adjacent samples.

Insular samples of *insularis* (Isla Ángel de la Guarda), *nudicauda* (Carmen), and *latirostra* (Danzante), and to a lesser extent *anthonyi* (Todos Santos), are sharply set off from their mainland neighbors on PC-1 but not on PC-2, although not always in the same direction of color change. For example, *anthonyi* is darker overall than its mainland counterparts, either *intermedia* (Group A) or *egressa* Orr (Group D), while *insularis* is paler than the sample of *gilva* Rhoads (Group I) from the adjacent coast, as are both *nudicauda* and *latirostra* relative to *ravida* (Group O). Only *abbreviata* (San Francisco) is markedly different from mainland samples (Group S of *arenacea* or Group R of *pretiosa*) on PC-2. All remaining insular taxa exhibit little, if any, differentiation between neighboring island populations or those closest on the peninsula.

Our analysis of color, both the univariate X-coefficients and the principal component summaries of these variables, excludes color pattern not reflected in these measurements. For example, both the insular *anthonyi* and *martinensis* Goldman the northwest Pacific coast were noted in their respective descriptions for the conspicuous blackish outer sides of the hind legs and inner sides of the ankles, a feature that is unique to these two insular taxa among all individuals and samples examined by us.

Morphology, mtDNA, and nuclear gene markers.—Our molecular data for Baja California populations of the desert woodrat are extensive for the mtDNA *cyt-b* gene, with sequences of 138 individuals from a total of 66 localities. These data include at least a single individual from all islands from which desert woodrats are, or were, known (see Figs. 6, 7, and 8). Our microsatellite data are much more limited, just 84 individuals representing 31 separate localities, including 29 specimens from four island populations (Isla San Marcos, Isla Danzante; Isla San José, and Isla San Francisco).

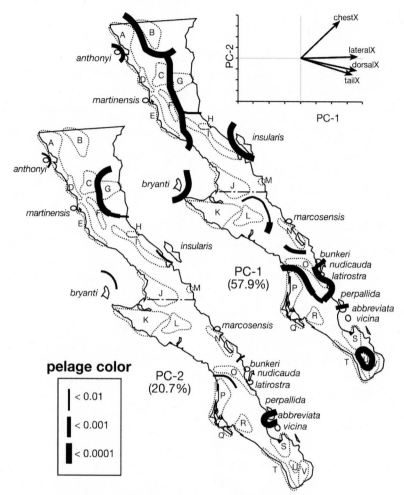

Figure 105. Color differentiation among peninsular and insular samples of the desert woodrat complex in Baja California, based on a principal components analysis of colorimetric X-coefficients. Separate maps detail differences in PC-1 and PC-2 axes, which combine to explain 78.6% of the variation. Line thickness (inset box in lower left; ANOVA, Fisher's PLSD pairwise test) indicates the level of significance between geographically adjacent samples. The inset in the upper right identifies character vectors on both axes.

Three *cyt-b* subclades are present in Baja California (Fig. 6). One, subclade 1A, is broadly distributed throughout the peninsula from the northern gulf coast to the cape region (Fig. 7). The second, subclade 1B, is limited to the northwestern coast of Baja and adjacent southern California (Fig. 8). The third, Subclade 1D, is restricted to the insular taxon *insularis* and is completely and uniquely defined by both phylogenetic position within the mtDNA tree (Fig. 5) and diagnosed by its strongly differentiated glans penis (Fig. 31) and craniodental characters (Figs. 96, 97, and 101). In this case, therefore, there is complete concordance across these respective character sets. However, the boundaries between subclades 1B and 1A along the peninsular are not completely concordant with the craniodental transitions in either the overall size or size-free trends depicted in Figs. 95 and 97. For example, subclade 1B includes the subspecies *intermedia* (morphological Groups A and C), *egressa* (Groups D, E, and F), and *gilva* (Groups B and I), as well as the adjacent Pacific coast insular taxa *anthonyi* (Todos Santos) and *martinensis* (San Martín). All remaining peninsular and insular sample groups (excepting *insularis*) belong to subclade 1A. The major morphological transitions among these groups, however, are positioned geographically within each of these two subclades rather than between them. For example, the boundary between *intermedia* and *egressa* lies within subclade 1B. Boundaries among *ravida*, *pretiosa*, and *arenacea* are within subclade 1A, which also includes some of the insular samples (e.g., *anthonyi*, *martinensis*, *marcosensis*) and their adjacent mainland samples (Fig. 97). Even in color attributes, only the sharp transition in the north (between *intermedia* + *egressa* and samples to the immediate east and south) is there a morphological boundary concordant with that between mtDNA subclades 1A and 1B (compare the geographic position of clade boundaries in Fig. 6 with the colorimetric PC-1 transition in Fig. 105).

Our microsatellite samples are restricted to the southern half of the peninsula, and thus of mtDNA subclade 1A, including four insular populations. Consequently, these data are inadequate to address the correspondence of nuclear gene phylogeographic structure and the geographic placement of the mtDNA subclades along the peninsula, although there is general concordance between sample membership in phyletic clusters based on both types of molecular data (compare the trees in Figs. 4, 5, and 14).

We provide data for 18 microsatellite loci, including sample size, mean number of alleles per locus, number of monomorphic loci, levels of heterozygosity, and deviations from Hardy-Weinberg equilibrium (Fis), in Table 38. We pooled samples from three geographic regions along the peninsula (Fig. 106). Two loci deviate from Hardy-Weinberg expectations in each of two pooled peninsular samples (northern and southern Gulf); all remaining samples are in equilibrium. There is a strong relationship between the number of alleles and sample size ($r = 0.885$, Z-value = 2.798, $p = 0.0051$) but the sizes of our peninsular and insular samples are equivalent, on average. Gene diversity is also similar across all samples, although the mean number of alleles is significantly lower for insular than peninsular ones (means = 3.4 and 8.6, respectively; ANOVA, $F_{(1,5)}$ = 46.206, p = 0.0092). Similarly, insular populations on average have higher numbers of monomorphic loci than do the three peninsular samples (4.8 versus 1.0), although this difference is not statistically significant (ANOVA, F = 3.680, p = 0.1132). The insular sample from Isla San José (*perpallida*) has higher values of each attribute than other insular populations, comparable to mainland population values.

We map localities and provide an unrooted network linking each of the seven microsatellite samples, based on a matrix of pairwise Fst values, in Fig. 106. The three peninsular samples are closely similar, with pairwise Fst < 0.022 and not significantly different from zero for each comparison. These three samples occupy the center of the network. Insular samples are linked to their geographically adjacent peninsular samples (*abbreviata* and *perpallida* both link the southern Gulf sample, *marcosensis* Burt to the northern Gulf). However, in all cases the average Fst between an insular population and its peninsular counterpart is substantial and significantly higher than zero (insular to peninsular Fst values range from 0.087 [San José to southern Gulf; p < 0.05] to 0.292 [San Francisco to southern Gulf; p < 0.001]). On average, therefore, *perpallida* from Isla San José is weakly distinct from the mainland although it is sharply differentiated from *abbreviata* from the adjacent Isla San Francisco, only 3 km distant (Fst = 0.360, p < 0.001). Indeed, *abbreviata* is strongly separated from all other samples, peninsular or insular (mean pairwise Fst value = 0.423, range 0.291 [to northern Gulf] to 0.654 [to Isla Danzante]). The samples of *marcosensis* from Isla San Marcos and *latirostra* from Isla Danzante are only slightly less differentiated relative to all others, with mean Fst values of 0.397 and 0.405, respectively.

Table 38. Allele frequency and genotypic diversity indices for 18 microsatellite loci for 3 pooled mainland and 4 insular populations of the desert woodrat in Baja California.

Sample (locality number)	Mean N	Mean # alleles	# Mono-morpic loci	H_e	H_o	F_{is}
Northern Gulf (BCS-12, BCS-15, BCS-16, BCS-21, BCS-22)	14.9	10.4	1	0. 819	0. 732	0. 11[1]
West Coast (BCS-39, BCS-40, BCS-41)	5.6	6.2	1	0. 782	0. 730	0. 03
Southern Gulf (BCS-73, BCS-84, BCS-90, BCS-91, BCS-108, BCS-120)	13.1	9.1	1	0. 785	0. 758	0. 13[1]
Isla San Marcos (BCS-30)	8.0	3.3	4	0. 426	0. 470	0. 07
Isla Danzante (BCS-54)	5.8	2.4	5	0. 329	0. 343	0. 05
Isla San José (BCS-78)	8.7	5.1	1	0. 637	0. 558	0. 04
Isla San Francisco (BCS-79)	4.8	2.7	9	0. 254	0. 233	0. 09

[1] $p < 0.001$, based on bootstrapping over loci with 1000 repetitions

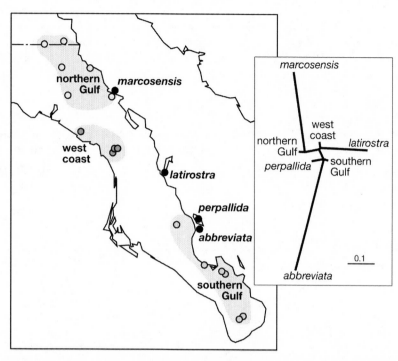

Figure 106. Map of pooled localities comprising three peninsular and four insular samples of desert woodrats for which data from 18 microsatellite loci are summarized in Table 38. The network on the right is an unrooted neighbor-joining tree linking each of these 7 samples based on a matrix of pairwise Fst values. Branch lengths are drawn proportional, with the scale provided in the lower right.

Taxonomic considerations.— The trends in both overall size and size-free cranial dimensions as well as color and color pattern among the insular and mainland taxa are complex. Some of these taxa appear diagnosable in both cranial size and shape axes (*anthonyi*, *insularis*, and *marcosensis*), but most exhibit either minor differentiation or none at all, especially in size-free shape, relative to other insular or adjacent mainland samples and taxa. At the molecular level, only *insularis* is phylogenetically outside all other peninsular or insular Baja California samples in its position in the mtDNA tree. Both *marcosensis* and *abbreviata* are well differentiated from their adjacent mainland samples in microsatellite loci and *perpallida* is weakly differentiated. Overall, there is little relationship between the current taxonomic designation, species or subspecies, of these insular forms and their degree of differentiation (Table 39). The "species" *N. bryanti* and *N. bunkeri* cannot be distinguished from mainland samples of *N. lepida* in cranial shape, even if they are different in size. In contrast, the "subspecies" *N. l. insularis* and *N. l. marcosensis* differ greatly from adjacent taxa in both size and shape. Clearly a review of the status of each taxon requires evaluation. We direct our attention to this review in the nomenclatural section below.

Table 39. Insular taxa of the *Neotoma lepida* group, including the taxonomic level of their original description and current allocation as well as whether nor not each taxon differs significantly from geographically adjacent samples in cranial size, size-free cranial shape, or color.

Taxon	Original designation	Current allocation[1]	Size[2]	Shape[2]	Color[3]
anthonyi	species	species	strong	strong	moderate
martinensis	species	species	no	strong	no
bryanti	species	species	strong	no to weak	strong
insularis	species	subspecies	no to weak	strong	strong
marcosensis	subspecies	subspecies	strong	moderate	no
bunkeri	species	species	strong	weak	no
nudicauda	species	subspecies	moderate	weak	strong
latirostra	subspecies	subspecies	strong	no	strong
perpallida	subspecies	subspecies	no	no	no
abbreviata	species	subspecies	strong	moderate	no
vicina	subspecies	subspecies	moderate	no	no

[1] Hall (1981); Álvarez-Castañeda and Cortés-Calva (1999); Musser and Carleton (2005).
[2] PC-1 (size) and CAN-1 (size-free) score comparisons to nearest other insular taxon (in the case of island clusters, such as the central and southeastern mid-rift groups) or to the immediately adjacent mainland sample (see Figs. 97 and 99 to 104); no = non-significant; weak = $p < 0.01$; moderate = $p < 0.001$; strong = $p < 0.0001$.
[3] Comparisons based on PC-1 colorimetric scores only (Fig. 105).

TRANSITIONS WITHIN THE "DESERT" MORPHOLOGICAL GROUP

Western Desert Transect

This analysis includes all samples of the desert mitochondrial DNA subclade 2A, distributed from southeastern Oregon and southern Idaho through western Utah, essentially all of Nevada, eastern California, and extreme northeastern Baja California in Mexico (Fig. 107). The area encompasses the type localities of five taxa, four of which are recognized as valid subspecies in the current literature (e.g., Hall, 1981): *lepida*, *desertorum* (listed as a junior synonym of *lepida* by Goldman, 1932, and subsequent authors), *nevadensis*, *marshalli*, and *grinnelli*. With the exception of Thomas's *lepida*, for which the type locality is unknown (see Goldman, 1932), we have examined the holotypes of each of these named forms and have molecular sequence data for topotypes of three (*desertorum*, *nevadensis*, and *marshalli*).

We grouped all specimens into currently designated subspecies based on the range map in Hall (1981). Within each subspecies, we then grouped localities somewhat arbitrarily by geographic proximity, which resulted in 16 taxonomically and geographically pooled samples upon which we based all statistical comparisons. With the exception of samples of *marshalli* (from Carrington and Stansbury islands in the Great Salt Lake, Utah; Group 14, map, Fig. 107), each of the recognized subspecies is divided into two or more separate groups to permit analysis of variation within these formally recognized taxa as well as among them. Our final samples include three separate groups for *nevadensis* in the northwestern part of the transect region (Groups 1-3), two groups for *grinnelli* along the western margins of the lower Colorado River (Groups 15-16), and 10 groups for *lepida* (including *desertorum*; Groups 4-13), as well as the single sample of *marshalli*.

Localities and sample sizes.—In the analyses below, we include craniodental measurement from 616 adult specimens and colorimetric variables from 827 individuals. A total of 170 of these were sequenced for the mtDNA *cyt-b* gene, with sequenced individuals present in 12 of the 16 geographic samples, including at least one sample from each recognized subspecies. Sequences available from 19 topotypes of *desertorum* from Furnace Creek, Death Valley (locality CA-405; Inyo Co., California) were taken from ear biopsies removed during a longitudinal population study. Groups 9 and 10 are the same as the W Mohave and E Mojave samples, respectively, examined in the Tehachapi Transect, and Group 16 includes all individuals of the Desert-e sample of the San Gorgonio Pass Transect. As in previous analyses, the list of specimens for each geographic sample includes the number of individuals for which we examined craniodental

(n_m), colorimetric (n_c), glans penis (n_g), and mtDNA sequences (n_{DNA}), in addition to the specific localities taken from specimen labels and museum catalog numbers. Localities are numbered as in the Appendix.

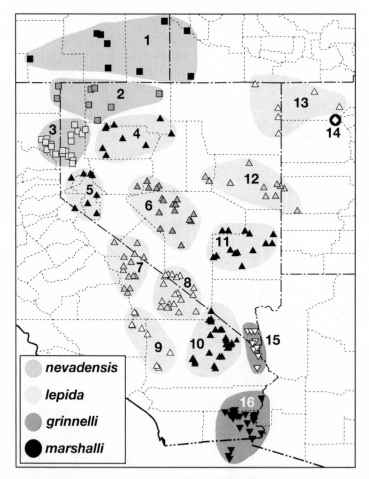

Figure 107. Map of the 16 grouped localities used in the analysis of craniodental and color characteristics of woodrats distributed through the western desert of the United States and adjacent northeastern Baja California, Mexico. Grouped localities for each of the four recognized subspecies are indicated by a different gray tone and individual localities by separate symbols. The type locality of *nevadensis* is within geographic Group 2; that of *desertorum* (currently a junior synonym of *N. l. lepida*) is in Group 8; that of *grinnelli* is in Group 16; and that of *marshalli* is in Group 14.

Group 1 [*nevadensis*] (total $n_m = 21$, $n_c = 34$)

IDAHO:– CANYON CO.: (1) ID-1; $n_m = 1$, $n_c = 2$; MVZ 67676, 67679; OWYHEE CO.: (2) ID-2; $n_m = 2$, $n_c = 6$; MVZ 79366-79370.

OREGON:– HARNEY CO.: (3) OR-2; $n_m = 1$, $n_c = 3$; MVZ 79356-79358; (4) OR-3; $n_m = 7$, $n_c = 8$; SDNHM 16810-16817; (5) OR-4; $n_m = 1$, $n_c = 3$; USNM 216031-216032, 222294; (6) OR-5; $n_m = 3$, $n_c = 3$; MVZ 79359-70361. LAKE CO.: (7) OR-1; $n_m = 1$, $n_c = 1$; USNM 247782. MALHEUR CO.: (8) OR-7; $n_m = 2$, $n_c = 2$; SDNHM 16819-16820; (9) OR-8; $n_m = 1$, $n_c = 2$; MVZ 79363-79364; (10) OR-8a (not found); $n_m = 1$, $n_c = 3$; USNM 208083-208085.

Group 2 [*nevadensis*] (total $n_m = 17$, $n_c = 30$, $n_g = 2$, $n_{DNA} = 8$)

CALIFORNIA:– MODOC CO.: (1) CA-424; $n_m = 6$, $n_c = 6$, $n_g = 1$, $n_{DNA} = 6$; MVZ 197159-197164; (2) CA-425; $n_c = 2$; MVZ 79371-79372.

NEVADA:– HUMBOLDT CO.: (3) NV-30; $n_m = 2$, $n_c = 2$, $n_g = 1$, $n_{DNA} = 2$; MVZ 197167-197168; (4) NV-31; $n_m = 2$, $n_c = 9$; MVZ 8280-8288 [MVZ 8282, holotype of *nevadensis* Taylor]; (5) NV-32; $n_c = 1$; MVZ 7888; (16) MV-34; $n_c = 1$; MVZ 74208; (7) NV-36; $n_c = 1$; MVZ 74210; (8) NV-37; $n_m = 2$; MVZ 96707, 96709; (9) NV-38; $n_m = 1$, $n_c = 2$; MVZ 79377-79378; (10) NV-39; $n_m = 2$, $n_c = 2$; MVZ 149770-149775; (11) NV-41; $n_m = 1$; MVZ 79380. WASHOE CO.: (12) NV-1; $n_m = 1$, $n_c = x$; MVZ 74176); (13) NV-2; $n_c = 2$; MVZ 74177-74178; (14) NV-3; $n_c = 1$; MVZ 74180; (15) NV-4; $n_c = 1$; NVZ 74179.

Group 3 [*nevadensis*] (total $n_m = 39$, $n_c = 61$, $n_g = 1$, $n_{DNA} = 2$)

CALIFORNIA:– LASSEN CO.: (1) CA-415; $n_m = 5$; MVZ 149356-149360; (2) CA-416; $n_c = 1$; MVZ 126466; (3) CA-417; $n_m = 1$, $n_c = 1$; USNM 67897; (4) CA-418; $n_m = 4$, $n_c = 6$; USNM 100866, 100868-100871, 100873; (5) CA-419; $n_m = 3$, $n_c = 3$; MVZ 77608-77609, 114334; (6) CA-420; $n_c = 1$; MVZ 183920; (7) CA-421; $n_m = 1$, $n_c = 1$; MVZ 114333; (8) CA-422; $n_c = 1$; USNM 67896; (9) CA-423; $n_m = 5$, $n_c = 14$; MVZ 39852-39858, 41321-41327.

NEVADA:– PERSHING CO.: (10) NV-46; $n_m = 1$, $n_c = 2$, $n_g = 1$, $n_{DNA} = 2$; MVZ 197165-197166. WASHOE CO.: (11) NV-5; $n_c = 2$; MVZ 31589, 31592; (12) NV-6; $n_m = 1$, $n_c = 1$; MVZ 74186; (13) NV-7; $n_c = 1$; MVZ 74190; (14) NV-8; $n_m = 5$, $n_c = 4$; USNM 78283-78287; (15) NV-9; $n_m = 1$, $n_c = 2$; MVZ 74192; (16) NV-10; $n_c = 1$; MVZ 79375; (17) NV-11; $n_c = 1$; USNM 94721; (18) NV-12; $n_c = 2$, $n_c = 1$; MVZ 74182-74183; (19) NV-13; $n_c = 1$; MVZ 74185; (20) NV-14; $n_m = 2$; MVZ 74188, 74190; (21) NV-15; $n_c = 1$; MVZ 74197; (22) NV-16; $n_m = 1$, $n_c = 1$; USNM 78280; (23) NV-17; $n_c = 1$; MVZ 74198; (24) NV-18; $n_m = 1$, $n_c = 3$; MVZ 74200-74202; (25) NV-19; $n_m = 3$, $n_c = 4$; MVZ 74193-74195, 74197; (26) NV-21; $n_c = 1$; MVZ 74205-75207; (27) NV-22; $n_m = 1$; MVZ 96700; (28)

NV-23; n_m = 3, n_c = 3; USNM 78029-78030, 78279, 44669; (29) NV-25; n_m = 1, n_c = 1; MVZ 88310.

Group 4 [*lepida*] (total n_m = 13, n_c = 31)
NEVADA:– HUMBOLDT CO.: (1) NV-40; n_m = 1, n_c = 2; MVZ 74216; (2) NV-42; n_m = 1, n_c = 1; USNM 78289. LANDER CO.: (3) NV-72; n_m = 5, n_c = 7; USNM 32361-32363, 32365-32368; (4) NV-73; n_c = 1; MVZ 71097. PERSHING CO.: (5) NV-43; n_m = 1, n_c = 2; MVZ 74218-74219; (6) NV-44; n_m = 1, n_c = 2; MVZ 74215, 74220; (7) NV-45; n_m = 1, n_c = 1; MVZ 74224; (8) NV-47; n_m = 1, n_c = 2; MVZ 74221, 74223; (9) NV-48; n_m = 1, n_c = 6; MVZ 68501-68503, 68505-68507; (10) NV-49; n_c = 1; MVZ 68508; (11) NV-50; n_c = 1; MVZ 68509; (12) NV-51; n_c = 1; MVZ 68500; (13) NV-52; n_c = 3; MVZ 74225-74227; (14) NV-53; n_m = 1; MVZ 95356; (15) NV-54; n_c = 1; MVZ 68492.

Group 5 [*lepida*] (total n_m = 11, n_c = 23, n_g = 2)
NEVADA:– CHURCHILL CO.: (1) NV-62; n_m = 1, n_c = 1; MVZ 88115; (2) NV-63; n_c = 1; MVZ 88312; (3) NV-63a; n_g = 2; CSULB 8101, 8104. DOUGLAS CO.: (4) NV-60; n_m = 1, n_c = 1; MVZ 64786; (5) NV-61; n_c = 1; MVZ 86569. LYON CO.: (6) NV-55; n_m = 1, n_c = 1; MVZ 64437; (7) NV-56; n_m = 1, n_c = 1; MVZ 143952; (8) NV-57; n_m = 4, n_c = 7; MVZ 64439-64445; (9) NV-58; n_c = 1; MVZ 159860; (10) NV-59; n_m = 1, n_c = 4; MVZ 64447, 64449-64450, 64452. MINERAL Co.: (11) NV-65; n_c = 1; MVZ 40456. STOREY CO.: (12) NV-29; n_m = 1, n_c = 1; MVZ 88311. WASHOE CO.: (13) NV-24; n_c = 2; USNM 244672, 244677; (14) NV-27; n_m = 1, n_c = 1; MVZ 71103.

Group 6 [*lepida*] (total n_m = 36, n_c = 47, n_g = 1, n_{DNA} = 4)
NEVADA:– CHURCHILL CO.: (1) NV-64; n_m = 1, n_c = 1; MVZ 85232. LANDER CO.: (2) NV-74; n_c = 1; USNM 24952; (3) NV-75; n_m = 1, n_c = 2; MVZ 64454-64455; (4) NV-76; n_c = 1; MVZ 64456; (87) NV-77; n_m = 2, n_c = 3; MVZ 64458-64459, 64461; (6) NV-78; n_m = 2, n_c = 3; MVZ 93417, 93649-93650; (7) NV-79; n_m = 1, n_c = 1; USNM 93648. NYE CO.: (8) NV-95; n_m = 1, n_c = 1; MVZ 45792; (9) NV-96; n_m = 8, n_c = 11; MVZ 45765-45767, 45770-45772, 45775-45776, 45778, 45782-45784; (10) NV-97; n_c = 1; MVZ 88313; (11) NV-98; n_m = 3, n_c = 3; MVZ 58433, 58435-58436; (12) NV-99; n_m = 2, n_c = 2; MVZ 58437-58438; (13) NV-100; $_m$ = 3, n_c = 3; USNM 94255, 94258, 211037; (14) NV-101; n_m = 1; USNM 93647; (15) NV-102; n_m = 4, n_c = 4; MVZ 58445, 58448-58450; (16) NV-103; n_m = 2, n_c = 2; MVZ 58440-58442; (17) NV-104; n_m = 1, n_c = 1; MVZ 58441; (18) NV-105; n_m = 1, n_c = 2; MVZ 49446-49447; (19) NV-106; n_m = 2, n_c = 3, n_g = 1, n_{DNA} = 4; MVZ 199364-199367; (20) NV-110; n_m = 1, n_c = 2; MVZ 49448-49449.

Group 7 [*lepida*] (total n_m = 46, n_c = 57, n_g = 17, n_{DNA} = 23)
CALIFORNIA:– INYO CO.: (1) CA-382; n_m = 1, n_c = 3; MVZ 15508-15510; (2) CA-382a; n_g = 1; LACM 43046; (2) CA-383; n_m = 1, n_c = 1; MVZ 26321; (3) CA-384; n_m = 5, n_c = 6; MVZ 16783-16788; (4) MC-385; n_m = 1; MVZ 38932; (5) CA-386; n_c = 1; MVZ 77726; (6) CA-387; n_m = 1, n_c = 1; MVZ 77727; (7) CA-388; n_m = 14, n_c = 17, n_{DNA} = 19; MVZ 195276-195284, 202448-204457; (8) CA-389; n_m = 4, n_c = 4, n_g = 8, n_{DNA}=4; MVZ 195286-195289; (9) CA-390; n_m = 1, n_c = 1; MVZ 77725; (10) CA-391; n_c = 1, n_g = 8; MVZ 121135; (11) CA-392; n_m = 5, n_c = 6; MVZ 26323-26326, 26330-26333. MONO CO.: (12) CA-412; n_m = 1, n_c = 1; MVZ 115385; (13) CA-413; n_c = 1; MVZ 26336; (14) CA-414; n_m = 1, n_c = 1; MVZ 16789.

NEVADA:– ESMERALDA CO.: (15) NV-121; n_c = 1; MVZ 40854; (16) NV-122; n_m = 8, n_c = 8; MVZ 40860-40863, 40868-40869, 40871-40872; (17) NV-123; n_c = 1; MVZ 59619; (18) NV-124; n_m = 3, n_c = 3; MVZ 38720-38722.

Group 8 [*lepida*, including *desertorum*] (total n_m = 87, n_c = 123, n_g = 11, n_{DNA} = 28)
CALIFORNIA:– INYO CO.: (1) CA-393; n_m = 1; MVZ 26304; (2) CA-394; n_m = 4, n_c = 6; MVZ 26295, 26297-26298, 26300-26301, 26303; (3) CA-395; n_m = 2, n_c = 6; MVZ 26313-26315, 26319-26320; (4) CA-396; n_m = 1, n_c = 1; MVZ 26305; (5) CA-397; n_m = 3, n_c = 4; MVZ 74426-74429; (6) CA-398; $_m$ = 6, n_c = 2, n_{DNA} = 4; MVZ 26291-26292, 192238-192241; (7) CA-399; n_m = 4, n_c = 5; MVZ 26286-26287, 26289-26290; 26294; (8) CA-400; n_m = 1, n_c = 1; MVZ 26285; (9) CA-401; n_c = 2; MVZ 61361-61362; (10) CA-402; n_m = 1, n_c = 1; MVZ 26278; (11) CA-403; n_c = 1; MVZ 26282; (12) CA-404; n_m = 1, n_c = 1; MVZ 26276; (13) CA-405; n_m = 42, n_c = 39, n_{DNA} = 19; MVZ 26239-26242, 26245-26247, 26249-26257, 26259-26274, 26279-26281, 26283-26284, 126002, USNM 25739/33139 [holotype of *desertorum* Merriam], USNM 33050, 33137-33138, 34097-34098, 34104-34105, 34109, 34111-34112, 34130, 34140, 34506-34507; (14) CA-405a; n_g = 1; LACM 32311; (14) CA-406; n_m = 3, n_c = 3; MVZ 26248, 61364-61365; (15) CA-407; n_c = 2; MVZ 161192-161193; (16) CA-408; n_c = 1; MVZ 161200; (17) CA-409; n_m = 1, n_c = 2; MVZ 93053-93054); (19) CA-410; n_c = 2; MVZ 161196-161197; (20) CA-411; n_c = 2; MVZ 161189-161190.

NEVADA:– NYE CO.: (21) NV-111; n_m = 1, n_c = 3; MVZ 161208-161209, 161211; (22) NV-112; n_m = 2, n_c = 6; MVZ 93075-93076, 93078-93081; (23) NV-113; n_c = 1; MVZ 93082; (24) NV-114; n_m = 2, n_c = 5; MVZ 86574-86575, 93067-93068, 93074; (25) NV-115; n_m = 2, n_c = 2; MVZ 161203-161204; (26) NV-116; n_m = 5, n_c = 18, n_g = 10, n_{DNA} = 5; MVZ 195290-195307; (27) NV-117; n_m = 2, n_c = 2; MVZ 48906, 48908, 59376; (28) NV-118; n_m = 1, n_c = 1; USNM 34502; (29) NV-119; n_m = 2, n_c = 4; USNM 26723, 27103, 34484, 34491.

Group 9 [*lepida*] (total n_m = 69, n_c = 78, n_g = 1, n_{DNA} = 37)
CALIFORNIA:– INYO CO.:   (1) CA-380; n_m = 2, n_c = 2; MVZ 28209-28210; (2) CA-380a; n_g = 1; LACM 75406; (2) CA-381; n_m = 22, n_c = 25, n_{DNA} = 25; MVZ 202459-202483; KERN CO.:   (3) CA-91; n_m = 2, n_c = 2, n_{DNA} = 2; MVZ 195264-195265; (4) CA-92; n_m = 3, n_c = 3; MVZ 143941, 143943-143944; (5) CA-93; n_c = 1; MVZ 134633; (6) CA-94; n_m = 10, n_c = 10, n_{DNA} = 10; MVZ 195266-195275.   SAN BERNARDINO CO.:   (7) CA-328; $_m$ = 18, n_c = 22; MVZ 5374, 5994, 6006-6007, 6075, 6077-6078, 6080-6092, 6827-6828; (8) CA-329; n_m = 8, n_c = 8; MVZ 28208-28210, 31434-31437; (9) CA-335; n_m = 3, n_c = 3; MVZ 21035-21037; (10) CA-336; n_m = 1, n_c = 1; MVZ 145684; (11) CA-337; n_c = 1; MVZ 158991.

Group 10 [*lepida*] (total n_m = 84, n_c = 98, n_{DNA} = 16)
CALIFORNIA:– SAN BERNARDINO CO.:   (1) CA-334; n_c = 2; MVZ 65594-65595; (2) CA-343; n_c = 2; MVZ 77231-77232; (3) CA-344; n_c = 1; MVZ 31420; (4) CA-345; n_c = 1; MVZ 31421; (5) CA-346; n_m = 2, n_c = 1; MVZ 31425, 31427; (6) CA-347; n_m = 2, n_c = 3; MVZ 31431-31433; (7) CA-349; n_m = 7, n_c = 10, n_{DNA} = 10; MVZ 195313-195319, 199349-199351; (8) CA-351; n_m = 1, n_c = 1; MVZ 121169; (9) CA-352; n_m = 1, n_c = 1, n_{DNA} = 1; MVZ 195320; (10) CA-353; n_m = 1, n_c = 3; MVZ 81957, 93063-93094; (11) CA-354; n_m = 2, n_c = 2; MVZ 196354-196356; (12) CA-355; n_m = 1, n_c = x; MVZ 81956; (13) CA-356; n_c = 1; MVZ 80250; (14) CA-357; n_m = 19, n_c = 19; MVZ 80251-80257, 80259-80270; (15) CA-358; n_m = 1, n_c = 1; MVZ 143950; (16) CA-359; n_m = 13, n_c = 11; MVZ 80236-80240, 80242-80249; (17) CA-360; n_m = 1, n_c = 1; MVZ 81946; (18) CA-361; n_m = 5, n_c = 6; MVZ 81950-81955; (19) CA-362; n_m = 2, n_c = 2; MVZ 81944-81945; (20) CA-363; n_m = 1, n_c = 1; MVZ 81942; (21) CA-364; n_m = 4, n_c = 6; MVZ 80230-80235; (22) CA-365; n_m = 1, n_c = 1; MVZ 31418; (23) CA-366; n_m = 3, n_c = 5, n_{DNA} = 5; MVZ 195308-195313; (24) CA-368; n_c = 1; MVZ 61182; (25) CA-369; n_c = 1; MVZ 86564; (26) CA-370; n_m = 3, n_c = 3; MVZ 86548, 86550, 86552; (27) CA-371; n_m = 1, n_c = 1; MVZ 86567; (28) CA-372; n_m = 3, n_c = 3; MVZ 86553-86554, 86558; (29) CA-373; n_m = 3, n_c = 3; MVZ 86545, 93060, 93062; (30) CA-374; n_m = 1, n_c = 1; MVZ 86546; (31) CA-375; n_m = 1, n_c = 1; MVZ 86544.
NEVADA:– NYE CO.:   (32) NV-120; n_m = 6, n_c = 3; USNM 25961, 26708, 26717, MVZ 33375, 33377, 33379, 34116, 34122, 34131.

Group 11 [*lepida*] (total n_m = 27, n_c = 33, n_g = 3, n_{DNA} = 11)
NEVADA:– LINCOLN CO.:   (1) NV-125; n_m = 1, n_c = 1; MVZ 86829; (2) NV-126; n_m = 1, n_c = 1; MVZ 59619; (3) NV-127; n_m = 1, n_c = 2; MVZ 48924-48925, 48927; (4) NV-128; n_m = 5, n_c = 7; MVZ 48931-48935, 48937-48938; (5)

NV-129; $n_m = 3$, $n_c = 3$; MVZ 48920, 48922-48923; (6) NV-130; $n_m = 1$; MVZ 53220; (7) NV-131; $n_m = 1$, $n_c = 2$; MVZ 53209, 53212; (8) NV-132; $n_m = 2$, $n_c = 2$; MVZ 48917-48918; (9) NV-133; $n_m = 1$, $n_c = 1$; MVZ 53213; (10) NV-135; $n_m = 5$, $n_c = 9$, $n_g = 3$, $n_{DNA} = 11$; MVZ 197130-197140; (11) NV-136; $n_m = 1$, $n_c = 1$; MVZ 53217. NYE CO.: (12) NV-107; $n_m = 1$, $n_c = 1$; MVZ 58453; (13) NV-108; $n_m = 2$, $n_c = 1$; MVZ 53204-53205; (14) NV-109; $n_m = 1$, $n_c = 2$; MVZ 53200-53201.

Group 12 [*lepida*] (total $n_m = 30$, $n_c = 35$)
NEVADA:– EUREKA CO.: (1) NV-80; $n_m = 1$, $n_c = 1$; MVZ 71102; (2) NV-81; $n_m = 1$, $n_c = 1$; MVZ 179585. WHITE PINE CO.: (3) NV-82; $n_m = 1$, $n_c = 1$; MVZ 46250; (4) NV-83; $n_m = 1$, $n_c = 1$; MVZ 46249; (5) NV-84; $n_c = 1$; MVZ 46248; (6) NV-85; $n_m = 3$, $n_c = 4$; MVZ 53195, 53195-53197; (7) NV-86; $n_m = 1$, $n_c = 2$; MVZ 79389-79390; (8) NV-87; $n_m = 3$, $n_c = 3$; MVZ 79391, 79398, 79400; (9) NV-89; $n_m = 6$, $n_c = 7$; MVZ 79401, 79382-79384, 79386-79388; (10) NV-90; $n_m = 8$, $n_c = 12$; MVZ 42024-42031, 42035-42036, 42040, 42042-42043; (11) NV-91; $n_m = 1$; MVZ 46427; (12) NV-92; $n_m = 1$, $n_c = 1$; MVZ 42044; (13) NV-93; $n_m = 1$, $n_c = 1$; MVZ 42022.
UTAH:– MILLARD CO.: (14) UT-9; $n_m = 2$; USNM 356958-356959.

Group 13 [*lepida*] (total $n_m = 14$, $n_c = 30$, $n_g = 2$, $n_{DNA} = 4$)
NEVADA:– ELKO CO.: (1) NV-66; $n_m = 1$, $n_c = 3$, $n_g = 2$, $n_{DNA} = 4$; MVZ 197126-197129; (2) NV-67; $n_c = 2$; MVZ 68486-68487; (3) NV-68; $n_m = 1$, $n_c = 2$; MVZ 68490-68491; (4) NV-69; $n_c = 1$; MVZ 68494; (5) NV-70; $n_m = 2$, $n_c = 3$; MVZ 68496-68498; (6) NV-71; $n_m = 4$, $n_c = 9$; MVZ 46251-46259.
UTAH:– BOX ELDER CO.: (7) UT-1; $n_m = 3$, $n_c = 5$; MVZ 44075-44078, USNM 43172; (8) UT-3; $n_m = 3$, $n_c = 3$; USNM 264310-264312; (9) UT-4; $n_c = 2$; USNM 133074-122075.

Group 14 [*marshalli*] (total $n_m = 6$, $n_c = 8$, $n_{DNA} = 1$)
UTAH:– TOOELE CO.: (1) UT-5; $n_m = 4$, $n_c = 6$; USNM 263979, 263981-263985 [USNM 263984, holotype of *marshalli* Goldman]); (2) UT-6; $n_m = 2$, $n_c = 2$, $n_{DNA} = 1$; USNM 263978-263979, BYU 18771.

Group 15 [*grinnelli*] (total $n_m = 68$, $n_c = 78$, $n_g = 24$, $n_{DNA} = 14$)
CALIFORNIA:– SAN BERNARDINO CO.: (1) CA-376; $n_c = 7$; MVZ 20974-20980; (2) CA-377; $n_m = 2$, $n_c = 2$; MVZ 95021-95022; (3) CA-377a; $n_g = 1$, CSULB 3984; (4) CA-378; $n_m = 1$, $n_c = 3$; MVZ 10424-10426; (5) CA-378a-b; $n_g = 1$; LACM 75561; (6) CA-379; $n_m = 3$, $n_c = 7$; MVZ 61839-61845.

NEVADA:– CLARK CO.: (5) NV-141; n_c = 1; MVZ 71801; (6) NV-142; n_m = 14, n_c = 12, n_g = 4, n_{DNA} = 14; MVZ 195245-195258; (7) NV-142a; n_g = 1; CSULB 8105; (7) NV-143; n_m = 20, n_c = 15, n_g = 7; MVZ 149285, 149287-149296, 149298-149300, 149320-149325; (8) NV-145; n_m = 1, n_c = 2; MVZ 61847-61848; (9) NV-146; n_c = 1; MVZ 96715; (10) NV-147; n_m = 18, n_c = 15, n_g = 9; MVZ 149265, 149267-149268, 149270-149272, 149274-139184, 149297; (11) NV-148; n_m = 6, n_c = 5; MVZ 61850-61851, 61858-61862; (12) NV-149; n_c = 3; MVZ 96716-96717, 102627; (13) NV-150; n_m = 3, n_c = 5; MVZ 61853-61857.

Group 16 [*grinnelli*] (total n_m = 63, n_c = 38, n_{DNA} = 27)
CALIFORNIA:– IMPERIAL CO.: (1) CA-195; n_m = 1; LACM 91731; (2) CA-196; n_m = 3; LACM 91642-91643, 91656; (3) CA-197; n_m = 1, n_c = 1; MVZ 84768; (4) CA-198; n_m = 2, n_c = 2; MVZ 84766-84767; (5) CA-199; n_m = 5; LACM 91654-91655, 91647-91649; (6) CA-201; n_m = 3; LACM 91651-91653; (7) CA-202; n_m = 1; LACM 91650; (8) CA-203; n_m = 1; LACM 72801; (9) CA-204; n_m = 4, n_c = 3; MVZ 65885-65888; (10) CA-205; n_m = 12, n_c = 5, n_{DNA} = 26; MVZ 122927-122928, 195259-195263, 215616-215640; (11) CA-206; n_m = 5, n_c = 8; MVZ 10446, 10448-10452, 10455-10456; (12) CA-207; n_m = 2, n_c = 2; MVZ 95023-95024; (13) CA-208; n_c = 1; MVZ 10429; (14) CA-209; n_m = 5; LACM 63700-63701, 63703, 63707, 63711; (15) CA-209; n_m = 2; LACM 75552, 75555; (16) CA-210; n_m = 5, n_c = 5; MVZ 10435, 10437-10439, 10717 [MVZ 10438, holotype of *grinnelli* Hall]; (17) CA-211; n_c = 1; MVZ 10441; (18) CA-212; n_m = 1, n_c = 1; MVZ 95025; (19) CA-213; n_c = 1; MVZ 10444. RIVERSIDE CO.: (20) CA-312; n_m = 5, n_c = 4; MVZ 149261-149264, 149266; (21) CA-313; n_c = 1, n_{DNA} = 1; MVZ 199817; (22) CA-315; n_c = 1; MVZ 10427.

MEXICO:– BAJA CALIFORNIA: (23) BCN-101; n_m = 3, n_c = 1; MVZ 111919-111921; (24) BCN-100; n_m = 1, n_c = 1; USNM 136996; (25) BCN-102; n_m = 1, n_c = 1; USNM 136648.

Habitat.— The habitats occupied by desert woodrats throughout the interior deserts of the United States vary considerably, especially among the geographic groups we assemble here for analysis. Animals build nests in crevices within rock outcrops of a variety of compositions, from basalt to limestone. They also construct nests at the base of shrubs, primarily either sagebrush (*Artemisia* sp.) or rabbitbrush (*Chrysothamnus* sp.) in the Great Basin Desert and yuccas (both Joshua Tree and Mojave Yucca) as well as Creosote Bush in the Mojave Desert, the two broad regions included in the Western Desert Transect. Typically, the range of habitats occupied becomes more restricted in northeastern California, Oregon, northern Nevada, and Idaho, where desert woodrats occur mostly in the massive flood basalt flows at lower elevations in the intermontane basins (Figs.

108 and 109), below the elevational range of the Bushy-tailed Woodrat, *Neotoma cinerea*. In eastern California and throughout most of Nevada, however, desert woodrats are found equally commonly in rock outcrops of a wide range of compositions (Figs. 110 and 111).

Figure 108. Basalt flow habitat of *N. l. nevadensis* east of Cedarville, Modoc Co., California (Group 2, locality CA-424). Photo taken in July 2001.

Figure 109. Basalt flows of the Virgin Valley, Humboldt Co., Nevada (Group 2, locality NV-30), near the type locality of *N. l. nevadensis* Taylor. Photo taken in July 2001.

Figure 110. Yellow-red rhyolite outcrops amid sagebrush and rabbitbrush, habitat of *N. l. lepida* near the northwestern margins of the range of this subspecies (Group 6, locality NV-106 – McKinney Tank, Nye Co., Nevada). Photo taken in August 2002.

Figure 111. Small andesite outcrops amid Utah Juniper, Joshua Tree, and sagebrush, habitat of *N. l. lepida* near the southeastern margins of the range of this subspecies (Group 11, locality NV-135 – Delamar Mts., near Caliente, Lincoln Co., Nevada). Photo taken in July 2001.

    Craniodental and colorimetric variation.—Standard descriptive statistics for all external and craniodental variables are given separately for each geographic Group in Table 40. Twenty of the 21 craniodental variables and three of the four external variables exhibit significant variation among the 16 geographic samples, with $p < 0.0001$ for 16 of these (ANOVA, $F_{(15,618)} > 2.950$ for craniodental variables, $F_{(15,479)} > 4.462$ for external variables). Only Total Length (TOL; $F_{(15,479)}$, $p = 0.2542$) and Mesopterygoid Fossa Width (MFW; $F_{(15,618)}$, $p = 0.0702$) do not exhibit significant differences. However, few variables differ significantly in comparisons between geographically adjacent groups, based on Fisher's PLSD posterior tests for paired samples. In all but three comparisons, no more than six of the total of 25 variables are statistically separable, suggesting that samples are weakly differentiated, if at all, across the entire sample range. The three pairwise comparisons for which substantial numbers of variables exhibit significant differences are those between the two samples of *N. l. lepida* adjacent to the single group locality of *N. l. marshalli* (Groups 12 and 13 versus 14, where 18 and 16 of the 21 craniodental variables, respectively, are significantly different), and between geographic Groups 6 and 11 of *N. l. lepida*, where 17 variables differ significantly (Fig. 107). Importantly, no more than four variables differ between any sample of *N. l. nevadensis* and an adjacent sample group of *N. l. lepida* (e.g., comparisons between Groups 1, 2, 3 versus Group 4 and/or 5; Fig. 107), and no more than five variables differ in comparison between any geographic sample of *N. l. lepida* and either sample of *N. l. grinnelli* (Groups 8, 10, and 11 versus Group 15 or 16). When samples are grouped by current subspecies, significant differences are also present, with *grinnelli* and *marshalli* uniformly most different but *nevadensis* only minimally distinguishable from *lepida*. For the latter pair of subspecies only six variables are significantly different, while the pooled geographic samples of *lepida* Thomas differ from those of *grinnelli* by 15 and from *marshalli* by 17.

    The geographic differences among grouped samples are due, in part, to a general latitudinal effect. Nineteen of the 25 external and craniodental variables exhibit a significant relationship with latitude but not with longitude in a multiple regression model using these two as independent variables. Regression coefficients are low, even if significant, ranging from $r = 0.100$ (AW vs. latitude) to $r = -0.346$ (TAL vs. latitude). For the most part, these coefficients are positive (CIL, Fig. 112, top), so that individual dimensions increase with latitude. Only for Rostral Width (RW) and tail length (TAL) is there a strong negative relationship with latitude (Fig. 112, bottom).

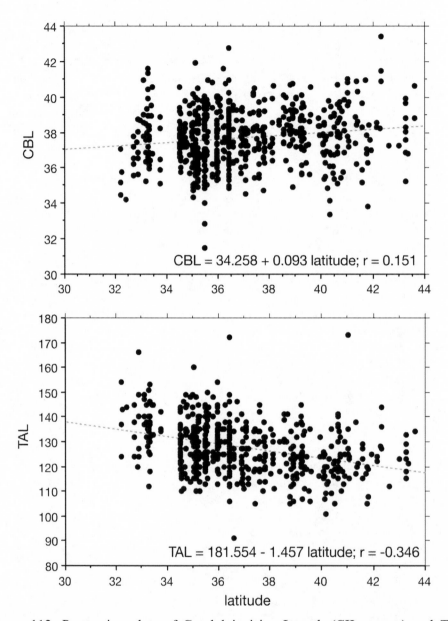

Figure 112. Regression plots of Condyloincisive Length (CIL; upper) and Tail Length (TAL; lower) and latitude for all individuals of the Western Desert Transect, illustrating general size trends across geography. Regression equations are given.

Table 40. External and cranial measurements of adult (age classes 1-5) specimens of the 16 geographic samples of *Neotoma lepida* along the Western Desert Transect in California, Oregon, Idaho, Nevada, and Utah; see Fig. 107). Means, standard error, sample size, and range are given with each designated geographic group.

Variable	Group 1 *nevadensis*	Group 2 *nevadensis*	Group 3 *nevadensis*	Group 4 *lepida*	Group 5 *lepida*	Group 6 *lepida*	Group 7 *lepida*	Group 8 *lepida*
external								
TOL	293.0±2.9	289.4±5.2	298.1±3.4	286.7±2.5	296.8±4.5	290.5±2.6	294.7±1.9	293.8±1.9
	20	11	12	36	10	34	39	73
	272-329	259-316	273-310	259-314	265-310	253-320	266-315	263-352
TAL	125.2±1.8	122.1±2.4	128.6±2.8	119.4±1.3	126.8±3.5	122.4±1.5	126.0±1.5	128.1±1.2
	20	111	12	36 28	10	34	39	73
	112-144	108-138	113-141	101-134	106-141	105-139	91-141	106-172
HF	31.4±0.2	31.2±0.3	31.9±0.3	30.7±0.3	31.1±0.5	30.7±0.2	31.5±0.2	30.5±0.2
	21	14	12	39 32	11	37	45	66
	30-35	20-33	30-34	24-32	29-35	28-35	29-33	21-34
E	29.2±0.5	28.8±0.4	28.3±0.7	27.9±0.4	29.3±0.8	27.5±0.5	28.5±0.4	26.9±0.4
	9	13	6	29	11	30	45	61
	27-31	26-31	27-31	24-34	24-33	22-32	24-32	21-35
cranial								
CIL	38.79±0.40	37.99±0.55	37.95±0.32	37.42±0.24	38.45±0.44	34.95±0.22	37.81±0.19	37.88±0.17
	21	14	12	43	11	37	45	84
	36.8-42.7	33.8-40.9	35.8-39.8	33.4-40.7	35.7-40.7	35.0-40.8	34.5-40.5	34.8-42.8
ZB	21.28±0.24	20.68±0.21	20.8±0.27	20.46±0.14	21.06±0.31	20.97±0.13	20.78±0.24	20.61±0.09
	21	14	12	43	11	37	45	84
	19.5-23.8	19.3-21.9	19.2-22.8	18.6-22.4	19.6-23.0	18.7-22.3	18.8-23.2	19.1-22.8

Table 40 (continued)

Variable	Group 1 nevadensis	Group 2 nevadensis	Group 3 nevadensis	Group 4 lepida	Group 5 lepida	Group 6 lepida	Group 7 lepida	Group 8 lepida
IOC	5.20±0.06	5.18±0.06	5.15±0.22	5.15±0.03	5.18±0.06	5.07±0.04	4.94±0.03	4.98±0.03
	21	14	12	43	11	37	45	84
	4.8-5.9	4.8-5.5	4.8-5.5	4.7-5.4	4.8-5.5	4.5-5.4	4.5-5.4	4.5-5.8
RL	15.85±0.23	15.62±0.23	15.59±0.22	15.36±0.14	15.75±0.19	15.67±0.13	15.41±0.11	15.51±0.09
	21	14	12	42	11	37	45	84
	14.3-19.1	13.8-17.0	14.4-16.8	12.9-17.2	14.6-16.8	13.8-17.3	14.0-17.1	13.9-17.9
NL	15.26±0.22	15.07±0.27	15.19±0.23	14.88±0.15	15.60±0.16	15.40±0.15	14.97±0.11	15.06±0.10
	21	14	12	43	1	37	45	84
	14.0-18.4	12.9-17.1	14.0-16.5	12.2-16.8	14.8-16.7	13.3-18.3	13.5-16.4	13.2-18.4
RW	6.24±0.07	6.08±0.09	6.11±0.07	6.11±0.03	6.25±0.09	6.14±0.04	6.14±0.04	6.24±0.03
	21	14	12	43 32	11	37	45	84
	5.8-6.9	5.7-6.8	5.7-6.6	5.5-6.5	5.8-6.7	5.5-6.5	5.5-6.8	6.7-7.0
OL	13.94±0.12	13.71±0.16	13.75±0.13	13.71±0.06	13.98±0.15	13.80±0.06	13.57±0.06	13.60±0.05
	21	14	12	43	11	37	45	84
	12.9-15.1	12.5-14.4	13.1-14.7	12.6-14.9	13.0-14.6	12.8-14.6	12.5-14.4	12.7-15.3
DL	11.33±0.18	11.09±0.24	10.92±0.13	10.86±0.11	11.40±0.22	11.33±0.09	10.98±0.10	11.07±0.07
	21	14	12	43	11	37	45	84
	10.2-13.5	9.6-12.6	10.2-11.8	9.1-11.9	10.2-12.4	9.8-12.5	9.3-12.3	9.8-12.8
MTRL	8.06±0.06	7.97±0.09	8.11±0.10	7.99±0.03	7.88±0.07	7.96±0.05	7.88±0.04	7.88±0.04
	21	14	12	43	11	37	45	84
	7.6-8.6	7.5-8.9	7.6-8.7	7.6-8.4	7.5-8.2	7.1-8.8	7.1-8.5	6.9-8.6
IFL	8.48±0.12	8.55±0.16	8.40±0.13	8.43±0.7	8.73±0.14	8.47±0.09	8.27±0.06	8.33±0.05
	21	14	12	43	11	37	45	84
	7.4-9.7	7.4-9.6	7.8-9.3	7.4-9.3	7.8-9.3	6.8-9.5	7.4-9.4	7.4-9.8

Table 40 (continued)

Variable	Group 1 nevadensis	Group 2 nevadensis	Group 3 nevadensis	Group 4 lepida	Group 5 lepida	Group 6 lepida	Group 7 lepida	Group 8 lepida
PBL	17.8±0.19	17.65±0.29	17.64±0.18	17.28±0.13	17.93±0.22	17.93±0.14	17.42±0.11	17.46±0.09
	16.1-19.7	15.8-19.7	16.7-19.0	15.2-18.9	16.9-19.1	15.4-19.5	15.8-19.5	15.6-19.9
	21	14	12	43	11	37	45	84
AW	7.05±0.06	6.98±0.05	7.01±0.08	6.94±0.04	6.95±0.04	7.07±0.04	7.00±0.04	6.99±0.03
	6.7-7.8	6.6-7.3	6.5-7.4	6.5-7.5	6.8-7.3	6.6-7.6	6.5-7.5	6.5-7.8
	21	14	12	43	11	37	45	84
OCW	9.02±0.07	8.89±0.06	9.01±0.09	9.03±0.04	9.24±0.10	9.04±0.05	8.96±0.04	8.97±0.03
	8.4-9.4	8.6-9.3	8.5-9.4	8.5-9.5	8.8-9.7	8.4-9.8	8.5-9.6	8.4-9.5
	21	14	12	43	11	37	45	84
MB	16.94±0.14	16.69±0.13	17.01±0.15	16.65±0.08	17.18±0.17	17.05±0.09	16.76±0.07	16.71±0.07
	15.8-18.3	15.9-17.4	16.1-17.9	15.6-17.9	16.2-18.3	16.0-18.2	15.7-17.5	15.3-18.5
	21	14	12	43	11	37	45	84
BOL	5.83±0.08	5.60±0.09	5.52±0.09	5.59±0.09	6.59±0.11	5.55±0.05	5.57±0.05	5.62±0.04
	5.4-6.6	5.1-6.2	5.1-6.2	4.5-8.8	5.2-6.4	4.8-6.1	5.0-6.4	4.7-6.6
	21	14	12	43	11	37	45	84
MFL	7.88±0.11	7.61±0.13	7.49±0.11	7.51±0.06	7.86±0.17	7.76±0.07	7.80±0.05	7.59±0.06
	7.1-9.3	6.5-8.2	7.1-8.3	6.8-8.3	6.7-8.8	7.0-8.7	7.0-8.3	5.7-8.8
	21	14	12	43	11	37	45	84
MFW	2.30±0.03	2.33±0.03	2.31±0.04	2.35±0.03	2.49±0.07	2.25±0.04	2.28±0.03	2.33±0.02
	2.06-2.7	2.2-2.6	2.1-2.5	2.1-2.7	2.1-2.9	1.7-2.8	1.8-2.7	1.8-2.8
	21	14	12	43	11	37	45	84
ZPW	4.22±0.07	4.00±0.05	3.99±0.07	4.01±0.07	4.19±0.13	4.08±0.05	4.10±0.03	4.16±0.3
	3.7-4.7	3.8-4.4	3.7-4.4	3.1-5.1	3.8-4.6	3.4-4.8	3.5-4.6	3.7-5.1
	21	14	12	43	11	37	45	84

Table 40 (continued)

Variable	Group 1 nevadensis	Group 2 nevadensis	Group 3 nevadensis	Group 4 lepida	Group 5 lepida	Group 6 lepida	Group 7 lepida	Group 8 lepida
CD	15.71±0.08	15.54±0.12	15.71±0.10	15.57±0.07	15.98±0.13	15.68±0.07	15.49±0.07	15.47±0.05
	21	14	12	43	11	37	45	84
	14.9-17.0	14.7-16.3	15.3-16.4	14.0-16.3	15.2-16.7	14.9-16.6	14.5-16.4	14.5-17.1
BUL	7.16±0.08	7.20±0.07	7.18±0.06	7.08±0.04	7.34±007	7.16±0.53	7.15±0.04	7.24±0.03
	21	14	12	43	11	37	45	84
	6.5-8.2	6.7-7.6	6.8-7.4	6.5-7.6	6.7-7.6	6.6-7.7	6.3-7.9	6.6-8.0
BUW	7.52±0.08	7.44±0.08	7.59±0.07	7.37±0.04	7.64±0.09	7.58±0.05	7.37±0.03	7.46±0.03
	21	14	12	43	11	37	45	84
	7.1-8.6	6.9-8.0	7.2-8.1	6.8-8.1	7.3-8.1	6.9-8.2	6.8-7.9	6.9-8.3

Variable	Group 9 lepida	Group 10 lepida	Group 11 lepida	Group 12 lepida	Group 13 lepida	Group 14 marshalli	Group 15 grinnelli	Group 16 grinnelli
external								
TOL	294.4±2.1	293.4±1.7	291.9±2.3	292.5±2.2	290.0±4.4	308-5±10.2	293.1±2.3	295.6±2.1
	60	81	24	27	12	4	59	51
	265-330	261-336	265-312	264-319	270-327	291-338	245-319	265-337
TAL	130.8±13	128.9±1.1	125.4±1.8	123.3±1.8	118.4±2.4	140.0±11.2	130.7±1.2	136.7±1.4
	60	81	24	27	12	4	59	51
	110-154	110-160	105-142	107-145	105-136	125-173	113-149	112-166
HF	31.7±0.2	30.5±0.2	30.2±0.3	30.4±0.2	31.7±0.3	32.8±0.6	30.7±0.1	30.5±0.2
	62	84	24	30	14	5	68	56
	28-34	27-36	27-35	28-33	30-33	31-34	28-34	27-33

Table 40 (continued)

Variable	Group 9 lepida	Group 10 lepida	Group 11 lepida	Group 12 lepida	Group 13 lepida	Group 14 marshalli	Group 15 grinnelli	Group 16 grinnelli
E	29.8±0.5	27.9±0.3	28.4±0.4	28.6±0.4	28.8±0.7	26.4±1.9	29.7±0.2	30.3±0.3
	44	78	27	30	12	5	68	53
	23-34	22-33	24-32	24-32	21-31	19-29	27-33	24-35
cranial								
CIL	37.79±0.16	30.51±0.17	37.40±0.24	38.01±0.20	37.71±0.41	39.89±0.58	37.19±0.20	37.60±0.22
	64	84	27	30	14	5	68	57
	35.0-41.0	34.3-41.9	34.8-39.9	35.7-40.0	35.1-40.9	37.8-41.0	31.5-40.5	34.2-41.6
ZB	20.82±0.09	20.41±0.10	20.42±0.14	20.78±0.13	20.75±0.26	21.71±0.36	20.36±012	20.45±0.13
	64	84	27	30	14	5	68	57
	19.1-22.7	18.6-22.8	19.0-21.9	19.5-22.3	19.0-22.6	20.7-22.6	17.5-22.1	18.4-23.1
IOC	5.05±0.03	5.06±0.02	4.94±0.05	5.03±0.06	5.17±0.04	5.33±0.08	5.07±0.03	5.00±0.03
	64	84	27	30	14	5	68	57
	4.7-5.6	4.6-5.5	4.3-5.6	4.1-5.6	5.0-5.5	5.2-5.6	4.6-5.7	4.6-5.4
RL	15.40±0.08	15.18±0.08	15.23±0.13	15.60±0.11	15.52±0.19	16.62±0.19	15.17±0.11	15.32±0.10
	64	84	27	30	14	5	68	57
	13.9-17.1	13.7-17.1	14.0-16.9	14.6-17.4	14.5-17.2	16.0-17.0	12.6-17.1	14.1-17.5
NL	14.98±0.08	14.88±0.07	14.85±0.14	15.23±0.12	14.91±0.16	15.86±0.24	14.66±0.10	14.89±0.12
	64	84	27	30	14	5	68	57
	13.7-16.9	13.4-16.2	13.4-16.6	14.0-17.2	14.2-16.1	15.2-16.5	11.7-16.5	13.5-17.0
RW	6.23±0.03	6.21±0.03	6.05±0.05	6.14±0.05	6.05±0.09	6.23±0.06	6.25±0.04	6.30±0.04
	64	84	27	30	14	5	68	57
	5.2-6.7	5.7-6.8	5.5-6.58	5.5-6.7	5.6-6.7	6.0-6.4	5.6-6.9	5.7-7.0

Table 40 (continued)

Variable	Group 9 *lepida*	Group 10 *lepida*	Group 11 *lepida*	Group 12 *lepida*	Group 13 *lepida*	Group 14 *marshalli*	Group 15 *grinnelli*	Group 16 *grinnelli*
OL	13.72±0.06	13.50±0.05	13.52±0.09	13.76±0.08	13.55±0.13	13.96±0.13	13.43±0.06	13.60±0.08
	64	84	27	30	14	5	68	57
	12.7-15.0	12.5-14.9	12.1-14.5	12.5-14.3	12.5-14.4	13.5-14.3	11.8-14.4	12.5-15.3
DL	10.91±0.07	10.84±0.08	10.83±0.08	11.20±0.09	10.92±0.18	11.62±0.34	10.76±0.09	10.92±0.10
	64	84	27	30	14	5	68	57
	9.8-12.4	9.4-13.1	10.2-11.6	10.1-12.2	9.9-12.3	10.4-12.3	8.8-12.7	9.5-12.9
MTRL	8.00±0.03	7.86±0.03	7.77±0.08	7.84±0.05	8.06±0.05	8.41±0.10	7.93±0.03	7.89±0.07
	64	84	27	30	14	5	68	57
	7.3-8.6	7.1-8.4	6.9-8.4	7.4-8.4	7.7-8.4	8.0-8.6	7.2-8.7	7.3-8.5
IFL	8.31±0.05	8.28±0.05	8.27±0.06	8.48±0.07	8.33±0.11	8.18±0.19	8.22±0.06	8.36±0.07
	64	84	27	30	14	5	68	57
	7.4-9.4	7.3-9.6	7.8-9.0	7.7-9.2	7.7-9.1	7.5-8.7	6.8-9.2	7.1-9.7
PBL	17.38±0.08	17.16±0.11	17.30±0.09	17.77±0.10	17.52±0.19	18.79±0.33	17.12±0.10	17.33±0.11
	64	84	27	30	14	5	68	57
	15.6-18.7	15.8-19.5	16.5-18.3	16.7-18.7	16.6-18.7	17.7-19.5	14.9-18.6	15.7-19.5
AW	7.00±0.03	6.95±0.03	6.81±0.05	6.96±0.05	6.99±0.09	7.56±0.06	6.90±0.03	6.93±0.03
	64	84	27	30	14	5	68	57
	6.5-7.6	6.4-7.9	6.0-7.2	6.4-7.5	6.4-7.5	7.4-7.7	6.4-7.6	6.5-7.5
OCW	9.10±0.04	8.94±0.03	8.89±0.05	8.99±0.06	9.16±0.08	9.45±0.10	8.94±0.04	8.85±0.05
	64	84	27	30	14	5	68	57
	8.5-9.8	8.4-9.6	8.4-9.5	8.4-9.5	8.9-9.7	9.2-9.7	8.0-9.7	8.0-9.6
MB	17.02±0.06	16.73±0.06	16.70±0.10	16.91±0.09	16.91±0.14	17.46±0.09	16.83±0.08	16.84±0.10
	64	84	27	30	14	5	68	57
	16.1-18.3	15.5-18.4	15.6-17.8	15.8-17.8	16.2-18.2	17.1-17.7	14.7-18.2	15.2-18.1

Table 40 (continued)

Variable	Group 9 lepida	Group 10 lepida	Group 11 lepida	Group 12 lepida	Group 13 lepida	Group 14 marshalli	Group 15 grinnelli	Group 16 grinnelli
BOL	5.55±0.04	5.48±0.04	5.31±0.08	5.63±0.06	5.54±0.09	5.98±0.05	5.51±0.05	5.47±0.05
	64	84	27	30	14	42	68	57
	5.0-6.5	4.7-6.3	4.7-6.2	4.9-6.1	5.1-6.1	5.3-6.8	4.5-6.3	4.8-6.6
MFL	7.69±0.05	7.72±0.05	7.50±0.11	7.52±0.07	7.46±0.15	7.77±0.05	7.54±0.05	7.59±0.05
	64	84	27	30	14	42	68	57
	7.2-8.8	6.1-9.2	6.5-8.8	6.6-8.3	6.5-8.8	6.8-8.4	6.2-8.6	6.7-8.7
MFW	2.33±0.03	2.36±0.02	2.28±0.04	2.30±0.04	2.29±0.04	2.59±0.03	2.36±0.02	2.34±0.03
	64	84	27	30	14	42	68	57
	1.9-2.8	1.9-3.0	1.9-2.8	1.8-2.6	2.0-2.5	2.2-3.0	2.0-2.8	1.8-2.8
ZPW	4.21±0.0.3	4.06±0.02	3.91±0.06	4.07±0.04	4.14±0.06	4.05±0.03	3.98±0.03	4.08±0.03
	64	84	27	30	14	42	68	57
	3.8-4.8	3.5-4.7	3.3-4.5	3.6-4.6	3.7-4.5	3.7-4.4	3.3-4.5	3.6-4.7
CD	15.64±0.05	15.39±0.05	15.34±0.09	15.47±0.09	15.53±0.09	15.48±0.06	15.42±0.06	15.49±0.07
	64	84	27	30	14	42	68	57
	14.5-16.7	14.6-16.6	14.4-16.4	14.8-16.7	14.8-16.0	14.9-16.4	14.4-17.0	14.1-17.0
BUL	7.21±0.03	7.19±0.03	7.07±0.05	7.13±0.04	7.25±0.08	6.68±0.02	7.28±0.04	7.31±0.04
	64	84	27	30	14	42	68	57
	6.5-8.0	6.6-7.9	6.6-7.6	6.6-7.6	6.8-7.6	6.1-7.1	6.3-8.0	6.5-7.8
BUW	7.48±0.03	7.45±0.03	7.42±0.05	7.58±0.05	7.57±0.07	6.92±0.03	7.50±0.04	7.57±0.04
	64	84	27	30	14	42	68	57
	7.0-8.0	7.0-8.1	6.9-7.9	6.9-8.0	7.2-8.0	6.4-7.3	6.7-8.3	7.0-8.2

The increase in size with latitude suggested by condyloincisive length (Fig. 112) and most other variables is mirrored by similar trends in the principal components vectors, especially those of the 1st PC axis. In this analysis, only the first three axes have eigenvalues greater than 1.0. The first axis explains 53.4% of the total pool of variation while remaining axes individually explain no more than 7.6%. All variables have relatively high and positive loadings on PC-1, all with highly significant correlation coefficients of individual variables versus PC-1 scores (p < 0.0001 in all cases; Z-value = 7.376 for logMTRL, the variable with the smallest loading [0.285]; Table 41). Taken together, these data support the interpretation that PC-1 represents general size. The single measure of skull length (Condyloincisive Length, logCIL) has the highest loading on PC-1 (0.949) and is thus the best univariate measure of overall size as well. It is not surprising, therefore, that PC-1 scores exhibit a significant relationship to latitude (r = 0.121, $F_{(1,630)}$ = 9.322, p = 0.0024). The position of individuals along the second PC axis is most heavily influenced by Maxillary Toothrow Length (logMTRL) and Alveolar Width (logAW) and less so by Interorbital Constriction (logIOC) opposed by Diastemal Length (logDL) (Table 41). The third PC axis is dominated by the width of the mesopterygoid fossa (logMFW) contrasted with the length of the toothrow (logMTRL). PC-2 scores are not related to the latitudinal position of a sample (r = 0.041, $F_{(1,630)}$ = 1.059, p = 0.3038), but those of PC-3 are, and in the opposite direction as PC-1 scores (r = 0.200, $F_{(1,630)}$ = 26.207, p < 0.0001).

Geographically grouped samples, either the numerical groups identified in Fig. 107 or those groups pooled by subspecies, exhibit significant differences in their respective PC scores on each of the first three axes (by group, ANOVA, $F_{(15,618)}$ = 3.848, p < 0.0001; by subspecies, ANOVA, $F_{(3,630)}$ = 5.377, p < 0.01). However, scatterplots of PC scores on combinations of the 1st, 2nd, and 3rd axes reveal broadly overlapping distributions of the geographic groups or subspecies and thus little obvious separation on any combination of axes (Fig. 113).

As with the univariate analyses, there is only limited differentiation among geographically adjacent groups when pairwise comparisons are made using Fisher's PLSD posterior test based on ANOVAs (Fig. 114). For example, in only four cases are adjacent groups significantly different in their mean PC-1 scores. Groups 5 and 7 as well as Groups 6 and 11, all of the subspecies *lepida*, differ with p = 0.015 and 0.0014, respectively. And, the single sample of *marshalli* (Group 14) differs from both adjacent samples of *lepida* (groups 12 and 13) with p = 0.0058 and 0.0064, respectively. Group 14 also differs from adjacent Groups 12 or 13 in their respective mean PC-2 scores (p = 0.0001 and 0.0168, respectively) although they do not differ in their mean PC-3 scores. Importantly, none of the samples of either *nevadensis* or *grinnelli* exhibit differences relative to adjacent ones of *lepida* on any PC axis (p > 0.18 in all pairwise comparisons).

Table 41. Principal component eigenvalues and factor loadings for log-transformed cranial variables of adult specimens of the Western Desert Transect.

Variable	PC-1	PC-2	PC-3
log CIL	0.949	-0.200	-0.083
log ZB	0.893	-0.048	-0.008
log IOC	0.486	0.371	0.033
log RL	0.881	-0.173	-0.110
log NL	0.841	-0.213	-0.094
log RW	0.655	0.285	0.140
log OL	0.815	-0.105	-0.008
log DL	0.871	-0.363	0.030
log MTRL	0.285	0.659	-0.346
log IFL	0.759	-0.284	0.072
log PBL	0.884	-0.181	-0.112
log AW	0.481	0.568	-0.043
log OCW	0.618	0.294	-0.095
log MB	0.855	0.065	-0.011
log BOL	0.785	-0.046	0.046
log MFL	0.671	-0.180	0.049
log MFW	0.159	0.176	0.892
log ZPW	0.676	0.156	-0.097
log CD	0.782	0.077	0.003
log BUL	0.670	0.179	0.153
log BUW	0.745	0.162	0.153
eigenvalue	11.220	1.595	1.054
% contribution	53.43	7.59	5.02

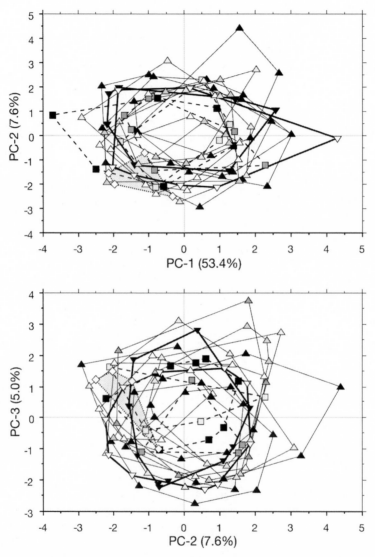

Figure 113. Ellipses encompassing all individual scores of each of the 16 geographic Groups of the Western Desert Transect on the first three principal components axes: Above – PC-1 versus PC-2; below – PC-2 versus PC-3. The three grouped localities of *nevadensis* are identified by squares, the 10 of *lepida* by upright triangles, the two of *grinnelli* by inverted triangles, and the single grouped locality of *marshalli* by diamonds and light gray infill. Components were extracted from the covariance matrix of log-transformed variables; factor loadings and eigenvalues are provided in Table 40.

Colorimetric variation across this transect is similar to the pattern described above for craniodental variables, although more geographically adjacent, paired groups exhibit significant differences for both univariate and multivariate variables. Each of the X-coefficient values for the four topographic regions of the skin is statistically heterogeneous among the 16 groups (ANOVA, where $F_{(15,829)}$ is always > 8.974 and p is always < 0.0001). Much of this level of difference is due to the three geographic samples of *nevadensis* in the northwestern part of the sampled range (Groups 1, 2, and 3; map, Fig. 107), which are considerably darker for all four variables than samples of other subspecies (Table 42). If these samples are excluded from the analysis, significant differences are still present among the 13 remaining groups, but the level of significance decreases an order of magnitude or more, to $0.001 > p < 0.01$ for each X-coefficient (ANOVA, $F_{(12,690)} > 2.758$).

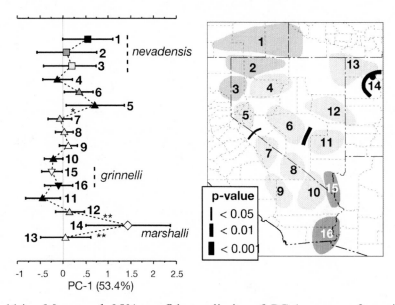

Figure 114. Mean and 95% confidence limits of PC-1 scores of craniodental variables for the 16 geographic samples of the Western Desert Transect (left) and map of general groupings (right). Symbols and sample numbers correspond to those in the transect map, Fig. 107. Grouped localities belonging to the subspecies *nevadensis*, *grinnelli*, and *marshalli* are identified; all remaining groups are of the nominate subspecies. As near as possible, samples are arranged geographically on the left; significance levels (based on Fisher's PLSD posterior comparison from ANOVA) between geographically adjacent areas are indicated: * p < 0.05, ** p < 0.01. Significance levels among geographic groups are also indicated by line width (as indicated in the inset) on the map, right.

Table 42. Colorimetric X-coefficients for the four topographic regions of the study skin for geographic samples of the Western Desert Transect. Mean, standard error, sample size, and range are given for each sample.

Sample	Dorsal-X	Tail-X	Lateral-X	Chest-X
1	8.881±0.396	5.418±0.482	25.169±0.714	41.104±0.934
(*nevadensis*)	34	34	34	34
	5.85-19.18	2.36-15.31	16.73-33.83	30.83-55.53
2	8.455±0.280	5.697±0.322	27.67±0.763	42.395±1.222
(*nevadensis*)	37	37	37	37
	4.87-11.99	2.98-121.58	17.88-38.75	24.84-59.17
3	8.39±0.252	4.253±0.192	24.769±0.620	41.174±0.724
(*nevadensis*)	69	69	69	69
	4.27-14.20	2.07-9.18	15.00-36.93	26.59-51.10
4	10.57±0.290	7.555±0.504	30.221±0.804	46.567±0.985
(*lepida*)	31	31	31	31
	5.45-12.99	2.88-13.51	17.86-40.63	33.58-55.73
5	11.565±0.565	7.921±0.624	31.376±0.698	46.151±1.270
(*lepida*)	23	23	23	23
	6.11-16.38	2.89-14.08	22.34-35.01	29.89-59.65
6	11.881±0.287	8.346±0.360	30.033±0.743	45.412±1.107
(*lepida*)	46	46	46	46
	7.44-16.01	4.11-12.93	17.73-40.47	23.46-62.46
7	12.231±0.383	9.047±0.429	32.958±0.757	47.788±0.795
(*lepida*)	56	56	56	56
	6.62-18.72	3.29-16.55	21.57-43.82	36.30-63.09
8	13.579±0.261	9.175±0.250	31.487±0.406	47.484±0.491
(*lepida*)	123	123	123	123
	7.63-25.25	3.76-17.03	19.33-45.43	31.94-61.66
9	12.766±0. 293	8.413±0.336	32.748±0.479	46.614±0.639
(*lepida*)	78	78	78	78
	7.89-19.78	3.14-15.76	24.66-43.67	36.21-61.80
10	12.313±0.238	8.775±0.263	32.013±0.467	47.058±0.580
(*lepida*)	99	99	99	99
	6.61-18.49	4.05-15.61	21.30-44.18	29.10-60.03
11	13.189±0.462	9.124±0.519	31.318±0.631	45.987±1.012
(*lepida*)	34	34	34	34
	8.43-18.32	4.87-16.93	22.15-37.35	35.46-56.91

Table 42 (continued)

Sample	Dorsal-X	Tail-X	Lateral-X	Chest-X
12	13.285±0.448	10.151±0.464	31.531±0.648	44.914±1.016
(*lepida*)	36	36	36	36
	7.07-17.22	4.02-15.11	24.96-40.90	31.41-55.63
13	13.036±0.523	8.392±0.616	30.160±1.207	47.884±1.446
(*lepida*)	30	30	30	30
	8.14-18.58	2.26-20.26	14.71-39.61	29.76-61.13
14	16.130±1.049	10.975±1.518	34.544±1.695	42.921±2.105
(*marshalli*)	8	8	8	8
	10.88-20.78	7.12-18.50	29.11-42.88	34.58-51.43
15	14.195±0.271	9.50±0.289	33.264±0.555	49.622±0.675
(*grinnelli*)	103	103	103	103
	7.10-19.87	3.34-18.69	15.67-48.11	33.46-67.38
16	14.238±0.409	9.244±0.389	30.098±0.762	46.955±0.858
(*grinnelli*)	38	38	38	38
	8.98-20.90	5.46-14.21	36.24-54.76	36.24-54.76

Both dorsal and tail color (Dorsal-X and Tail-X) vary significantly with the geographic position of the individual samples, based on a multiple regression model using both latitude and longitude as independent variables ($r = 0.274$, $F_{(2,700)} = 28.358$, $p < 0.0001$ and $r = 0.134$, $F_{(2,700)} = 6.405$, $p = 0.0018$, respectively). Dorsal-X is significantly related to both latitude and longitude ($p = 0.0001$ and < 0.0001, respectively), but Tail-X is only related to longitude ($p = 0.0011$). The dorsum overall and dorsal tail stripe become darker both to the north and west across the sampled range of the Western Desert Transect. Neither Lateral-X nor Chest-X exhibits similar trends.

Significant correlations exist between the X-coefficients of most topographic regions of the skin, with r-values ranging from a high of 0.310 (Dorsal-X vs. Tail-X, $p < 0.0001$) to a low of 0.106 (Tail-X vs. Lateral-X, $p = 0.0051$). Dorsal-X and Lateral-X, however, are not significantly correlated ($r = 0.068$, $p = 0.0709$). In comparison to the other transects, there appears to be greater independence in the coloration of the four different topographic regions of the skin among Western Desert Transect samples.

To examine the degree of concordance in colorimetric variables across the transect, we again employed a principal components analysis using the X-coefficients for each topographic region. Of the four axes extracted, only PC-1 has an eigenvalue greater than 1.0; it explains 48.4% of the variation (Table 43). All four X-coefficients load positively and equally highly on this axis, suggesting that it expresses primarily the degree of darkness or paleness around the entire body. Hence, PC-1 in this analysis mirrors the pattern observed for other transects. PC-2, which explains 21.1% of the variation, contrasts Dorsal-X and Tail-X, on the one hand, with Lateral-X and Chest-X, on the other. PC-3 (18.0% of the variation) primarily contrasts Lateral-X and Chest-X, and PC-4 (12.6%) contrasts Dorsal-X and Tail-X.

Table 43. Principal component factor loadings for colorimetric variables of adult specimens of the Western Desert Transect.

Variable	PC-1	PC-2	PC-3	PC-4
Dorsal-X	0.769	-0.362	-0.147	0.506
Tail-X	0.743	-0.464	0.071	-0.477
Lateral-X	0.636	0.444	0.628	0.066
Chest-X	0.622	0.549	-0.545	-0.123
eigenvalue	1.935	0.844	0.718	0.502
% contribution	48.4	21.1	18.0	12.6

Scores for each of the four PC axes are significantly heterogeneous among the sampled populations, with p-values ranging from < 0.0001 (PC-1 and PC-2) to 0.0081 (PC-3) and 0.0060 (PC-4). Bivariate combinations of any pair of these axes, however, fail to show substantive differences among the 16 geographic groups, with the exception that the samples of *N. l. nevadensis* have significantly higher scores on PC-1 (i.e., are darker) than do those of other subspecies samples (Fig. 115). Minor ($p < 0.05$) differences do exist in comparisons between geographically adjacent sample groups other than those between subspecies, and samples of both *N. l. grinnelli* and *N. l. marshalli* tend to be the palest (Fig. 115). Given the color differences between subspecies and the geographic position of their sample groups, it is not surprising that PC-1 scores are highly correlated with geography, as indexed by latitude and longitude. In a multiple regression analysis using both latitude and longitude as independent variables, $r = 0.580$ (ANOVA, $F_{(2,835)} = 211.464$, $p < 0.0001$), with latitude and longitude individually significant at $p < 0.0001$. As with the univariate Dorsal-X and Tail-X coefficients, overall color becomes darker to the north and to the west among the 16 geographic samples, with a rather sharp distinction between samples of *N. l. nevadensis* and samples of the other subspecies (Fig. 115). Strong clinal trends are evident even if the very dark *N. l. nevadensis* samples are removed from the analysis ($r = 0.219$, ANOVA, $F_{(2,702)} = 17.609$, $p < 0.0001$), with p-values for both latitude and longitude remaining at or below 0.0001. Bi-directional clinal variation remains even within the exclusive set of samples assignable to *N. l. lepida* ($r = 0.164$, ANOVA, $F_{(2,553)} = 7.656$, $p = 0.0005$), with $p = 0.0006$ and 0.0198, respectively, for both latitude and longitude. Clearly, color is paler in the south and east and becomes progressively darker to the north and west, culminating in a step in this clinal pattern in the shift between northwestern-most samples of *N. l. lepida* (Groups 4, 5, and 13) and those of *N. l. nevadensis* (Groups 1-3).

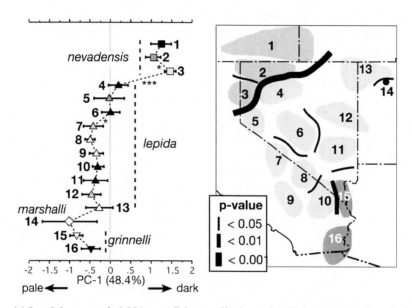

Figure 115. Mean and 95% confidence limits of PC-1 scores of colorimetric variables for the 16 geographic samples of the Western Desert Transect (left) and map of general groupings (right). Symbols and sample numbers correspond to those in the transect map, Fig. 107. Geographic groups are identified by currently recognized subspecies (*N. l. nevadensis* [Groups 1-3], *N. l. lepida* [Groups 4-13], *N. l. grinnelli* [Groups 15-16], and *N. l. marshalli* [Group 14]) and are arranged geographically on the left; significance levels between geographically adjacent areas are indicated: * $p < 0.05$, ** $p < 0.01$. Significance among geographic groups is also indicated by line width (as indicated in the inset) on the map, right.

Taxonomic considerations.—The analyses presented above involve comparisons among multiple geographic samples of four currently recognized subspecies of the desert woodrat: *lepida* (including *desertorum*), *nevadensis*, *marshalli*, and *grinnelli*. In this section, we reiterate the basis for each of these taxa, as delineated in their original descriptions, and evaluate those differences based on the data we have gathered and analyzed. We conclude with our perceptions on the validity of each.

The actual locality where the holotype of *lepida* was obtained is not known. Thomas (1893, p. 235) gave it only as "Utah," which Goldman (1932, p. 61) later emended to "somewhere on 'Simpson's route' between Camp Floyd (a few miles west of Utah Lake), Utah, and Carson City, Nevada." Goldman's designation places the type locality most likely in our geographic Groups 12 or 13. Neither of these grouped localities exhibits significant differences for univariate external, craniodental, or colorimetric variables, or for multivariate PC scores, in comparison to geographic Group 8, which contains the type locality of *desertorum* (Figs. 114 and 115). Thus, Goldman's (1932, p. 61) action to place *desertorum* in the synonymy of Thomas' *lepida* seems fully justified.

Goldman (1932, p. 62) also placed Taylor's *nevadensis*, originally described as a species separate from both *desertorum* and *lepida*, in synonymy of *N. lepida*, and, moreover, as a synonym of the nominate subspecies. All subsequent authors have followed Goldman's first action, but these same authors have retained *nevadensis* as a valid subspecies (Grinnell, 1933; Hall, 1946, 1981). In his description of *nevadensis*, Taylor compared his specimens to series of *desertorum* in the Museum of Vertebrate Zoology collection from both the type locality in Death Valley and the Mojave Desert of southern California as well as to Thomas' description of *lepida*. Taylor stated that his *nevadensis* was uniformly smaller in all external measurements, most notably in the larger ratio of hind foot to total length and smaller ratio of tail to total length (both due to an absolutely longer tail in *desertorum*). He also noted that the skull of *nevadensis* was smaller and "differently shaped," as the "frontal profile is not flattened on the same plane" (p. 292). Finally, in color, *nevadensis* was said to average darker, especially on the mid-dorsum and dorsal tail stripe, with the ventral color more whitish and less buffy than *desertorum* Merriam from the Mojave Desert. As our analyses included all of those specimens used by Taylor in his study, as well additional materials collected subsequently, we can directly evaluate these stated differences.

Of the four external measurements, none exhibit significant differences when comparison is made between geographically adjacent samples of *nevadensis* and *lepida* (Groups 2 and 3 versus Groups 4, 5, or 13). However, in comparison to the samples that Taylor used in his diagnosis, *nevadensis* from Group 2 does have a slightly longer hind foot (ANOVA, $F_{(4,237)}$ = 2.684, Fisher's PLSD, p = 0.0198 to Group 8 and 0.0205 to Group 10) and longer tail (but only in comparison to Group 10, $F_{(15,555)}$ = 8.468, p = 0.0254, not to Group 8, p = 0.0524). Contrary to Taylor's description, cranial dimensions are not smaller in his *nevadensis* than in *lepida* (his samples of *desertorum*), as only six of the 21 craniodental variables are significantly different (by ANOVA comparing pooled samples of both subspecies; see above), and for five of these (IOC, OL, MTRL, IFL, and BOL) *nevadensis* is actually larger (only in BUL is *nevadensis* is smaller than samples of *lepida*). Taylor's conclusion that *nevadensis* has a smaller skull is based on his inclusion of subadult specimens in his comparisons, which is both indicated by his description of the rounded frontal profile (cited above; a standard pattern of skull growth is the flattening of the cranial vault with increasing age [e.g., Myers and Carleton, 1981]) and by our age designations. Only one of the nine specimens from the type locality used by Taylor in his description is an adult by our criteria, and even the holotype (MVZ 8282) is age class "6", or a subadult not included in our statistical summaries.

If Taylor's conclusions of the distinctness of his *nevadensis* and *lepida* (including *desertorum*) in both external and cranial features were wrong, his eye did properly discern the differences in color tones, particularly of the dorsum and dorsal tail stripe. Both of these characters are significantly darker in all of our samples of *nevadensis* than any sample of *lepida*, either those geographically adjacent or the ones further south that Taylor used in his comparisons. This distinctness is sharply defined geographically, as is evident in the principal components analysis presented in Fig. 115. Furthermore, in a canonical variates analysis comparing the grouped samples of all four subspecies using the four X-coefficients, *nevadensis* is significantly darker than the other three (ANOVA, Fisher's PLSD posterior test; p < 0.0001 in comparison to *lepida*, 0.0038 to *marshalli*, and 0.0043 to *grinnelli*), while the other three subspecies are indistinguishable from each another (p > 0.5474 in all comparisons). The darker coloration exhibited by *nevadensis* is likely an adaptation for concealing coloration, because desert woodrats in the geographic range of this taxon (northeastern California, southeastern Oregon, northwestern Nevada, and southwestern Idaho) characteristically, and nearly exclusively, inhabit the extensive flood basalt flows found throughout the region (Figs. 108 and 109). Populations of *lepida* further to the east or south occupy a much wider range of habitats and soil conditions. Here, melanic or overly dark individuals are limited to the small, localized lava fields (those near Big Pine in the Owens Valley [locality CA-388] or Little Lake [CA-381] and the Pisgah and Amboy lave flows of eastern California [CA-349, CA-351]) that are juxtaposed among broad expanses of other, paler substrates. Taylor's *nevadensis*, therefore, is distinguishable from other woodrats in the Great Basin primarily by darker overall color tones, not by either molecular (mtDNA sequences) or external or craniodental measurements.

In his description of *marshalli* from the islands in the Great Salt Lake, Utah, Goldman (1939) distinguished this taxon by its larger size, "more noticeable in the skull than in other parts" (p. 357) and overall paler coloration. Both of these general statements do, in fact, legitimately diagnose *marshalli* relative to geographically adjacent samples of *lepida*. For example, *marshalli* is significantly larger in overall cranial size, as indexed by PC-1 scores (Fig. 114). We further illustrate the relative distinctness of *marshalli* with respect to the other recognized subspecies included in Western Desert Transect by comparing the Mahalanobis D^2 (MD^2) distances within and among *lepida*, *nevadensis*, and *grinnelli* to that between these three taxa and *marshalli*, based on the CVA of the 21 log-transformed craniodental variables (Fig. 116). Excluding *marshalli*, the average MD^2 within each subspecies is only slightly lower (mean = 3.685) than it is between them (mean = 4.355) and thus only marginally significantly different (ANOVA, $F_{(1,79)}$ = 4.957, p = 0.0288). Moreover, the mean MD^2 distances do not differ among any pair of these three subspecies (ANOVA, Fisher's PLSD posterior test for pairwise comparisons, p-values range from a low of 0.2836 to a high of 0.8263). Importantly, however, the mean MD^2 distances between *lepida*, *nevadensis*, and *grinnelli* relative to *marshalli* are each significantly different (p < 0.0001 in all comparisons).

Goldman's observation that the overall color at all four topographic regions of the skin of his *marshalli* is paler is also true, although the mean scores on PC-1 are not significantly different (Fig. 115, and above). The combination of color and craniodental size, therefore, could be used to justify recognition of *marshalli* as a valid subspecies, even though based on mtDNA results the taxon is clearly very closely related, genealogically, to population samples of the nominate subspecies of *N. lepida*. These genetic data suggest that the morphological distinctness of *marshalli* is likely of recent origin, either due to drift in a small insular population or to local adaptation.

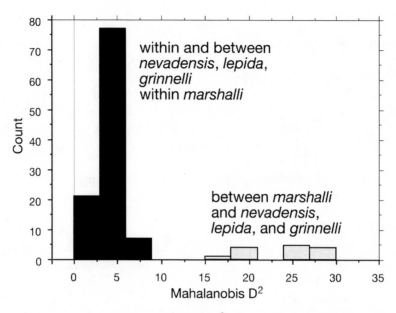

Figure 116. Histogram of Mahalanobis D^2 values within and between the 16 geographic groups of the Western Desert Transect. Note the limited morphometric divergence within and between samples of the subspecies *lepida* Thomas, *grinnelli* Hall, and *nevadensis* Taylor but the extreme difference between these three subspecies samples and the single sample of *marshalli* Goldman from islands in the Great Salt Lake.

Hall (1942), in his description of *grinnelli*, noted that this taxon differed from *N. l. lepida* "in actually and relatively longer tail, slightly lesser average size in most other parts measured, and lighter color" (p. 369). Our sample of *grinnelli* from Group 16, which includes the type locality, does have a significantly longer tail (mean 136.7mm; Table 39) than any sample of *N. l. lepida*, but this tail length is also significantly longer than that of the other *grinnelli* sample, Group 15 (mean 129.6). And, the latter sample is not different from any of the geographically adjacent samples of *N. l. lepida*, such as Group 10 (mean 128.9). Tail length, however, exhibits a weak clinal trend with respect to latitude ($r = -0.346$, $F_{(1,567)} = 76.996$, $p < 0.0001$) across the entire sampled range, such that the long-tailed *grinnelli* of Group 16 is only at the southern terminus of this cline (Fig. 112, bottom). For most craniodental variables, either sample of *grinnelli* falls within the range exhibited by the collective samples of *N. l. lepida* (Table 39). The "lighter color" is minimally true for our Group 15 sample, from the northern end of the range of *grinnelli* as mapped by Hall (1942), but not for Group 16, which includes the holotype (Fig. 115). The latter group cannot be distinguished from any of the 11 samples of *N. l. lepida*, and the former differs only at a p < 0.05) from Group 10 of *N. l. lepida*, to the immediate west (Fig. 115). Since color has a strong latitudinal component, our samples of *grinnelli* are clearly only at the southern end of a gradual cline that begins in northern Nevada and ends in northeastern Baja California. When samples are grouped by subspecies, even a multivariate canonical analysis of the colorimetric variables fails to distinguish *grinnelli* from *lepida* (ANOVA, Fisher's PLSD posterior test, p = 0.547).

Finally, as noted above, mtDNA haplotypes of samples within the mapped range of *grinnelli* are all part of the desert subclade 2A group and, in some cases, are broadly shared across hundreds of kilometers within this overall range. And, as we describe in the next section, our geographic samples of *lepida* and *grinnelli* cannot be distinguished by their respective allelic arrays in the 18 microsatellite loci. Consequently, there appears little valid reason for the continued recognition of *grinnelli* as a valid geographic entity within *N. lepida*, from either a morphological or molecular standpoint.

Eastern Desert Transect
(Transitions across the Colorado River)

We include in this final transect those samples of the desert mtDNA Clade 2 that occur on both sides of the Colorado River. This area encompasses the eastern-most samples of *lepida* examined in the previous analysis, those of *grinnelli* from the west side of the lower Colorado River in southern California and northeastern Baja California, all samples from north of the Colorado River in northern Arizona, Utah, and Colorado (subspecies *monstrabilis* and *sanrafaeli*), and all samples from Arizona and Sonora from south and east of the Colorado River, which collectively include the named forms *devia, auripila, bensoni, flava, aureotunicata,* and *harteri*. This set of samples also includes those belonging to mtDNA subclade 2A (*lepida* and *grinnelli*), subclade 2B (*monstrabilis* and *sanrafaeli*), subclade 2C (*devia*), subclade 2D (*auripila*), and subclade 2E (*auripila, aureotunicata, bensoni, flava,* and *harteri*).

Samples from east and south of the Colorado River in Arizona have been considered a species (*Neotoma devia*, including *auripila, aureotunicata, bensoni, flava,* and *harteri*) distinct from *N. lepida* (including *grinnelli*) to the west and north in California, Nevada, Utah, and Arizona (Mascarello, 1978; Musser and Carleton, 1993, 2005). Musser and Carleton (1993) initially placed *monstrabilis* and *sanrafaeli* in synonymy of the species *N. devia*, noting that this allocation required confirmation, but later reversed their opinion by listing these two taxa in the synonymy of *N. lepida* (Musser and Carleton, 2005, p. 1056). Hoffmeister (1986) disagreed with Mascarello's opinion on the specific status of *N. devia* and regarded all Arizona taxa, including *monstrabilis*, as only subspecifically distinct from *N. lepida* to the west of the Colorado River. He also recognized only two subspecies in Arizona, regarding *monstrabilis* as a junior synonym of *devia* and *flava* and *harteri* as synonyms of *auripila*. Finally, he suggested that both *bensoni* and *aureotunicata* from northwestern Sonora were only "color variants and may be referable" to *auripila* (p. 421). He placed the geographic dividing line between *devia* and *auripila* at about the La Paz-Mohave Co. line (= the Bill Williams River) in west-central Arizona.

We pooled localities into 13 geographic groups for analysis (Fig. 117). These include both samples of *grinnelli* (Groups 15 and 16) and four of the set of *lepida* samples (Groups 10-13) used in the Western Desert Transect analysis. The remaining groups contain a single sample of the subspecies *sanrafaeli* (Group 17, which includes the type locality [UT-34]), two of *monstrabilis* (Groups 18 and 19, divided approximately east and west of the Kaibab Plateau), and four samples encompassing Arizona and Sonora south and east of the Colorado River. Group 18 contains the type locality of *monstrabilis* [AZ-21], Group 20 includes the type locality of *devia* [AZ-50], and Group 23 contains the type localities of the other five subspecies (*auripila* [AZ-84], *bensoni* [S-2], *flava* [AZ-79], *aureotunicata* [S-6], and *harteri* [AZ-69]). The Arizona and Sonora samples are defined primarily by their separate mtDNA clade membership, with the exception of mtDNA subclade 2C, which is divided into one group from the Colorado Desert north of Flagstaff, Arizona (Group 20) and another along the eastern side of the lower Colorado River south of Boulder Dam (Group 21). As with previous analyses, we have included the holotypes or topotypes for each of these taxa in our analyses, obtaining DNA sequences from topotypes, or near topotypes, of *devia*, *monstrabilis*, *sanrafaeli*, *auripila*, *bensoni*, *aureotunicata*, and *harteri*.

Individuals from a few of the localities are unassigned to any geographic group; these are considered as "unknowns" and are excluded from the summaries of group statistics and both univariate and multivariate comparisons that follow. Most of these localities lie along the Virgin River in southeastern Nevada and adjacent Arizona and Utah, the approximate boundary between mtDNA subclades 2A and 2B. Three other localities lie either between the ranges of *lepida*, *monstrabilis*, and *sanrafaeli* in central Utah (localities NV-138 and NV-139) or represent the rather geographically isolated locality northwest of Phoenix in south-central Arizona (locality AZ-68). We use canonical discriminant analysis to determine the morphometric assignment of individuals from these localities to the pre-defined geographic groups. In the case of the Virgin River samples, we are specifically interested in the degree of morphological and molecular concordance, and we use the posterior probabilities of membership in the predefined reference samples to assess this concordance.

Localities and sample sizes.—The Eastern Desert group of samples includes craniodental measurements from 693 adult woodrats and colorimetric variables from 768 individuals. We sequenced the mtDNA *cyt-b* gene for 163 individuals representing 11 of the 13 geographic samples and 20 individuals from three of the "unknown" localities along the Virgin River in Nevada and Arizona. We also have data from 18 microsatellite loci for 285 individuals from 31 localities. Samples from west and north of the Colorado River in California, Nevada, and western Utah are the same as those used in the Western Desert Transect (Groups 10-13 of *lepida* and 15-16 of *grinnelli*). Sample details for each of these groups are in the Western Desert Transect account, above. Other pooled samples, including the number of individuals for which craniodental (n_m), colorimetric (n_c), glans penis (n_g), or mtDNA sequences (n_{DNA}) were examined, and specific localities (numbered as in the Appendix) we list here.

Group 17 [*sanrafaeli*] (total $n_m = 14$, $n_c = 24$, $n_g = 11$, $n_{DNA} = 20$)
COLORADO:– RIO BLANCO CO.: (1) CO-1; $n_m = 2$, $n_c = 4$; USNM 148012-148015.
UTAH:– EMERY CO.: (2) UT-32; $n_{DNA} = 2$; BYU 18153-18154; (2) UT-33; $n_m = 3$, $n_c = 8$, $n_g = 5$, $n_{DNA} = 8$; MVZ 199391-199398. GARFIELD CO.: (3) UT-28; $n_{DNA} = 1$; BYU 18300; (3) UT-29; $n_m = 2$, $n_c = 4$; USNM 158536-158359. GRAND CO.: (4) UT-34; $n_m = 2$, $n_c = 2$, $n_g = 1$, $n_{DNA} = 3$; [type locality of *sanrafaeli* Kelson], MVZ 199388-199390; (5) UT-34a; $n_g = 1$; CSULB 11118. WAYNE CO.: (6) UT-31; $n_m = 5$, $n_c = 6$, $n_g = 4$, $n_{DNA} = 6$; MVZ 199399-199404.

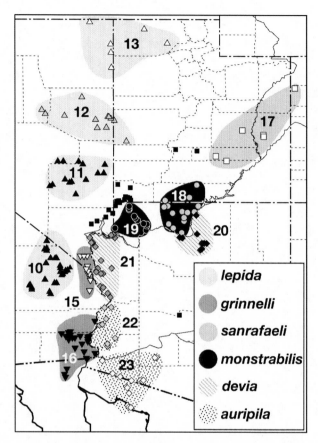

Figure 117. Map of the 13 grouped localities of the Eastern Desert Transect used in the analysis of craniodental and color characteristics of woodrats. Grouped localities for each of the six recognized subspecies (Hall, 1981; Hoffmeister, 1986) are indicated by different patterns and individual localities by separate symbols. Data for Groups 10-13 and 15-16 are given in the Western Desert Transect, above. The type locality of *sanrafaeli* is within geographic Group 17; that of *monstrabilis* in Group 18; that of *devia* in Group 20; and that of *auripila* in Group 23 (in which also are the type localities of *aureotunicata*, *bensoni*, *flava*, and *harteri*, all currently listed as synonyms). Solid squares identify those localities that are collectively considered as group "unknown."

Group 18 [*monstrabilis*] (total $n_m = 39$, $n_c = 59$, $n_g = 3$, $n_{DNA} = 11$)
ARIZONA:– COCONINO CO.: (1) AZ-19; $n_m = 1$; MVZ 56493; (2) AZ-20; $n_c = 1$; USNM 161173; (3) AZ-21; $n_m = 12$, $n_c = 20$, $n_{DNA} = 2$; MVZ 56494-

56510, 197153-197154; USNM 243123 [holotype of *monstrabilis* Goldman]; (4) AZ-22; n_m = 2, n_c = 2; UNSM 243443-243444; (5) AZ-23; n_m = 2, n_c = 5; USNM 161166-161170; (6) AZ-24; n_m = 1, n_c = 1; USNM 161171; (7) AZ-25; n_m = 1, n_c = 1; USNM 161172; (8) AZ-26; n_m = 2, n_c = 2; USNM 248998, 250014; (9) AZ-27; n_m = 1, n_c = 1; USNM 243446; (10) AZ-28; n_m = 1, n_c = 2, n_{DNA} = 4; MVZ 197155-197158; (11) AZ-29; n_c = 1; USNM 529790; (12) AZ-30; n_m = 3, n_c = 7; USNM 215637-215639, 215542-215543, 216048; (13) AZ-31; n_m = 2, n_c = 2; USNM 248782-248783; (14) AZ-34; n_c = 2; MVZ 61188-61189; (15) AZ-36; n_c = 1; MVZ 58799; (16) AZ-39; n_c = 2; USNM 161174-161175; (17) AZ-50a; n_m = 1, n_c = 1; MVZ 51710.

UTAH:– KANE CO.: (18) UT-26; n_{DNA} = 1; BYU 18017; (19) UT-27; n_m = 2, n_c = 2; USNM 578066-578067; (20) UT-24; n_m = 4, n_c = 6; MVZ 56490-56492; USNM 161176-161177, 190301. SAN JUAN CO.; (21) UT-25; n_m = 4, n_g = 3, n_{DNA} = 4; MVZ 199374-199377.

Group 19 [*monstrabilis*] (total n_m = 44, n_c = 75, n_{DNA} = 18)

ARIZONA:– MOHAVE CO.: (1) AZ-4; n_c = 2; USNM 243131-243132; (2) AZ-6; n_m = 2, n_c = 2; USNM 243133-243134; (3) AZ-7; n_m = 7, n_c = 12, n_{DNA} = 12; MVZ 197141-197152; (4) AZ-8; n_m = 2, n_c = 7; MVZ 58798, 61185-61187, USNM 263121-263123; (5) AZ-9; n_m = 2, n_c = 5; MVZ 56511-56513; SDNHM 12684-12685; (6) AZ-10; n_m = 3, n_c = 5; SDNHM 12661-12662, 12679; USNM 263015, 263113, 263115; (7) AZ-11; n_m = 3, n_c = 3; USNM 262068-262070; (8) AZ-12; n_m = 2, n_c = 3; USNM 243127-243129; (9) AZ-13; n_m = 2, n_c = 3; USNM 243124-243126; (10) AZ-14; n_m = 5, n_c = 9; MVZ 56514-56518, SDNHM 12789, 12692-12693, 12796, USNM 161642; (11) AZ-15; n_m = 5, n_c = 9; (MVZ 56519-56527; (12) AZ-16; n_m = 6, n_c = 6, n_{DNA} = 6; MVZ 199368-199373; (13) AZ-17; n_m = 5, n_c = 8; MVZ 56528-56531, USNM 263017-263020; (14) AZ-18; n_c = 1; USNM 532731.

Group 20 [*devia*] (total n_m = 41, n_c = 72, n_{DNA} = 20)

ARIZONA:– COCONINO CO.: (1) AZ-32; n_m = 1, n_c = 1; USNM 532728; (2) AZ-33; n_c = 1; USNM 529792; (3) AZ-35; n_c = 1; MVZ 56542; (4) AZ-37; n_m = 3, n_c = 3, n_{DNA} = 3; MVZ 199378-199380; (5) AZ-38; n_m = 1, n_c = 1; USNM 529789; (6) AZ-40; n_m = 1, n_c = 1; USNM 202466; (7) AZ-41; n_m = 1, n_c = 1; USNM 202463; (8) AZ-42; n_c = 1; USNM 250098; (9) AZ-43; n_m = 5, n_c = 11; USNM 215545-215449, 215640-215641, 215816-215819; (10) AZ-45; n_m = 6, n_c = 8; LACM 5598, 5599; MVZ 56544-56549,; USNM 244153; (11) AZ-46; n_c = 2; USNM 251014-251015; (12) AZ-47; n_m = 1, n_c = 3; USNM 244099-244100, 244148; (13) AZ-48; n_m = 6, n_c = 6, n_{DNA} = 7; MVZ 199381-199387; (14) AZ-49; n_m = 7, n_c = 19, n_{DNA} = 10; MVZ 197093-197102; (15) AZ-50; n_m = 9, n_c = 13;

USNM 226374-226375, 226376 [holotype of *devia* Goldman], 226377-226378, 226380, 226390-226391, 226394, 226398-226401.

Group 21 [*devia*] (total n_m = 97, n_c = 71, n_{DNA} = 13)

ARIZONA:– MOHAVE CO.: (1) AZ-51; n_m = 3, n_c = 2; USNM 270666-270668; (2) AZ-52; n_m = 1; USNM 270661; (3) AZ-53; n_m = 4; USNM 270663-270665; (4) AZ-54; n_m = 1, n_c = 2; MVZ 56535-56536; (5) AZ-55; n_m = 5, n_c = 9; MVZ 128176-128180; SDNHM 22667-22669; (6) AZ-56; n_m = 2, n_c = 8, n_{DNA} = 10; MVZ 197103-197112; (7) AZ-57; n_m = 3, n_c = 3; MVZ 56539-56541; (8) AZ-58; n_m = 1, n_c = 2; USNM 227815, 227818; (9) AZ-59; n_m = 22, n_c = 3; MVZ 149190, 149192-149207, 149211-149215; (10) AZ-60; n_m = 5, n_c = 5; MVZ 61863-61867; (11) AZ-61; n_m = 1, n_c = 2, n_{DNA} = 2; MVZ 197114-197114; (12) AZ-62; n_m = 7; MVZ 149216, 149219-149224; (13) AZ-63; n_m = 1, n_c = 1; MVZ 10457; (14) AZ-64; n_m = 21, n_c = 21; MVZ 149230-149251; (15) AZ-66; n_m = 19, n_c = 12; SDNHM 12996,-12997, 13002-13003, 13017-13022, 13025, 13030-13032, 13043-13044, 13049, 13143, 13156; (16) AZ-67; n_m = 1, n_c = 1, n_{DNA} = 1; MVZ 199820.

Group 22 [*auripila*] (total n_m = 27, n_c = 33, n_{DNA} = 16)

ARIZONA:– LA PAZ CO.: (1) AZ-70; n_m = 2, n_c = 2; USNM 181040, 181121; (2) AZ-71; n_m = 1, n_c = 1, n_{DNA} = 1; MVZ 199818; (3) AZ-72; n_m = 1, n_c = 2; MVZ 6265162652; (4) AZ-73; n_m = 9, n_c = 8; MVZ 149252-149260; (5) AZ-74; n_m = 6, n_c = 6, n_{DNA} = 6; MVZ 195235-195240. YUMA CO.: (6) AZ-75; n_m = 2, n_c = 2; USNM 525880-525881; (7) AZ-76; n_m = 3, n_c = 3; MVZ 62653-62655; (8) AZ-77; n_m = 3, n_c = 9, n_{DNA} = 9; MVZ 197115-197123.

Group 23 [*auripila*] (total n_m = 75, n_c = 81, n_g = 3, n_{DNA} = 18)

ARIZONA:– YUMA CO.: (1) AZ-78; n_c = 1; USNM 505202; (2) AZ-79; n_m = 6, n_c = 6; MVZ 62657 [holotype of *flava* Benson], 62656; SDNHM 12245, 12253-12255; (3) AZ-80; n_m = 5, n_c = 5; MVZ 62658-62662; (4) AZ-81; n_m = 1, n_c = 1; MVZ 76178. PIMA CO.: (5) AZ-82; n_m = 3, n_c = 3; SDNHM 12410, 12427, 12446; (6) AZ-83; n_m = 1, n_c = 1; USNM 251322; (7) AZ-84; n_m = 7, n_c = 8, n_{DNA} = 1; MVZ 62663-62664, 62666-62669, 202447 [type locality of *auripila* Blossom]. MARICOPA CO.: (8) AZ-69; n_m = 12, n_c = 12, n_{DNA} = 4; MVZ 199819, 200713-200714, JLP 19737 [uncataloged, tail only], SDNHM 11433-11435, 11450-11451, 11458, 11462 [holotype of *harteri* Huey], 11463-11464.

MEXICO:– SONORA: (9) S-1; n_m = 1; MVZ 83237; (10) S-2; n_m = 11, n_c = 13, n_g = 3, n_{DNA} = 5; MVZ 76179-76180, 83238-83240, 83242-83245, 200709-200712; CIB 4561 [type locality of *bensoni* Blossom]; (11) S-3; n_m = 4, n_c = 4; MVZ 83246-83249; (12) S-4; n_m = 7, n_c = 10; MVZ 83258-83267; (13) S-5; n_m =

8, n_c = 8; MVZ 83250-83257; (14) S-5; n_m = 3, n_c = 3, n_{DNA} = 8; MVZ 200705-200708, CIB; (15) S-6; n_m = 3, n_c = 3; SDNHM 10852, 10907 [holotype of *aureotunicata* Huey], 10934; (16) S-6; n_m = 2, n_c = 3; MVZ 83269-83271.

unknown (total n_m = 47, n_c = 56, n_g = 18, n_{DNA} = 22)
ARIZONA:– MARICOPA CO.: (1) AZ-68; n_m = 4, n_c = 4; MVZ 62670-62673. MOHAVE Co.: (2) AZ-1; n_m = 4, n_c = 4, n_{DNA} = 4; MVZ 199360-199363; (3) AZ-2; n_m = 2; USNM 261978, 262072; (4) AZ-3; n_m = 1, n_c = 1; USNM 243130; (5) AZ-5; n_m = 2, n_c = 2; USNM 261977, 262071; (6) AZ-65; n_{DNA} = 2; BYU 18947-18948.

NEVADA:– CLARK CO.: (7) NV-137; n_m = 1; USNM 40388; (8) NV-138; (n_m = 8, n_c = 8, n_g = 4, n_{DNA} = 8; MVZ 199352-199359; (9) NV-139; n_m = 3, n_c = 8, n_g = 2, n_{DNA} = 8; MVZ 202484-202491; (10) NV-140; n_m = 19, n_c = 18, n_g = 10; MVZ 149301-149319.

UTAH:– WASHINGTON CO.: (11) UT-11; n_m = 1, n_c = 1; USNM 167511; (12) UT-11a; n_g = 1; CSULB 1475; (13) UT-12; n_m = 1, n_c = 2; USNM 167461, 167508; (14) UT-12a; n_g = 1; CSULB 1473; (15) UT-13; n_m = 1, n_c = 1; USNM 513356; (16) UT-14; n_m = 2; USNM 190302-190303; (17) UT-15; n_m = 1, n_c = 1; USNM 327185; (18) UT-16; n_m = 3, n_c = 3, n_g = 1; MVZ 45397; USNM 40389, 167510; (19) UT-17; n_m = 2, n_c = 2; MVZ 61183-61184; (20) UT-18; n_m = 1, n_c = 1; MVZ 149537; (21) UT-19; n_m = 1, n_{DNA} = 2; LACM 90500, BYU15034-15035; (22) UT-21; n_m = 2; USNM 327182, 327186; (23) UT-22; n_m = 2; USNM 327183-327184; (24) UT-21a; n_g = 1; CSULB 1479; (25) UT-22a-b; n_g = 2; CSULB 1461, 1468. PIUTE CO.: (26) UT-23; n_m = 4; USNM 157877, 157879-157881. WAYNE CO.: (27) UT-30; n_m = 1; USNM 175882.

Habitat.—Throughout the area covered by the Eastern Desert Transect in southern Utah, southeastern Nevada, and Arizona, desert woodrats are largely restricted to rock outcrops or rimrock shelves of basalt or other lavas, sandstones, limestones, or on occasion mudstones—virtually any solid rock type that fractures or otherwise contains a structure where nests can be constructed with some security. In areas where other woodrat species are not present, desert woodrats also construct nests in clumps of dense vegetation, primarily patches of yucca (*Yucca* sp.), either prickly-pear or cholla cactus (*Opuntia* sp.), and occasionally at the base of a variety of shrubs. In western Arizona and northwestern Sonora, desert woodrats are restricted to rocky habitats where they co-occur with the White-throated Woodrat, *Neotoma albigula*, which is almost always the only woodrat occupying vegetated areas on the desert pavement or along shallow arroyo courses. In southern Utah, desert woodrats occur at elevations below the Bushy-tailed Woodrat, *Neotoma cinerea*, but north of Flagstaff in northern Arizona, they co-occur with the Stephen's (*Neotoma stephensi*), Bushy-tailed (*Neotoma cinerea*), and White-throated woodrats in the basalt mesas within piñon-juniper woodlands (Dial and Czaplewski, 1990).

We illustrate the range of habitats occupied by desert woodrats through the transect area with photographs of type localities, or nearby sites, for six of the formally named taxa that span the north to south extent of the Eastern Desert Transect (Figs. 118 through 123).

Figure 118. Outcrops of Estrada Sandstone near the type locality of *N. l. sanrafaeli* at Rock Canyon Corral, Grand Co., Utah (Group 17, locality UT-34). Utah Juniper, sagebrush (*Atremisia* sp.), rabbitbrush (*Chrysothamnus* sp.), and saltbush (*Atriplex* sp.). Photo taken in August 2002.

Figure 119. Kaibab Limestone at Ryan, Coconino Co., Arizona (Group 18, locality AZ-21), the type locality of *N. l. monstrabilis*. Piñon-juniper woodland; understory of sagebrush and yucca. Photo taken in July 2001.

Figure 120. Exposures of Moenkopi sandstone, about 1 mi E of Tanner Tank, Coconino Co., Arizona, the type locality of *N. l. devia*; Gray Mountain is in the background (Group 20, locality AZ-49). Vegetation is dominated by Shadscale (*Atriplex confertifolia*), Spiny Hop Sage (*Grayia spinosa*), Snakeweed (*Gutierrezia* sp.), and Banana Yucca (*Yucca bacata*). Photo taken in July 2001.

Figure 121. Basalt hills near the type locality of *N. l. harteri*, 1.2 mi E Black Gap, Maricopa Co., Arizona (Group 23, locality AZ-69). Dominant vegetation in foreground is Creosote Bush (*Larrea tridentata*). Photo taken in March 2003.

Figure 122. Agua Dulce Mountains, approximately 7 mi E Papago Well, Pima Co., Arizona, the type locality of *N. l. auripila* (Group 23, locality AZ-84). Desert woodrats are rare but found exclusively on the small granite hillocks exposed above the Creosote Bush flats. Photo taken in October 2003.

Figure 123. Basalt flows at Tanque de los Papagos, Sonora, Mexico, the type locality of *N. l. bensoni* (Group 23, locality S-2). Vegetation primarily Velvet Mesquite, Foothill Paloverde, and Whitethorn Acacia in the sandy washes with Creosote Bush, Limberbush, Ocotillo, Saguaro, and cholla dominating the basalt and pumice stone exposures. Photo taken in March 2003.

Craniodental and colorimetric variation.—We provide standard descriptive statistics for the four external and 21 craniodental variables for each geographic group in Table 44 (Groups 17 through 23) and Table 39 (Groups 10-13 and 15-16). All variables, both external and craniodental, exhibit significant mean differences among the 13 samples (p < 0.0001, ANOVA, $F_{(12,592)}$) for each measurement except total length (ANOVA, TOL, $F_{(12,525)}$ = 2.127, p = 0.0141). The pattern of differences is strongly structured geographically, with the majority of significant ones present between paired samples along the Colorado River as opposed to other regions included in this analysis. We summarize the number of univariate differences among all geographically adjacent samples in Fig. 124. Few variables (five or less, except for Groups 13 and 17) differentiate all pairwise samples except those that are for the most part on opposite sides of the Colorado River in California, Nevada, and Arizona. For example, a mode of 15 characters (range 11-19) differ significantly (p < 0.01) between the two samples of *grinnelli* from southeastern California (Groups 15 and 16) and those of *devia* (Group 21) and *auripila* (Groups 22 and 23) on the Arizona side of the river. The degree of cross-river differentiation becomes progressively less to the north and east, however, as the western sample of *monstrabilis* (Group 19) differs from that of *devia* to its south (Group 21) by 10 variables and to the eastern sample of *devia* (Group 20) by six. Similarly, the eastern sample of *monstrabilis* (Group 18) differs from western *devia* by 13 characters and from the sample of *devia* to its immediate south (Group 20) by five variables. Interestingly, there are as many, or more, differences (13) between the two samples of *devia* (Groups 20 and 21), which are on the same side of the river, as there are between either of these samples and those of *monstrabilis* on the north side (Groups 18 and 19, respectively). The fewer differences between samples of eastern *devia* (Group 20) and those of *monstrabilis* from north of the Colorado River (Group 18) is consistent with Hoffmeister's (1986) results, upon which he argued that *devia* is only subspecifically distinct from *N. lepida* rather than a distinct species (Mascarello, 1978).

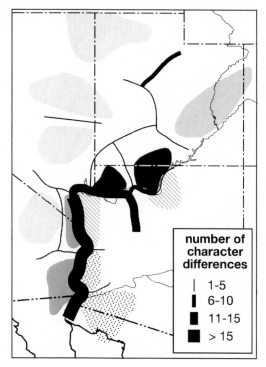

Figure 124. Summary map of the number of significant differences (ANOVA, Fisher's PLSD pairwise comparisons, p < 0.01) among the four external and 21 craniodental characters between geographically paired samples of the Eastern Desert Transect. The width of the line indicates the number of variables (inset in the lower right). Geographic samples are pattern-coded as in the map, Fig. 117.

Table 44. External and cranial measurements of adult (age classes 1-5) specimens of 7 geographic samples of *Neotoma lepida* along the Eastern Desert Transect in Utah and Arizona (Groups 17-23, see Fig. 117). Data for Groups 10-13 and 15-16 are given in Table 39, above. Means, standard error, sample size, and range are provided for each sample. Current subspecies allocations are given with each designated geographic group.

Variable	Group 17 sanrafaeli	Group 18 monstrabilis	Group 19 monstrabilis	Group 20 devia	Group 21 devia	Group 22 auripila	Group 23 auripila
external							
TOL	284.7±6.6	293.5±2.3	291.6±2.1	288.9±1.6	286.4±1.45	287.6±2.8	291.3±1.8
	13	35	41	37	88	18	50
	274-294	261-326	267-324	267-305	255-325	266-310	260-313
TAL	125.5±1.6	129.9±1.6	129.7±1.3	131.1±1.2	130.8±0.9	135.8±2.1	142.3±1.2
	13	35	41	37	88	18	50
	111-132	108-148	113-149	115-146	108-149	121-154	126-166
HF	30.7±03	31.2±0.2	31.5±0.2	30.5±0.5	29.6±0.1	29.59±0.3	29.7±0.1
	16	40	43	42	97	25	59
	29-33	27-34	29-33.5	24-33	27-32	27-32	28-32
E	31.2±0.3	27.3±0.6	28.4±0.6	29.8±0.5	20.1±0.3	31.91±0.4	30.4±0.3
	12	26	29	22	89	23	57
	29-33	24-33	24-33	24-32	24-24	25-35	25-34
craniodental							
CIL	37.18±0.19	37.48±0.19	37.14±0.18	37.02±0.17	36.42±0.13	35.92±0.11	35.75±0.16
	16	40	43	42	98	25	59
	36.16-38.85	35.0-40.1	35.2-39.6	33.4-40.7	35.7-40.7	34.0-38.4	32.7-38.6
ZB	20.24±0.13	20.43±0.13	20.32±0.11	20.52±0.12	20.05±0.07	19.83±0.11	19.70±0.09
	16	40	43	42	98	25	59
	19.6-21.5	19.1-22.3	19.1-21.7	18.6-22.2	18.3-22.3	18.8-21.3	18.1-21.9

Table 44 (continued)

Variable	Group 17 *sanrafaeli*	Group 18 *monstrabilis*	Group 19 *monstrabilis*	Group 20 *devia*	Group 21 *devia*	Group 22 *auripila*	Group 23 *auripila*
IOC	4.86±0.13	5.10±0.04	5.00±0.02	5.06±0.02	5.10±0.02	5.00±0.05	5.06±0.02
	16	40	43	42	98	25	59
	4.7-5.1	4.6-5.8	4.6-5.3	4.7-5.5	4.6-5.6	4.3-5.5	4.7-5.3
RL	14.94±0.11	15.00±0.10	14.94±0.10	14.70±0.09	14.85±0.06	14.41±0.09	14.19±0.07
	16	40	43	42	98	25	59
	14.0-15.7	13.6-16.8	13.8-16.3	13.2-16.9	13.6-17.0	13.5-15.5	12.9-15.7
NL	14.38±0.12	14.49±0.10	14.36±0.10	14.47±0.10	14.33±0.06	13.91±0.11	14.00±0.07
	16	40	43	42	98	25	59
	13.2-15.4	13.3-15.9	12.8-15.9	13.0-16.3	13.0-16.7	12.8-15.6	12.5-15.1
RW	6.14±0.06	6.26±0.03	6.18±0.04	6.15±0.04	6.04±0.03	6.10±0.04	5.89±0.03
	16	40	43	42	98	25	59
	5.7-6.7	5.8-6.7	5.5-6.6	5.9-7.2	5.6-7.5	5.7-6.5	5.4-6.3
OL	13.81±0.08	13.81±0.06	13.79±0.06	13.78±0.06	13.41±0.04	13.31±0.08	13.28±0.05
	16	40	43	42	98	25	59
	13.3-14.5	12.9-14.7	13.2-16.3	13.0-14.5	12.4-152	12.4-14.1	12.3-14.2
DL	10.56±0.10	10.70±0.08	10.60±0.09	10.53±0.07	10.40±0.06	10.31±0.11	9.94±0.07
	16	40	43	42	98	25	59
	9.8-11.4	9.6-11.8	9.5-11.9	9.2-12.0	9.1-13.0	9.2-11.7	8.6-11.0
MTRL	8.25±0.07	8.07±0.04	7.91±0.04	7.99±0.03	7.82±0.03	7.68±0.04	7.68±0.03
	16	40	43	42	98	25	59
	7.7-8.6	7.3-8.7	7.5-8.5	7.6-8.5	7.0-8.8	7.4-8.1	7.2-8.4
IFL	8.19±0.08	8.24±0.06	8.09±0.06	8.09±0.06	8.01±0.04	7.78±0.06	7.68±0.04
	16	40	43	42	98	25	59
	7.5-8.8	7.5-9.0	7.4-8.9	7.0-9.5	7.1-9.5	7.2-8.3	6.9-8.3

Table 44 (continued)

Variable	Group 17 sanrafaeli	Group 18 monstrabilis	Group 19 monstrabilis	Group 20 devia	Group 21 devia	Group 22 auripila	Group 23 auripila
PBL	17.32±0.10	17.33±0.10	17.12±0.10	17.23±0.08	16.91±0.06	16.75±0.11	16.27±0.07
	16.7-18.1	16.2-18.8	15.8-18.2	16.2-19.1	15.7-19.4	15.8-18.3	14.9-17.2
	16	40	43	42	98	25	59
AW	7.35±0.05	7.21±0.04	7.09±0.04	7.16±0.04	7.00±0.03	6.91±0.04	6.86±0.03
	7.1-7.7	6.6-7.7	6.6-7.5	6.6-7.8	6.5-7.9	6.6-7.3	6.3-7.4
	16	40	43	42	98	25	59
OCW	8.79±0.06	8.97±0.04	9.04±0.04	9.00±0.04	8.80±0.03	8.74±0.05	8.64±0.02
	8.4-9.2	8.4-9.6	8.6-9.6	8.7-10.5	8.3-9.9	8.4-9.5	8.2-9.1
	16	40	43	42	98	25	59
MB	16.73±0.11	17.80±0.09	16.82±0.08	16.50±0.08	16.37±0.05	16.32±0.10	15.98±0.05
	16.1-17.5	15.6-18.1	15.8-17.7	14.7-17.5	15.1-18.1	15.3-17.5	15.1-17.0
	16	40	43	42	98	25	59
BOL	5.49±0.05	5.60±0.05	5.55±0.04	5.66±0.05	5.50±0.03	5.43±0.05	5.36±0.04
	5.2-5.9	5.0-6.5	5.0-6.1	4.9-6.8	4.7-6.5	5.1-6.0	4.6-6.4
	16	40	43	42	98	25	59
MFL	7.04±0.08	7.24±0.06	7.32±0.05	7.08±0.05	6.82±0.04	6.58±0.06	7.03±0.04
	6.5-7.8	6.4-8.1	6.7-7.9	6.5-8.1	6.1-8.4	6.1-7.5	6.3-7.9
	16	40	43	42	98	25	59
MFW	2.33±0.04	2.42±0.03	2.39±0.03	7.08±0.03	2.25±0.02	2.16±0.04	2.32±0.03
	2.1-2.6	2.1-3.0	2.0-2.9	2.1-2.9	1.9-2.9	1.8-2.6	2.0-2.8
	16	40	43	42	98	25	59
ZPW	4.22±0.05	4.06±0.04	3.96±0.03	4.28±0.03	4.00±0.02	3.89±0.04	3.92±0.03
	3.8-4.7	3.6-4.6	3.5-4.3	3.9-4.6	3.6-4.4	3.6-4.4	3.5-4.4

Table 44 (continued)

Variable	Group 17 samrafaeli	Group 18 monstrabilis	Group 19 monstrabilis	Group 20 devia	Group 21 devia	Group 22 auripila	Group 23 auripila
CD	15.07±0.08	15.35±0.06	15.26±0.06	15.14±0.06	15.07±0.04	15.00±0.08	14.84±0.04
	16	40	43	42	98	25	59
	14.5-15.8	14.7-16.3	14.3-16.3	14.4-16.5	14.2-17.1	14.2-16.2	14.1-15.5
BUL	7.02±0.04	7.18±0.03	7.20±0.03	7.07±0.03	7.18±0.04	7.29±0.04	7.15±0.03
	16	40	43	42	98	25	59
	6.7-7.3	6.7-7.6	6.8-7.7	6.5-7.4	6.8-7.8	6.9-7.7	6.5-7.5
BUW	7.65±0.07	7.68±0.04	7.55±0.04	7.39±0.03	7.38±0.02	7.48±0.03	7.40±0.03
	16	40	43	42	98	25	59
	7.1-8.2	7.1-8.1	7.1-8.2	6.8-7.7	6.7-7.8	7.1-7.8	6.9-7.9

The pattern of increasing differentiation from north to south along the Colorado River is, at least partially, a function of differences in clinal variation of samples west of the river and those to the east and north. There is no relationship between character trends with geographic position when samples are restricted to those of the western part of this transect, the four samples of *lepida* (Groups 10-13) and the two of *grinnelli* Hall (Groups 15 and 16). The regression coefficient of condyloincisive length (CIL), for example, on both longitude and latitude is 0.088, with a slope non-significantly different from zero ($F_{(1,280)} = 2.178$, $p = 0.1412$ and $F_{(1,280)} = 2.163$, $p = 0.1425$, respectively). In sharp contrast, however, CIL is significantly clinal relative to both longitude and latitude for eastern samples along the Colorado River, from Group 17 (*sanrafaeli*) in the northeast to Group 23 in southwestern Arizona and northwestern Sonora (*auripila*): longitude, $r = 0.311$, $F_{(1,315)} = 33.839$, $p < 0.0001$; latitude, $r = 0.440$, $F_{(1,315)} = 75.828$, $p < 0.0001$. Each of the other univariate characters that exhibit significant differences across the sampled area mirror this difference in clinal trends.

This pattern is also revealed by multivariate analyses of these same characters. Mean scores on extracted principal components axes are significantly different when all of the 13 geographic groups are compared (ANOVA, $F_{(12,631)} > 10.648$, $p < 0.0001$ for each of the first three extracted axes). Only the first two axes have eigenvalues greater than 1.0 (Table 45), with the first explaining 49.4% of the variation and the remaining axes individually explaining no more than 8.5%. As in previous analyses, all variables have relatively high and positive factor loadings on PC-1, all with highly significant correlation coefficients of individual variables versus PC-1 scores ($p < 0.0001$ in all cases; Z-value = 9.144 for logMFW, the variable with the smallest loading of 0.337; Table 45). Thus, PC-1 can be interpreted as a general size axis, and the univariate measure of skull length (condyloincisive length, logCIL), with its highest loading on PC-1 (0.924), is the best univariate estimate of overall size. Given the geographic trends in CIL described above, therefore, it is not surprising that PC-1 scores exhibit a significant relationship with latitude when all samples are included together ($r = 0.310$, $F_{(1,701)} = 74.328$, $p < 0.0001$). Moreover, PC-1 scores of western samples (Groups 10-13 of *lepida* and Groups 15-16 of *grinnelli*) do not exhibit a clinal pattern with latitude ($r = 0.059$; $F_{(1,305)} = 0.971$, $p = 0.3252$) while eastern samples (Groups 17-23) are strongly clinal ($r = 0.498$, $F_{(1,329)} = 108.756$, $p < 0.0001$). The uniformity of western samples and the cline among eastern ones in PC-1 scores is apparent in Fig. 125, where there are no significant differences among adjacent samples of the western set of geographic groups and a clear trend towards decreasing size from eastern Utah to southwestern Arizona in the eastern group, although most of the cline is among Arizona samples south and east of the Colorado River (Groups 20 to 23).

Table 45. Coefficients of principal components (PC) analysis and canonical discriminant (CAN) analysis of 21 craniodental log-transformed variables for 13 geographic samples of the desert woodrat complex of the Eastern Desert Transect.

Variable	PC-1	PC-2	CAN-1	CAN-2
log CIL	0.924	-0.260	-0.817	-0.394
log ZB	0.863	-0.068	0.086	-0.550
log IOC	0.452	0.478	-0.225	-0.562
log RL	0.766	-0.301	0.131	0.657
log NL	0.819	-0.306	0.275	-0.597
log RW	0.748	0.219	0.172	0.175
log OL	0.805	-0.028	-0.654	0.405
log DL	0.842	-0.401	0.391	0.312
log MTRL	0.409	0.537	0.195	0.342
log IFL	0.819	-0.168	0.202	0.193
log PBL	0.814	-0.250	0.201	0.249
log AW	0.475	0.572	-0.336	0.385
log OCW	0.565	0.338	0.032	-0.040
log MB	0.598	-0.009	0.401	0.768
log BOL	0.739	0.062	-0.414	0.064
log MFL	0.667	-0.249	0.994	-0.353
log MFW	0.337	0.281	0.016	-0.056
log ZPW	0.563	0.193	-0.129	-0.053
log CD	0.810	0.117	0.292	-0.071
log BUL	0.637	0.141	-0.130	-0.619
log BUW	0.719	0.217	-0.221	0.149
eigenvalue	10.379	1.778	1.866	0.765
% contribution	49.42	8.47	42.82	17.56

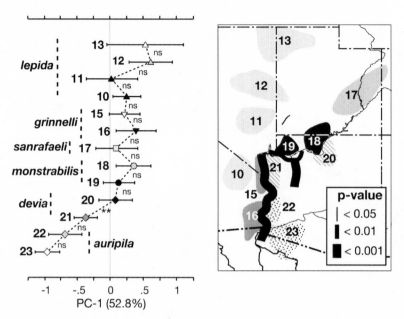

Figure 125. Mean and 95% confidence limits of PC-1 scores of craniodental variables for the 13 geographic samples of the Eastern Desert Transect (left) and map of general groupings (right). Symbols and sample numbers correspond to those in the transect map, Fig. 117. Current subspecific allocation of samples is indicated on the left. Samples are arranged geographically on the left, from north to south west of the Colorado River (localities 10-16) followed by those largely east of this river (Groups 17-23). Significance levels (based on Fisher's PLSD posterior comparison from ANOVA) between geographically adjacent areas are indicated: ns = non-significant; * $p < 0.05$, ** $p < 0.01$, *** $p < 0.001$. Significance between adjacent geographic groups is indicated by thickness of the lines on the map, right, as indicated in the inset.

The position of specimens on PC-2 is mostly influenced by the combination of Alveolar Width (AW), Maxillary Toothrow Length (MTRL), and Interorbital Constriction (IOC) contrasting with Diastemal Length (DL), Mesopterygoid Fossa Length (MFL), Nasal Length (NL), and Rostral Length (RL; Table 45). However, ellipses encompassing all individuals from each geographic group in scatterplots of PC scores are broadly overlapping, without clear spatial separation among any (Fig. 126). Thus, while mean scores are significantly different among the geographic groups included in this analysis, none of the groups is markedly different from others along any pair of planes formed by various PC axes, certainly not in the fashion that "coastal" and "desert" samples can be distinguished by PCA in the Tehachapi, San Gorgonio Pass, or San Diego Transects (Figs. 40, 63, and 79).

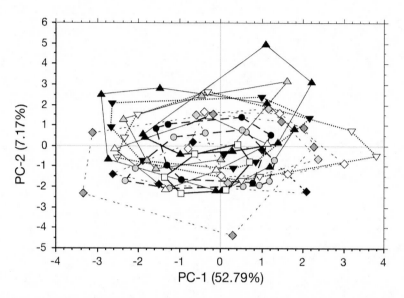

Figure 126. Ellipses encompassing all individual scores of each of the 13 geographic groups of the Eastern Desert Transect on the first two principal components axes. The four grouped localities of *N. l. lepida* are identified by upright triangles, the two of *N. l. grinnelli* by inverted triangles, the single group of *N. l. sanrafaeli* by open squares, the two groups of *N. l. monstrabilis* by circles, and the two samples each of *N. l. devia* and *N. l. auripila* by diamonds. Components were extracted from the covariance matrix of log-transformed variables; factor loadings and eigenvalues are provided in Table 44.

Despite the extensive overlap among all 13 samples of the Eastern Desert Transect in PCA scores (Fig. 126), some geographic structure on the second axis is apparent. Samples of *lepida* (Groups 10-13) are variable but overlap with those of *grinnelli* (Group 15), particularly the geographically adjacent Group 10 (*lepida*) and Groups 15 and 16 (*grinnelli*; Fig. 127). All *lepida* samples are also separable from both *monstrabilis* (Groups 18 and 19) and *sanrafaeli* (Group 17). Samples of *grinnelli* are separable from those of *devia* and *auripila* on the eastern side of the lower Colorado River, although less so for the southern set of grouped localities

(Groups 16 versus 22 and 23). The two samples of *devia* from south of the Colorado River (Groups 20 and 21) are sharply differentiated from one another, but there is no difference, or only slight differences, between either of these and their geographic counterparts of *monstrabilis* on the north side of the river. The second PC axis largely mirrors the molecular distinction of mtDNA subclade 2A (*lepida* and *grinnelli*, Groups 10-16) from subclades 2B (Groups 17-19), 2C (Groups 20-21), 2D (Group 22), and 2E (Group 23).

We investigated further the distinction among the samples of the Eastern Desert Transect by canonical variates analysis, which also allowed us to ascertain the placement of the "unknown" individuals with regard to geographically adjacent reference samples. We provide the standardized canonical coefficients for the first two axes in Table 44. The same general pattern of group relationship is apparent regardless of whether analyses are based on the 13 geographic groups as pre-defined units or whether a priori groups are based on subspecies or mtDNA subclade assignments. The geographically western samples of *lepida* (Groups 10-13) and *grinnelli* (Groups 15 and 16) overlap broadly and are virtually indistinguishable from one another, with only minimal statistical differences between any adjacent pair on either the first or second CAN axes (Fig. 128), which combine to explain 60.4% of the variation. Indeed, of the 273 specimens of these two sets of samples, 250 (or 91.6%) are correctly classified to one or the other sample with posterior probabilities above 0.90. These samples, in return, are strongly separable from all of those to the east from the upper Colorado River basin in Utah, both sides of this river in northern Arizona, and those east of the river in Arizona and Sonora. These two larger groups, which are completely separable on CAN-1 (Fig. 128, upper and lower left), correspond to the mtDNA subclade 2A (western samples) and the combination of subclades 2B through 2E (eastern samples). The eastern groups of samples, members of the two mtDNA subclades, are mostly separable from one another on CAN-2 (Fig. 128, upper, lower right), and are arranged from the northeast (Group 17, mtDNA clade 2B) to southwest (Group 23, mtDNA subclade 2E). The sample of *sanrafaeli* (Group 17) is not separable from either the geographically adjacent *monstrabilis* or from *devia* on CAN-1 ($p > 0.05$) but is strongly differentiated from both on CAN-2 ($p < 0.001$, Fisher's PLSD posterior tests). There are moderate ($p < 0.01$) to strong ($p < 0.001$) differences between each group of *monstrabilis* (Groups 18 and 19) and their respective adjacent groups of *devia* on the south side of the Grand Canyon (Group 20 and 21) on both canonical axes. Moderate differentiation ($p < 0.01$) on CAN-1 is also found between geographic Groups 21 and 22, which Hoffmeister (1986) allocated to *devia* and *auripila*, respectively, and which correspond to mtDNA subclades 2C and 2D. Finally, the two geographic groups of *auripila*, Group 22 (mtDNA subclade 2D) and Group 23 (mtDNA subclade 2E), are weakly ($p < 0.05$)

to sharply (p < 0.001) differentiated across the lower Gila River in southwestern Arizona on CAN-1 and CAN-2, respectively.

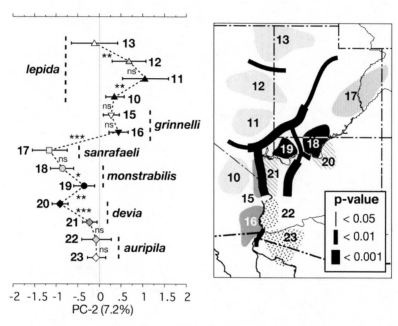

Figure 127. Mean and 95% confidence limits of PC-2 scores of craniodental variables for the 13 geographic samples of the Eastern Desert Transect (left) and map of general groupings (right). Symbols and sample numbers correspond to those in the transect map, Fig. 117. Current subspecific allocations of samples are indicated. Samples from west of the Virgin and Colorado rivers are arranged from north to south (localities 10-16) followed by the those from between these two rivers (Groups 17-19) and then those east of the Colorado River (Groups 20-23). Significance levels (based on Fisher's PLSD posterior comparison from ANOVA) between geographically adjacent areas are indicated: ns = non-significant; * p < 0.05, ** p < 0.01, ***p < 0.001.

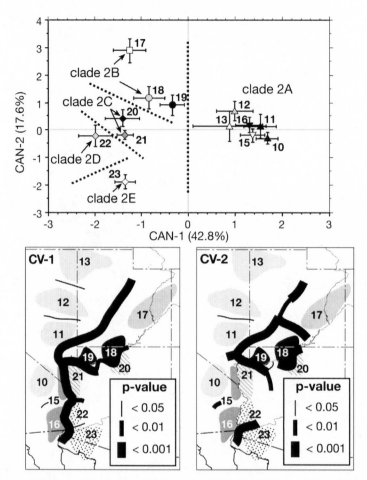

Figure 128. Above: Means and 95% confidence limits of the 13 geographic groups of the Eastern Desert Transect on the first two canonical variates axes, which combine to explain 60.4% of the total pool of variation in an analysis of 21 craniodental variables. Numbers identify sample groups (Fig. 117); samples of each mtDNA subclade are identified. Below: Patterns of differentiation among groups for CAN-1 (left) and CAN-2 (right) scores (ANOVA, Fisher's PLSD posterior test). Significance between adjacent geographic groups is indicated by line thickness separating groups, as indicated in the inset.

Overall, western desert samples (those of *lepida* and *grinnelli*) exhibit substantial uniformity in craniodental characters while those in the eastern part of this transect are markedly differentiated. The average Mahalanobis distance between western sets of samples is less than half that between eastern ones (4.38 versus 10.49, ANOVA, Fisher's PLSD critical difference = 2.907, p < 0.0001), with the mean of the latter almost as great as that between western and eastern groups (10.49 versus 13.42, Fisher's PLSD critical difference = 2.317, p = 0.0141). Hence, there is a quantitative difference in the degree of differentiation among the samples included in either of these broad geographic areas. Samples also become progressively more differentiated from north to the south, culminating with the strongly delineated groups along the lower Colorado River, in univariate craniodental variables (Fig. 118) and both multivariate principal components (Figs. 125-127) or canonical variates (Fig. 128) axes.

As with the Western Desert Transect, colorimetric variation among samples to the east is similar to the pattern described above for craniodental variables. We provide character means for the seven eastern samples of this transect in Table 46; similar data for Groups 10-13 (*lepida*) and Groups 15-16 (*grinnelli*) can be found in Table 42. Each of the four X-coefficients display highly significant differences when all 13 samples are compared (ANOVA, where $F_{(12,719)}$ > 6.686 and p < 0.0001), but with a pattern of relative homogeneity among western samples (Groups 10-13 of *lepida* plus Groups 15-16 of *grinnelli*) and substantial differentiation among these samples and the eastern groups (Group 17 to Group 23) as well as among the latter themselves. As with previous analyses, significant correlations are present between the X-coefficients for each topographic region of the skin, with Pearson correlation coefficients ranging from r = 0.488 (Dorsal-X and Tail-X; Z-value = 14.409, p < 0.0001) to r = 0.259 (Tail-X and Lateral-X; Z-value = 7.151, p < 0.0001). In general, Lateral-X exhibits the lowest correlations with other topographic regions. Not surprisingly, therefore, all four variables exhibit significant trends with geography, and with the same pattern. Color darkens coordinately around the body both from north to south and from west to east, with multiple regression coefficients using both latitude and longitude ranging from a low of 0.160 (Tail-X, $F_{(2,274)}$ = 9.546, p < 0.0001) to a high of 0.336 (Chest-X, $F_{(2,274)}$ = 46.067, p < 0.0001).

Table 46. Colorimetric X-coefficients for the four topographic regions of the study skin for geographic groups 17-23 of the Eastern Desert Transect (values for Groups 10-13 and 15-16 are provided in Table 42). Mean, standard error, sample size, and range are given for each sample.

Sample	Dorsal-X	Tail-X	Lateral-X	Chest-X
17 (*sanrafaeli*)	14.149±0.502 24 8.58-18.56	14.265±0.720 24 7.36-21.19	29.886±1.498 24 16.72-40.83	50.034±0.947 24 37.06-57.67
18 (*monstrabilis*)	11.897±0.367 58 6.24-18.96	10.448±0.376 58 4.84-18.97	31.296±0.809 58 14.75-41.65	47.756±786 58 34.41-61.71
19 (*monstrabilis*)	8.898±0.331 76 4.22-15.67	7.240±0.394 76 3.08-16.10	28.866±0.825 76 10.40-43.63	41.499±0.840 76 27.21-56.98
20 (*devia*)	10.492±0.310 65 6.35-20.03	5.668±0.252 65 2.54-11.69	27.218±0.790 65 14.05-41.27	38.853±1.015 65 23.12-53.40
21 (*devia*)	11.544±0.253 71 6.15-15.64	6.236±0.170 71 3.02-10.30	33.076±0.618 71 16.87-44.17	45.662±0.751 71 27.93-57.14
22 (*auripila*)	11.375±0.304 33 8.28-16.15	6.672±0.244 33 4.39-10.44	27.733±1.045 33 15.75-39.04	40.357±1.325 33 26.88-55.35
23 (*auripila*)	9.754±0.471 82 3.40-23.47	7.960±0.440 82 2.36-18.97	29.159±0.665 82 16.74-40.43	38.810±0.659 82 24.44-52.59

We summarize the overall, among-sample trends in color by a principal components analysis using the four X-coefficients. Of the four axes extracted, only PC-1 has an eigenvalue greater than 1.0; this axis explains 54.1% of the total pool of variation (Table 47). The pattern of character loading on each PC axis is the same as that for the analyses of other geographic areas, where all four variables load positively and nearly equally on PC-1, again suggesting that this axis expresses the degree of darkness or paleness around the entire body. PC-2, which explains 20.4% of the variation, contrasts Dorsal-X and Tail-X with Lateral-X and

Chest-X; PC-3 contrasts Tail-X and Lateral-X with Dorsal-X and Chest-X; and, finally, PC-4 pairs Dorsal-X and Lateral-X relative to Chest-X and Tail-X.

Table 47. Principal component factor loadings for colorimetric variables of adult specimens of the Eastern Desert Transect.

Variable	PC-1	PC-2	PC-3	PC-4
Dorsal-X	0.777	-0.341	-0.271	0.455
Tail-X	0.722	-0.463	0.460	-0.231
Lateral-X	0.647	0.664	0.314	0.202
Chest-X	0.789	0.213	-0.413	-0.403
eigenvalue	2.165	0.817	0.554	0.464
% contribution	54.12	20.42	13.86	11.61

The scores for each of the four PC axes are significantly heterogeneous among the sampled populations, with $p < 0.0001$ for each axis. PC-1 exhibits a highly significant relationship with the geographic position of each sample, with $r = 0.315$ (multiple regression with latitude and longitude as independent and PC-1 scores as dependent variables; $F_{(2,724)} = 36.004$, $p < 0.0001$, with the p-value for both latitude and longitude individually < 0.0001). This pattern is expected, because each X-coefficient alone exhibits similar relationships with geography and all four coefficients load equally on PC-1. Fig. 129 (left) illustrates the mean and 95% confidence limits for PC-1 scores of the 13 samples, arranged largely from north to south. Western-most samples (*lepida*, Groups 10-13, and *grinnelli*, Groups 15-16) as well as the northeastern *sanrafaeli* (Group 17) are the palest, while those from northwestern Arizona (*monstrabilis*, Group 19) and south and east of the Colorado River in Arizona (*devia*, Groups 20-21, *auripila*, Groups 22-23) are darkest. All of the latter areas, however, contain basalt flows where melanic woodrats are common (for example, the lava fields of Mt. Trumbull and the Toroweap Valley, Group 19; those north of the San Francisco Peaks, Group 20; and the Pinacate field in northwestern Sonora and adjacent Arizona, Group 23). There are no measurable color differences between any western desert samples of *lepida* (Fig. 129, right) and those of *grinnelli* along the California side of the lower Colorado River, except for Group 15, which is slightly paler ($p < 0.01$ in comparison to Group 10 of *lepida*. The sharpest differences in overall color exist between adjacent geographic groups along the Colorado River from east of the

Grand Canyon to its mouth in Mexico and secondarily along a north-south axis formed by the Kaibab Plateau.

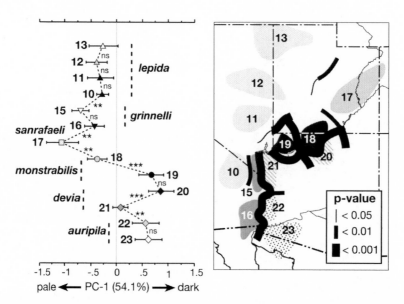

Figure 129. Mean and 95% confidence limits of PC-1 scores of colorimetric variables for the 13 geographic samples of the Eastern Desert Transect (left) and map of general groupings (right). Symbols and sample numbers correspond to those in the transect map, Fig. 117. Current subspecific allocation of samples is indicated on the left. Samples are arranged geographically on the left; significance levels (based on Fisher's PLSD posterior comparison from ANOVA) between geographically adjacent areas are indicated: ns = non-significant; ** $p < 0.01$, *** $p < 0.001$. Significance among geographic groups is also indicated by line width on the map, right, with the width equivalent to p-level (inset).

Morphology, mtDNA, and nuclear gene markers.—Each of the five subclades of the "desert" mtDNA Clade 2 are among the samples making up the Eastern Desert Transect. Two of these (subclades 2A and 2B) occur west and north of the Colorado River in California, Nevada, Utah, and northern Arizona, while the other three (subclades 2C, 2D, and 2E) are only found south and east of the Colorado River in northern and western Arizona (Fig. 6). These two sets of mtDNA groups also correspond to the chromosomal and electrophoretic groups defined by Mascarello (1978), and thus to the species *N. lepida* and *N. devia* as recognized by him and some subsequent authors (Musser and Carleton, 2005). However, as we described in the preceding section, both craniodental and colorimetric variation among pairs of these mtDNA haplotype subclades is complex, with substantial differences present for each of these datasets in some comparisons but not in others. Moreover, the distribution of tip types of the glans penis among these subclades is discordant with both molecular and other morphological traits (Fig. 30). The overall pattern of among-group differences and similarities is unlike the uniform concordance of differences in all character types observed between the "coastal" and "desert" groups and more akin to that observed in the analyses of transects within either of these broad geographic units. We have attempted to summarize these differences, or lack there of, in Table 48, using the same criteria and designations as we did above for the samples of the Baja California Transect (Table 39).

The synopsis presented in Table 48 is in general accord with the current species boundaries of these woodrats, wherein *N. lepida* (from west of the Colorado River) and *N. devia* (from east of that river) are regarded as distinct (Musser and Carleton, 2005, following Mascarello, 1978). In particular, samples from both sides of the lower Colorado River (*grinnelli* and *devia* in Table 48) are uniformly sharply divergent in craniodental size (PC-1, "size"), craniodental characters (CAN-1), and overall color (color PC-1). However, the differences between *monstrabilis* (from north of the Grand Canyon) and *devia* (to the south) are not as sharply defined, as overall size and craniodental shape either do not differ substantially or do so only at a moderate level. Only in color are there strong differences between these taxa. Indeed, the overall morphological differences between *monstrabilis* and *devia* are less than those between our two geographic samples of the latter (e.g., Groups 20 [2C-east in Table 48] and 21 [2C-west]). Again, the less-strongly defined transition between *devia* and *monstrabilis* was interpreted by Hoffmeister (1986) as evidence for intergradation and thus formed his rationale for arguing for the conspecificity of *N. lepida* and *N. devia* (sensu Mascarello, 1978). A question remaining, however, is if this apparent decrease in across-river sample discrimination in the area of the Grand Canyon, as opposed to the lower Colorado River, results simply from the clinal pattern of character variation exhibited in eastern samples but not in western ones (see above).

Mascarello, in his original 1978 study using a variety of genetic methods, did not include any samples from the crucial "transition" area between *monstrabilis* and *devia* in northern Arizona. This is particularly unfortunate because of the discordance in glans penis and limited differentiation in morphological traits described above. Mascarello's earlier study thus do not tell us if the type of chromosomal and allozymic differences that delineate samples from opposite sides of the lower Colorado River apply further to the north. We have tried to compensate for this lack of genetic data from samples across Grand Canyon region through analysis of our 18 microsatellite loci.

Table 48. Summary of morphological character differences between pairs of taxa (and mtDNA subclades) arranged as two species, *N. lepida* and *N. devia*, from west and east of the Colorado River, respectively. "no' = p > 0.05; "weak" = p < 0.05; "moderate" = p < 0.01; "strong = p < 0.001. Data from Figs. 118, 120, 121, and 122.

Comparison	Cranial PC-1 "size"	Cranial PC-2 "shape"	Cranial CAN-1	Color PC-1
within *lepida*				
lepida – monstrabilis	no	strong	strong	weak - strong
lepida – grinnelli	no	no	no	moderate
monstrabilis – sanrafaeli	no	no	no	no - moderate
between *lepida* and *devia*				
monstrabilis – devia	no - weak	no - weak	moderate - strong	strong
grinnelli – devia	strong	no - strong	strong	strong
within *devia*				
2C-east vs 2C-west	strong	strong	moderate	strong
2C – 2D	no	no	moderate	moderate
2D – 2E	no	no	weak	no

We provide the basic data for these microsatellite loci, including sample size, mean number of alleles per locus, gene diversity, levels of heterozygosity, and deviations from Hardy-Weinberg equilibrium (Fis), in Table 49. We pooled samples from eight geographic regions and mtDNA subclades, two each corresponding to subclade 2A (representing *lepida* and *grinnelli*), subclade 2B (*monstrabilis* and *sanrafaeli*), subclade 2C (*devia*, corresponding to morphological Groups 21 and 22), with single pooled samples for subclades 2D and 2E. One locus deviates from Hardy-Weinberg expectations in each of two pooled Arizona samples (2C-*devia* east and 2E); all remaining samples are in equilibrium. As in other analyses, there is a general relationship between the number of alleles and sample size (r = 0.988, Z-value = 5.733, p < 0.0001), but there is no difference in mean sample size when the samples are segregated by their geographic position relative to the Colorado River (e.g., all subclade 2A and 2B samples versus subclade 2C, 2D, and 2E; ANOVA, $F_{(1,6)}$ = 4.590, p = 0.0759). Nor is there any difference between the numbers of alleles (p = 0.0565) or gene diversity (p = 0.2695) between these broader geographic groups, although the subclade 2A and 2B samples (= *N. lepida*) are uniformly higher in all values than are those of subclades 2C to 2E (= *N. devia*).

We map the grouped sample localities listed in Table 49 in Fig. 130 and provide an unrooted neighbor-joining tree linking these 8 samples, based on a matrix of pairwise Fst values. The four samples from west of the Colorado River (*N. lepida*, mtDNA subclades 2A and 2B) differ significantly from those to the east (*N. devia*, mtDNA subclades 2C, 2D, and 2E) in average Fst value (mean Fst = 0.203; ANOVA, $F_{(1,5)}$ = 10.067, p = 0.0006), although the amount of diversity in this measure is as great within the four samples of *N. devia* (mean Fst = 0.163) as it is between the western (*N. lepida*) and eastern (*N. devia*) groups (ANOVA, Fisher's PLSD, p = 0.1607). This difference in within and between-group Fst measures is apparent in the proportional branch lengths depicted in the tree in Fig. 130. The difference, however, is due mostly to the very long-branch leading to the 2C-*devia* east sample. Overall, therefore, there is strong concordance between boundaries defined by the pooled microsatellite samples and the same boundaries delimited by mtDNA subclade membership or by morphological characters.

Table 49. Measures of diversity in 18 microsatellite loci for 8 samples of the "desert" morphological and mitochondrial groups; see Fig. 130) of the Eastern Desert Transect. Samples are identified by their mtDNA subclade, current subspecies allocation, and locality number(s) (see Appendix). Only samples identified by black circles in Fig. 130 are included in the summary statistics.

Sample (clade, locality number)	Mean N	Mean # alleles	Gene diversity	H_o	H_e	F_{is}
2A - *lepida* (CA-367, 405, NV-135, NV-142)	78.1	15.11	0.802	0.799	0.741	0.073
2A - *grinnelli* (CA-205, CA-300, CA-314)	69.6	14.22	0.759	0.819	0.777	0.052
2B - *monstrabilis* (AZ-7, AZ-16, AZ-21, AZ-28, UT-25)	16.6	6.22	0.736	0.627	0.599	0.045
2B - *sanrafaeli* (UT-31, UT-33, UT-34)	34.7	9.78	0.554	0.776	0.732	0.057
2C - *devia* east (AZ-37, AZ-47, AZ-49)	19.9	5.57	0.462	0.466	0.419	0.103[1]
2C - *devia* west (AZ-56, AZ-61, AZ-67)	12.9	5.63	0.575	0.589	0.589	0.000
2D - *auripila* (AZ-71, AZ-74, AZ-77)	25.4	7.72	0.655	0.687	0.628	0.087
2E - *auripila* (AZ-69, AZ-82, S-2, S-5)	14.1	6.38	0.756	0.655	0.546	0.171[1]

[1] significantly different from 0 at $p < 0.05$, based on bootstrapping over loci with 1000 repetitions.

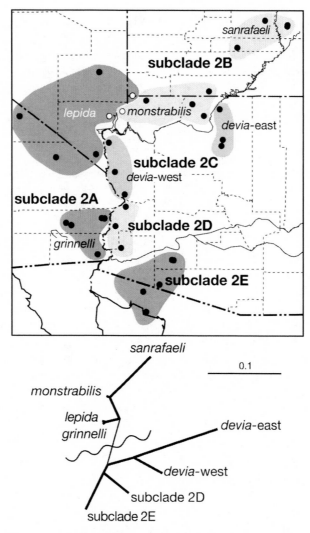

Figure 130. Map of pooled localities comprising 8 samples of desert woodrats of the Eastern Desert Transect for which data from 18 microsatellite loci are summarized in Table 49. The network on the bottom is an unrooted neighbor-joining tree linking each of these samples based on a matrix of pairwise Fst values. Branch lengths are drawn proportionally, with the scale provided in the upper right. The wavy line separates samples from west and east of the Colorado River. The three samples identified by white circles are "unknowns" (see Specimens examined, above) and were not included either in the summary statistics in Table 49 or in the construction of the tree.

We examine the concordance between geographic boundaries defined by microsatellite Fst measures and mtDNA and morphology groups more closely through the application of the assignment test option in Arlequin3 (Excoffier et al., 2005). This analysis is based on the methods of Paetkau et al. (1995) and Waser and Strobeck (1998), which determine the log-likelihood that individual multi-locus genotypes in each population actually come from that particular sample. In this case, we use the two "species", *N. lepida* (samples from west of the Colorado River) and *N. devia* (those from east of the river), as our sample groups. We plot these log-likelihood values in Fig. 131. As is clearly evident, there are two, non-overlapping clouds of points, and none of the specimens either lies on or close to the diagonal, along which an individual is equally likely to belong to either group. Importantly, if this analysis is limited solely to the samples of *monstrabilis* and *devia* on opposite sides of the Grand Canyon (Fig. 131, inset), the area where the morphological separation of samples is less sharp, individual assignments are still unambiguous. All specimens of subclade 2B from north of the Canyon cluster strongly relative to those of subclade 2C, with a greater overall degree of separation than for the total sample pool. These data are completely concordant with the mtDNA subclade membership of each of these same specimens, and the two datasets together provide no evidence of any kind for genetic intergradation, or gene flow, across the Colorado River in the Grand Canyon and Marble Canyon area, as well as further to the south. Consequently, the somewhat greater morphological similarity between *monstrabilis* and *devia* than between *grinnelli* and *devia* or *auripila*, for example, apparently does not result from differential gene flow. Rather, it most likely results from the differences in craniodental character clines as described above. The microsatellite assignments thus provide no support for Hoffmeister's (1986) hypothesis of intergradation across the Colorado River or for his disagreement with Mascarello's (1978) argument of separate species designations for desert woodrats on opposite sides of this river.

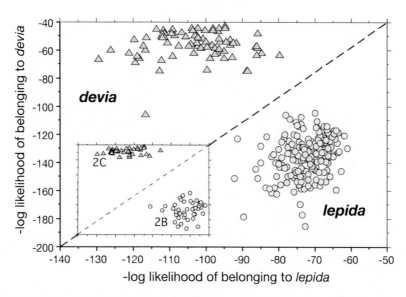

Figure 131.   Results of the assignment test for samples on both sides of the Colorado River (*devia* = mtDNA subclades 2C, 2D, and 2E; *lepida* = mtDNA subclades 2A and 2B), based on allele frequencies at 18 microsatellite loci. The negative log likelihood of each individual belonging to its own species / mtDNA subclade group is plotted against the negative log likelihood of it belonging to the other species / mtDNA subclade group.

Comparison within east or west sides of the Colorado River.—This transect encompasses the entire range of nine formally described taxa (*grinnelli, sanrafaeli, monstrabilis, devia, auripila, bensoni, flava, aureotunicata,* and *harteri*) and part of the range of a 10th (*lepida*). Here, we address the status of each of these through separate analyses that focus on paired transitions. Because we have reviewed evidence for the alternative hypotheses of the species status of those taxa divided by the Colorado River immediately above, here we focus solely on the uniqueness of these infraspecific taxa in four separate analyses: (1) between mtDNA subclades 2A (*lepida* and *grinnelli*) and 2B (*monstrabilis*); (2) between *sanrafaeli* in the upper Colorado River basin and *monstrabilis* to the immediate south (comparisons between two samples of mtDNA subclade 2B); and (3) between samples of *devia* and *auripila* on the east side of the Colorado River (between mtDNA subclades 2C, 2D, and 2E). Because Group 23 of this latter set of comparisons includes samples of five named forms (*auripila, bensoni, flava, aureotunicata*, and *harteri*), we also include an analysis (4) that involves these sets of taxa as a final comparison.

(1) subclades 2A and 2B: *lepida* and *grinnelli* versus *monstrabilis*.—These subclades contact one another along the Virgin River in southeastern Nevada, northwestern Arizona, and southwestern Utah (Fig. 132). To examine this transition, we performed two separate canonical analyses, one using the three groups geographically adjacent to the Virgin River (Group 11 [*lepida*] and Group 15 [*grinnelli*] west of the Virgin River and Group 19 [*monstrabilis*] to the east; Fig. 117) and a second using mtDNA subclades 2A or 2B as reference samples. We assigned each "unknown" specimen (localities AZ-5 to UT-14) to these a priori groups by their respective posterior probabilities. The results of both analyses were identical, although the analysis based on mtDNA subclade with only two reference samples resulted in higher posterior probabilities of group membership for each "unknown" specimen. Localities on either side of the Virgin River align almost perfectly with the geographic group(s) or mtDNA subclade on that side, as most individuals from north and west of the Virgin River were assigned to either Group 11 or Group 15 (mtDNA subclade 2A) and those from east and south of the river to Group 19 (mtDNA subclade 2B; Fig. 132). There are two exceptions to this overall pattern. The first of these are localities from the vicinity of St. George, where individuals from some sites immediately north of the river (e.g., localities UT-16 and UT-17) were assigned to Group 19 to the south while a single individual from Fort Pierce Wash (locality UT-19) on the south side was assigned to Group 11. The second exception is locality NV-139 (west slope Virgin Mts., Clark Co., Nevada), where six of the eight individuals were placed in Group 19 while two were assigned to Group 15. Each of these eight specimens is of the eastern mtDNA subclade 2B, which characterizes Group 19. Thus, despite limited discordance in an individual's morphological and molecular characteristics, the Virgin River apparently does mark a real boundary between both morphological and molecular geographic units of desert woodrats, although the separation is not absolute.

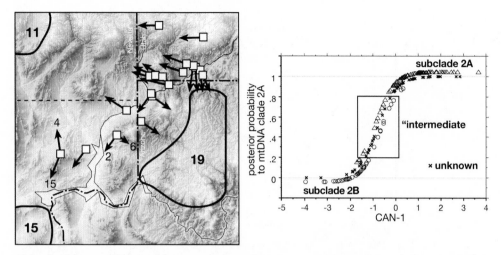

Figure 132. Left – Assignments of "unknown" samples (arrows) from localities along the Virgin River (open boxes) to Group 11 (*lepida*), Group 15 (*grinnelli*), and Group 19 (*monstrabilis*), based on posterior probabilities from the discriminant analysis of 21 craniodental variables. Individuals from two localities are assigned to more than one reference group; the numbers so assigned are indicated. Right -- Plot of the probability of membership to subclade 2A (*lepida*) and 2B (*monstrabilis*) relative to the score of that individual on the first CV axis. The box encloses "intermediate" individuals (those with probabilities between 0.2 and 0.8). The reference samples have been offset above and below the 1 and 0 probability lines, respectively, to improve visibility of their respective separation.

The pattern of posterior probabilities for individuals of each reference sample and the "unknowns," however, also documents a level of morphological intermediacy that suggests gene flow between western and eastern groups in the transition area represented by the Virgin River basin. We illustrate the degree of intermediacy of both reference samples and "unknown" individuals by plotting an individual's probability of assignment to mtDNA subclade 2A to its CAN-1 score (Fig. 132, right). Most (42 of 57, or 73.7%) of the "unknown" specimens have relatively high posterior probabilities of group assignments (p > 0.80), with the remaining 15 individuals exhibiting intermediate probabilities. Moreover, 23 of the 138 (16.7%) of those specimens from the reference samples also exhibit intermediate probabilities (between 0.2 and 0.8), whether the reference groups are defined a priori as the geographic Groups 11, 15 and 19 or as the two mtDNA subclades. This pattern of overlap in reference samples and the large number of intermediate "unknown" individuals in this analysis is markedly different from that of the contact areas between the "coastal" and "desert" morphological groups in California (Tehachapi [Fig. 45], Cajon Pass [Fig. 54], and San Gorgonio Pass [Fig. 64] transects) where there is both no intermediacy among reference sample individuals and no, or very few, intermediate "unknowns." The morphological suggestion of gene flow between subclade 2A and subclade 2B across the Virgin River is supported by the complete lack of separation of these two groups in the limited pool of microsatellite data we have for this region. The assignment test we performed above for the pooled samples on either side of the Colorado River, for example, fails to differentiate subclades 2A and 2B (ANOVA, $F_{(1, 178)} = 1.684$, p = 0.1961).

(2) upper and middle Colorado River: *monstrabilis* versus *sanrafaeli*.—In this analysis we include only geographic Group 17 (*sanrafaeli*) and Groups 18 and 19 (*monstrabilis*) from northern Arizona and southern Utah. Kelson (1949), in his description of *sanrafaeli*, compared that taxon to *monstrabilis*, which he considered the "nearest subspecies, geographically and morphologically" (p. 418). He noted that *sanrafaeli* averaged much lighter in overall color, although some specimens of *monstrabilis* were as pale. Both of these statements are certainly true (Fig. 129). Even though our two samples of *monstrabilis* are sharply different, the eastern Group 18 is intermediate in color between *sanrafaeli* and the darker western Group 19. Moreover, the degree of overall darkness (as indexed by PC-1 scores; Table 47) varies strongly from west to east from Group 17 to Group 19 individual localities (r = 0.502, $F_{(2,155)} = 19.745$, p < 0.0001, in a multiple regression of PC-1 scores against geographic position based on the independent locality latitude and longitude). This largely clinal shift likely results from background matching as lava fields become progressively more common west of the Kaibab Plateau in northwestern Arizona, culminating in the very dark individuals from the Toroweap Valley, vicinity of Mt. Trumbull, and Mokaac Wash in Mohave Co. (localities AZ-7, AZ-13 to AZ-16).

Cranially, Kelson described *sanrafaeli* as being larger in all dimensions, except braincase breadth, with a longer palatal bridge and both a longer and wider maxillary toothrow. Among our sample Groups 17, 18, and 19, however, Palatal Bridge Length (PBL) does not differ significantly (ANOVA, $F_{(2,96)} = 1.348$, p = 0.2647), although these samples do differ in both MTRL ($F_{(2,96)} = 11.233$, p < 0.0001) and AW ($F_{(2,96)} = 7.049$, p = 0.0014), with *sanrafaeli* (Group 17) larger than either sample of *monstrabilis* (Groups 18 and 19; Table 44). These differences are, again, clinal in nature, as the regression of individual values on both latitude and longitude of sample localities is significant (MTRL: r = 0.433, $F_{(2,95)} = 10.984$, p < 0.0001; AW: r = 0.393, $F_{(2,95)} = 8.674$, p = 0.0003).

The strong clinal pattern of differences among samples of *monstrabilis* and *sanrafaeli* is apparent from the distribution of individual scores resulting from a canonical variates analysis. For example, although mean CAN-1 scores for *sanrafaeli* specimens are significantly different from those of either *monstrabilis* sample (ANOVA, $F_{(2,96)}$ = 50.178, p < 0.0001), these scores overlap broadly with those of the geographically adjacent Group 18 of *monstrabilis* (Fig. 133). Moreover, these CAN-1 scores are strongly correlated with the geographical position of each individual locality (r = 0.669, $F_{(2,95)}$ = 41.432, p < 0.0001, in a multiple regression of CAN-1 scores against geographic position based on locality latitude and longitude). There is also no apparent step in this cline between Group 18 (*monstrabilis*) and Group 17 (*sanrafaeli*), as fitting non-linear curves to the relationship between geographic position and CAN-1 scores does not provide any significant increase to the relationship defined by linear analyses (p > 0.05 of slopes in all curvilinear to linear comparisons).

Overall, therefore, *sanrafaeli* Kelson appears only weakly differentiated from the samples of *monstrabilis* Goldman, with the few craniodental and color character differences varying along a relatively smooth cline from northeast to southwest across their respective ranges.

Finally, we comment on the specimens from Marysvale (locality UT-23) and Loa (locality UT-30) in central Utah (Fig. 133), which we placed in the "unknown" category. As these localities are intermediate in their geographic positions between our Group 17 (*sanrafaeli* Kelson), Group 18 (*monstrabilis* Goldman), and Group 12 (*lepida* Thomas), we performed a canonical variates analysis restricted to these three geographic groups as reference samples and assigned each of the "unknown" specimens accordingly. All five specimens (four from locality UT-23 and one from locality UT-30) are assigned to Group 18 (*monstrabilis* Goldman), with posterior probabilities > 0.908 in four cases and more intermediate values in the other two (0.624 and 0.680). Secondary assignments of these two specimens are to Group 12 (*lepida* Thomas). Importantly, all six specimens are excluded from membership in Group 17 (*sanrafaeli* Kelson), which is geographically closest, as the highest posterior probability to that group exhibited by any one specimen was only 0.0004. If one were to recognize *monstrabilis* Goldman as a taxon separate from *lepida* Thomas then its range should be expanded north to include Marysvale and Loa.

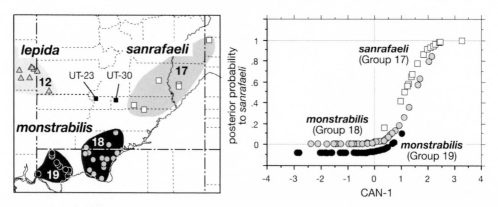

Figure 133. Left – Group samples of *N. l. lepida* (Group 12), *N. l. monstrabilis* (Groups 18 and 19), and *N. l. sanrafaeli* (Group 17) in eastern Nevada, southern Utah, and northern Arizona. Specimens from localities UT-23 and UT-30 are unassigned to subspecies. Right – Plot of the posterior probability of membership to *N. l. sanrafaeli* (Group 17) for each specimen of this subspecies and that of *N. l. monstrabilis* (Groups 18 and 19) relative to the score of that individual on the first CV axis. Symbols for Group 19 have been offset from the "0" line to improve visibility of the distribution of specimens in each group.

(3) east side of lower Colorado River: *devia* versus *auripila*, including mtDNA subclade 2C, 2D, and 2E.—Hoffmeister (1986) concluded that specimens from approximately south of the Bill Williams River in western Arizona belonged to a subspecies (*auripila*) separate from those to the north of that river (*devia*), distinguishing these two races primarily by size. The geographic break between Hoffmeister's subspecies generally corresponds to that between our mtDNA subclades 2C (to the north) and 2D + 2E (to the south). Here, we evaluate the degree of differentiation from north to south among our samples from western Arizona (Groups 21, 22, and 23), grouping these by subspecies (Group 21 = *devia*; Groups 22 + 23 = *auripila*) and separately by the three mtDNA subclades (Group 21 = subclade 2C, Group 22 = subclade 2D, and Group 23 = subclade 2E).

Analyses using all 13 samples of the Eastern Desert Transect detected only minimal divergence among these samples in western Arizona, in either univariate (Fig. 124) or multivariate principal components or canonical variates analyses (Figs. 125-128) of craniodental variables or PC analysis of color (Fig. 129). This overall pattern is confirmed by analyses restricted to the three samples on the east side of the lower Colorado River. Specimens from the Group 21 sample (*devia*) in the north (Fig. 117) are larger in size than those of Group 23 (*auripila*) in the south, but there is a gradual cline in size through this sample area (Fig. 134) without any obvious break, or step, at either the approximate boundary between the two races as drawn by Hoffmeister (1986, Map 5.90, p. 410) or between each of the three mtDNA subclades. There is a highly significant relationship between individual PC-1 scores (overall size) and latitudinal position along the lower Colorado River ($r = 0.353$, $F_{(1,177)} = 22.126$, $p < 0.0001$). Thus, although the southern-most and northern-most samples are significantly different in overall size, as indexed by PC-1 scores, there is no obvious break in size at the boundary between Hoffmeister's mapped ranges of *devia* and *auripila* (between our Groups 21 and 22, Fig. 128).

Nevertheless, the three samples on the Arizona side of the lower Colorado River are separable from one another in a canonical variates analysis of craniodental characters (Fig. 135). Group 23 (southern *auripila*) is nearly non-overlapping with Groups 21 and 22 on CAN-1 (ANOVA, Fisher's PLSD, $p < 0.0001$ in comparison to both Groups 22 and 21). This axis explains 76.7% of the variation and the placement of individuals is influenced most by Condyloincisive Length (logCIL), Rostral Length (logRL), and Braincase Breadth (logMB). Groups 21 (*devia*) and 22 (*auripila*, north) are not separable from one another on CAN-1 ($p = 0.0771$), but all three groups differ in CAN-2 scores ($p < 0.0001$ in each pairwise comparison). Consequently, although the majority of the variation in the canonical analysis is not concordant with Hoffmeister's placement of a subspecies boundary at approximately the Bill Williams River (between Groups 21 and 22), our three geographic groups, and thus the three mtDNA subclades, can be distinguished from one another in craniodental multivariate space when comparisons are limited to western Arizona alone. We included in our analysis the four "unknown" specimens from locality AZ-68 (New River Valley, 30 mi NW Phoenix), which represent the eastern-most limit in the range of desert woodrats in Arizona, south of the Mogollon Rim. This locality is well separated geographically from all localities further to the west (Fig. 135). Hoffmeister (1986) allocated these specimens to his concept of *auripila* (our Groups 22 and 23, combined). In our analysis, however, these fall in an intermediate position between all three of our geographic Groups (Fig. 135), with their posterior probabilities assigning two specimens to *devia* (Group 21) and one each to the two *auripila* Groups 22 and 23. It is, thus, not possible to assign the sample from New River Valley as a whole to either of the subspecies recognized by Hoffmeister.

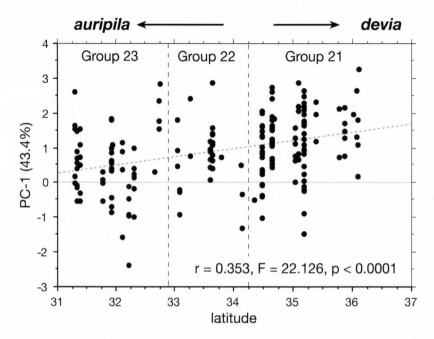

Figure 134. Linear regression of individual craniodental PC-1 scores against the latitudinal position of each separate locality for the pooled samples Group 21, Group 22, and Group 23 along the Arizona side of the lower Colorado River. Regression coefficients, F-value, and probability for each relationship are given. Vertical dashed lines separate scores for each geographic group, and the allocation to subspecies, as defined by Hoffmeister (1986), is indicated above each set of plots.

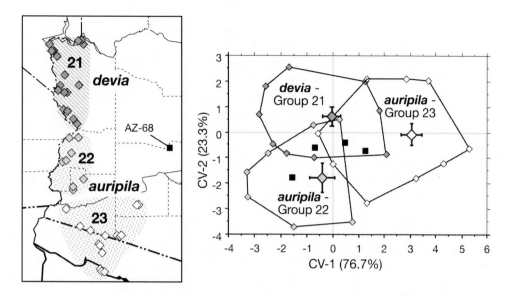

Figure 135. Left – Group samples of *devia* (Group 21) and *auripila* (Groups 22 and 23) from western Arizona. Specimens from locality AZ-68 are unassigned to subspecies. Subspecies designations follow Hoffmeister (1986). Right -- Scatterplot of CAN-1 and CAN-2 scores in a canonical analysis comparing *devia* (Group 21) and *auripila* (Group 22 and 23). Ellipses enclose all points for each group/subspecies, with the group centroid and 95% confidence limits along both planes indicated. The solid squares represent the positions of the four "unknown" specimens from locality AZ-68 (New River Valley, 30 mi NW Phoenix).

(4) *auripila, bensoni, flava, aureotunicata,* and *harteri* (Group 23).—The analyses we summarized above involved comparisons among the major geographic groups (= taxa) and mtDNA subclades currently recognized for the eastern desert portion of the range of the *Neotoma lepida* group. Since our analyses required grouping individual localities into larger geographic units for comparison, to this point we have provided no evaluation of formally named entities that might be included within any one of our sample groups. In most cases, each of our groups contains only a single taxon, based on the inclusion of type localities (Group 15 [*grinnelli*, locality CA-210, Western Desert Transect], Group 17 [*sanrafaeli*, locality UT-34], Group 18 [*monstrabilis*, locality AZ-21], and Group 20 [*devia*, locality AZ-50]). However, Group 23 includes the type localities of five taxa (*auripila* [locality AZ-84], *bensoni* [locality S-2], *flava* [locality AZ-79], *aureotunicata* [locality S-6], and *harteri* [locality AZ-69]), although Hoffmeister

(1986) had placed the latter four as synonyms of *auripila*. We conclude this section by comparing these geographic components of our Group 23, as we have for comparisons between the pooled geographic groups themselves. Our samples are restricted to the type series and/or topotypes of each subspecies (Fig. 136).

Each of these five races was diagnosed primarily on the basis of overall color, with few mean differences in craniodental characters mentioned in any description. Blossom (1933) noted that *auripila* had a smaller skull than *devia*, with a narrower braincase and relatively larger bullae, and he later (1935) defined *bensoni* as darker than *auripila* with a narrower Interorbital constriction and shorter maxillary toothrow. Benson (1935) contrasted his *flava* with Blossom's *auripila*, noting its more pallid color and smaller size. Finally, Huey (1937) characterized his *aureotunicata* by its bright buffy color and a slightly longer molar toothrow than either *auripila* or *flava*, and his *harteri* by its darker coloration and overall larger size, although he regarded it cranially as close to both *auripila* and *flava*.

Craniodental differences are slight among these five taxa. From a principal components analysis, *harteri* is larger than the other taxa, with *aureotunicata* next in size and *auripila* the smallest (PC-1 explains 43.6% of the variation; all variables load positively and highly, with the factor loading for logCIL = 0.936; data not shown). CIL or PC-1, as measures of overall size, are also strongly related to geographic position, using the latitude and longitude of each locality in a multiple regression as independent variables (CIL: $r = 0.467$, $F_{(2,56)} = 7.820$, $p = 0.0010$; PC-1: $r = 0.434$, $F_{(2,56)} = 6.504$, $p = 0.0029$). Samples are smaller in the west (*flava*) and become larger to the east (*harteri*). Subsequent PC axes individually explain no more than 9% of the total variation, and although significant differences do exist among taxa along PC-2 and PC-3, in no pairwise taxon comparison (using Fisher's PLSD posterior comparison) is the significance of difference less than $p = 0.05$. A clinal pattern is also apparent in a canonical analysis, when CAN-1 scores are regressed on the geographic position of localities in a multiple comparison using both latitude and longitude ($r = 0.671$, $F_{(2, 56)} = 64.263$, $p < 0.0001$). We conclude, therefore, that craniodental variation is slight and that the differences present are largely clinal.

The color differences noted by earlier authors are certainly correct, however, as simple visual comparisons of study skins or the quantitative measurements of color at the four topographic points on the body (dorsum, tail, lateral, and chest) attest. In a principal components analysis of the colorimetric X-coefficients for each topographic area, the first axis was equally influenced by each variable (individual loadings ranged from 0.717 to 0.886) and explained 68.3% of the total amount of variation present. Specimens aligned along this axis from palest (*flava*) to darkest (*bensoni*), with the other taxon samples intermediate (Fig. 136). These two taxon samples are significantly different from each other and from the other three in all pairwise comparisons; *auripila, aureotunicata,* and *harteri,* however, are not. Note in particular that some individuals of each race, with the exception of *flava,* are as dark as the melanic *bensoni* from the Pinacate lava flows. This is particularly true for our specimens of *aureotunicata,* which Huey (1937:349) characterized by its "very bright buff color." In fact, specimens from the vicinity of Puerto Peñasco on the Sonoran Gulf coast exhibit the broadest range of color of any taxon, from very pale to quite dark. Thus, while *flava* is paler than the others, as Benson (1935) noted in his description of this race, *harteri* is not darker than *auripila* to which Huey compared it. Curiously, although *harteri* and *aureotunicata* were described on successive pages in the same publication, Huey (1937) made no comparison between them.

The pattern of morphological variation expressed by this group of taxa is one of minimal craniodental differentiation, largely expressed as a cline from west to east. Color does vary substantially from pale to dark, but this variation is to be expected given the propensity for desert woodrats in this area to occupy slopes that are composed of either very pale granites and conglomerates or dark basaltic lavas (Figs. 121, 122, and 123). Restriction to rocky outcrops separated by intervening alluvial valleys of sandy desert harboring White-throated Woodrats (*Neotoma albigula*) also creates an insular distribution among local desert woodrat populations that likely promoted local differentiation of the kind we observe here. We address the taxonomic consequences of the patterns of variation below.

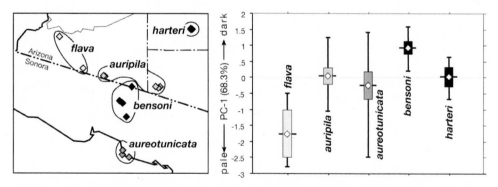

Figure 136. Upper – map of southwestern Arizona and northwestern Sonora, with individual localities of each of the five subspecies of desert woodrats described for the region plotted. Bottom – plot of the mean (horizontal lines and diamonds), range (vertical lines), and 95% confidence limits (boxes) of PC-1 scores based on the four X-coefficient colorimetric variables. Samples are arranged quasi-geographically, from west to east, and are distributed along the PC axis from palest (*flava*) to darkest (the melanic *bensoni*).

EVOLUTIONARY HISTORY OF THE *Neotoma lepida* GROUP

AGE OF LINEAGE DIVERSIFICATION

Woodrats of the *Neotoma lepida* group are known from a large number of localities of late and post-Pleistocene age from the midden record of western North America (reviewed in Betancourt et al., 1990). The FAUNMAP database (http://www.museum.state.il.us/research/faunmap/), for example, includes more than 40 late Wisconsin and Holocene sites from the United States. In contrast to this rich and relatively recent record, however, there are very few data for *Neotoma lepida* from earlier periods in the Pleistocene. The Paleobiology Database maintained by John Alroy (http://paleodb.org/) lists a single record, the Emory Burrow Pit locality in Orange Co., California dated as Irvingtonian to Rancholabrean in age (1.8 to 0.011 Ma); the University of California Museum of Paleontology database (http://bscit.berkeley.edu/ucmp/) lists a second Rancholabrean record, the Sternberg Pit locality in Kern Co, California (0.3 to 0.011 Ma); and the San Diego Natural History Museum's paleontology collection database (http://www.sdnhm.org/) likewise includes a single specimen, from an unnamed stream terrace in San Diego Co., also of Rancholabrean age. Given the paucity of fossils with ages older than about 40 Ka and the lack of precise dating for the few older records, the fossil record itself is of little use in establishing the timing of lineage diversification within the complex that we have uncovered by both mtDNA and nucDNA sequences (see above). We have thus employed a molecular-based approach to generate a hypothesis of lineage ages.

Arbogast et al. (2002) review the multiple difficulties in estimating divergence times from DNA sequences, on both phylogenetic and population genetic time scales. We acknowledge that two of the most critical issues regarding these estimations (single sequence data and recently separated taxa) apply to our data for the *Neotoma lepida* group and thus caution the reader that the hypotheses presented below are given only as a most general approximation and that they await confirmation by additional multiple sequence data.

We used the program RRTree, version 1.0 (Robinson-Rachavi and Huchon, 2000) to perform relative rate tests between each pair of mtDNA *cyt-b* clades and subclades to determine if the pattern of base substitution has behaved in a clock-like fashion. Constancy of rates is a necessary requirement if molecular dating is to be based on a molecular clock, regardless of the actual rate at which the clock ticks. In our case, we performed tests using the 1143 bp *cyt-b* dataset on the number of synonymous substitutions per synonymous site, the number of non-synonymous substitutions per non-synonymous site, and the number of synonymous transversions per fourfold degenerate site among all sequences. We included sequences of other species of *Neotoma* as an outgroup in the analysis (see above). In each analysis, a hypothesis of clock-like behavior during sequence diversification among these lineages of woodrats could not be rejected: synonymous sites (p = 0.839), non-synonymous sites (p = 0.954), and totally degenerate sites (p = 0.655).

Given that clock-like behavior of the mtDNA *cyt-b* sequences for the *Neotoma lepida* group could not be rejected, we estimated divergence dates based on 3rd position transversions, following the arguments of Irwin et al. (1991) and Smith and Patton (1993). However, as proposed by Edwards (1997; see also Edwards and Beerli, 2000), we corrected the genetic distance between each pair of clades for ancestral polymorphism using the formula $P_{net} = P_{AB} - \frac{1}{2} (P_A + P_B)$, where P_{net} is the corrected distance between clade A and clade B, P_{AB} is the mean genetic distances in pairwise comparisons of individuals from A versus B, and P_A and P_B are the mean genetic distances among individuals within each of these two clades. We used the Kimura 2-parameter distance and a rate of 1.7% per Ma sequence divergence based on 3rd position transversions (Smith and Patton, 1993). This rate estimate is derived from the split between *Mus* and *Rattus* estimated at 10 Ma, a date that is at the deeper end of the 10.3 to 8.8 Ma range of divergence dates for this taxon pair estimated from multiple nuclear genes by Steppan et al. (2004). The more recent divergence date of 8.8 Ma estimated for the *Mus-Rattus* split would yield a rate of 1.93%.

We provide estimated divergence dates and their standard errors (based on 500 bootstrap replicates implemented in MEGA3; Kumar et al., 2004) for each internal node in the clade phylogeny in Table 50. Because the relationships among subclades 2C, 2D, and 2E (from Arizona east and south of the Colorado River) are uncertain (Fig. 5), we treat these three as an unresolved trichotomy and report the average distances and time based on each possible set of comparison (e.g., 2C vs. 2D + 2E, 2D vs 2C + 2E, and 2E vs 2C + 2D). Note that the estimates given in Table 50 would be decreased by about 12% if a rate of 1.93%, based on a *Mus-Rattus* divergence of 8.8 Ma, were used for the calculation.

Table 50. Kimura 2-parameter distances (based on 3rd position transversions only) at internal nodes of mtDNA *cyt-b* tree, corrected for ancestral polymorphism (see text) and estimates of divergence dates (in Ma, or millions of years) derived from an estimated *Rattus-Mus* divergence of 1.7% per million years (see Smith and Patton, 1993).

Node	mean K2p ± SE	time (Ma) ± SE
Clade 1 vs Clade 2	2.7126 ± 0.8278	1.596 ± 0.487
within Clade 1		
1D vs 1A+1B+1C	1.3692 ± 0.5684	0.805 ± 0.334
1A vs 1B+1C	0.3978 ± 0.1484	0.234 ± 0.087
1B vs 1C	0.3141 ± 0.1908	0.185 ± 0.112
within Clade 2		
2A+2B vs 2C+2D+2E	0.8518 ± 0.3883	0.501 ± 0.228
2A vs 2B	0.0945 ± 0.0911	0.061 ± 0.033
2C vs 2D vs 2E	0.5039 ± 0.2206	0.296 ± 0.130

We illustrate the pattern and timing of diversification of the major lineages of the *Neotoma lepida* group in Fig. 137. Estimated divergence times (Table 50) range from an average of 1.6 Ma for the separation between Clade 1 and Clade 2, to approximately 61Ka for the Clade 2 subclades 2A and 2B. Not surprisingly, the estimated errors around each node are substantial, but at least the mean dates are consistent with the very limited fossil record and support the hypothesis that the base of the *Neotoma lepida* group is within the early Pleistocene. Importantly, the derivation of the subclade 1D (*insularis*) from Isla Ángel de la Guarda in the north-central Gulf of California, at approximately 800 Ka, is older than other subclade divergences within the complex. The differentiation between the two subclade clusters within Clade 2 that are separated by the Colorado River (e.g., subclades 2A and 2B versus subclades 2C, 2D, and 2E, or *N. lepida* versus *N. devia*) is estimated at 500 Ka, at a time well before divergences among the three continental subclades of Clade 1. Finally, with the exception of the desert subclades 2A and 2B, which appear to have had a quite recent divergence (ca. 61 Ka), all other diversification events, on average, are positioned within the Middle Pleistocene or within the Rancholabrean land mammal age (Fig. 137).

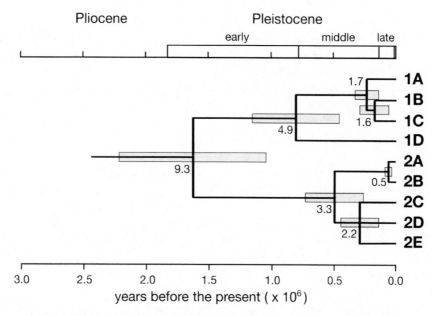

Figure 137. Neighbor-joining tree of mtDNA clades based on Kimura 2-parameter distances derived from 3rd position transversions from the complete (1143 bp) *cyt-b* dataset. Nodes are drawn consistent with the mean estimates of divergence dates given in Table 50, with one standard error estimates on either side of the mean indicated by the gray boxes. Arizona subclades 2C, 2D, and 2E are shown as an unresolved trichotomy, since the MP and Bayesian analyses (Fig. 5) hypothesize alternate topologies for their relationships. The mean numbers of 3rd position transversions among terminal branches stemming from each node are indicated.

COALESCENT HISTORY WITHIN CLADES

Times of divergence for internal nodes in the phylogenetic diversification of the lineages of woodrats of the *Neotoma lepida* group estimated by the application of a molecular clock ticking at a constant rate can be verified, in part, by examining the temporal depth of the sets of haplotypes contained within each clade and subclade. To do this, we employ coalescent methodology that is independent of the phylogenetic dating above. These analyses also allow us to examine the demographic history of each subclade, distinguishing long-term geographic and temporal stability from either population expansion or contraction. These types of analyses, however, are not without inherent problems. Although the smaller effective population size of mtDNA as compared to nuclear loci makes mtDNA a particularly useful marker for reconstructing species histories within a statistical phylogeographic framework (Templeton, 1998; Knowles and Maddison, 2002; Knowles, 2004), multiple loci are essential to determine the evolutionary significance of past demographic and biogeographic events. Thus, we offer the following analyses and interpretations under the important caveat that they are based on mtDNA sequences alone, our *cyt-b* data.

Coalescent Approaches

The earlier assessment of the history of mtDNA clades within the *Neotoma lepida* group (Patton and Álvarez-Castañeda, 2005), based on Tajima's (1989) test of selective neutrality and the pattern of pairwise haplotype differences (the "mismatch distribution" [Rogers and Harpending, 1992]), suggested temporal stability of each subclade of the coastal Clade 1 as well as the desert Clade 2 subclade 2C, but a history of recent expansion for the desert subclades 2A and 2B. Because we have substantially expanded the current dataset both with respect to numbers of sampled localities and sample sizes for each of these subclades, and recovered additional subclades within the desert Clade 2, we revisit the population history of each clade and subclade using both qualitative and quantitative analytical methods. As we detail below, this expanded dataset provides a slightly revised view of the respective clade histories. In the analyses that follow, all computations were performed using Arlequin3 (Excoffier et al., 2005) and the 801 bp dataset for the *cyt-b* gene and thus include all specimens and localities we have examined.

We employ both Tajima's D (Tajima, 1989) and Fu's Fs (Fu, 1997) to determine if there were deviations from neutral expectations in any of the samples analyzed, with significance assessed by using 1000 random permutations. The latter method is particularly sensitive to demographic expansion. Non-significant values suggest that evolution has been relatively independent of positive selection, heterogeneity of mutation rates, or major population perturbations during the coalescent history of the included sequences in the particular sample. Alternatively, significantly negative values suggest either a recent selective sweep (or other deviations from strict neutrality) or recent population expansion (see, for example, Aris-Brosou and Excoffier, 1996, with regard to Tajima's D).

We provide D and Fs values for clades and subclades in Table 51. All three subclades in Clade 1 and subclades 2A and 2B in Clade 2 exhibit significantly negative values for both measures. None of the Arizona subclades of Clade 2 (subclade 2C, 2D, or 2E), however, are significant for either measure. These results suggest that subclades 1A, 1B, 1C, 2A, and 2B have experienced either a recent selective sweep or population expansion, while subclades 2C, 2D, and 2E have been stable over their respective coalescent histories. However, because each of the latter three subclades is poorly sampled, additional data might indicate a different history. For example, the greatly expanded sampling of each subclade in Clade 1 now supports population expansion, a finding contrary to the stability originally posited by Patton and Álvarez-Castañeda (2005).

Although we cannot ignore the possibility of a selective sweep underlying those values for Tajima's *D* or Fu's *Fs* that are significantly negative, such measures are usually interpreted as indicating a history of population expansion (Hein et al., 2005). We thus examined the qualitative pattern of the coalescent history of each mtDNA clade and subclade through use of the mismatch distribution, the distribution of pairwise sequence differences among all haplotypes being compared. This distribution is expected to be multimodal for populations at demographic equilibrium, due to the highly stochastic nature of gene trees, but unimodal in those that have experienced a recent expansion. Moreover, expansion itself can result from several different historical processes, such as demographic expansion within populations (Rogers and Harpending, 1992; Slatkin and Hudson, 1991; reviewed in Harpending and Rogers, 2000) or range expansion with high levels of migration between neighboring demes (Excoffier, 2004; Ray et al., 2003). The type of expansion at the population level (demographic or range), the rate of expansion (explosive or exponential), and the size of the ancestral population prior to expansion all affect the pattern of haplotype diversity observed at any one time subsequent to the expansion. We used the program Arlequin3 (Excoffier et al., 2005) to calculate mismatch distributions under two demographic models, one of pure demographic expansion, wherein a stationary haploid population suddenly undergoes an increase, and spatial expansion in a 2-dimensional stepping-stone model, wherein the range of a population increases over time and over space. We employ the goodness-of fit-test, based on 500 bootstrap replicates, to assess the adequacy with which either expansion model can explain the empirical mismatch distribution.

We illustrate mismatch distributions for Clade 1 and subclades 1A, 1B, and 1C, in Fig. 138, and those for Clade 2 and subclades 2A and 2B, in Fig. 139. We present the distribution of pairwise differences for each group at the same scale to simplify comparative visualization. The distributions for both Clade 1 and Clade 2 are multimodal, as expected as each pools separate reciprocally monophyletic subclades. Note, however, that the main peak in the Clade 1 distribution (Fig. 138) is positioned well to the right of that of Clade 2 (Fig. 139), reinforcing the differences between the two clades in average pairwise divergence (Table 51) and the greater depth of among-subclade divergences in Clade 1 (Table 3). Overall, sudden expansion models provide a relatively poor fit to the empirical pairwise distribution for Clade 1 (p-value of goodness of fit = 0.322) and the spatial expansion model can be rejected (p = 0.026). This contrasts with Clade 2, where neither expansion model can be rejected (spatial expansion, p = 0.678; sudden expansion, p = 0.400).

The mismatch distributions for each subclade in Clade 1 are erratic, with a left-side 'shoulder' to the distribution for subclade 1B and a tendency for a similar shoulder for subclade 1A (Fig. 139). Only a sudden expansion model fits the observed data for subclades 1A and 1B (p = 0.906 and 0.890, respectively) while a spatial expansion model best fits the distribution for subclade 1C, although weakly (p = 0.396). Thus, subclades 1A and 1B appear to have had a different history of expansion than subclade 1C and, based on mean pairwise differences for each subclade (Table 51), expansion in all three has been at different times in the past (see below, and Table 52).

Mismatch distributions for subclades 2A and 2B, geographically distributed to the west and north of the Colorado River (Fig. 6), are strongly unimodal with their respective peaks at a low average pairwise difference (Table 51, Fig. 139). This pattern is expected under a model of range expansion, rather than purely demographic expansion, with a relatively low migration rate (i.e., $Nm <$ 50) between colonized demes (Excoffier, 2004). Nevertheless, neither model can be rejected for both subclades (p = 0.918 versus 0.704, respectively, for subclade 2A; p = 0.894 versus 0.592 for subclade 2B). Thus, the coalescent history of these two subclades has apparently been different than that of subclades of the coastal Clade 1. Moreover, the low average pairwise difference exhibited by both subclades 2A and 2B suggests a relatively recent expansion history, perhaps one still in progress. Neither subclade, however, appears to have expanded from a relictual population, as both exhibit high haplotype diversities (Table 4), not the low values expected if either expanded out of a refuge with a small effective number of females.

Table 51. Mean pairwise difference among all haplotypes, estimates of Tajima's *D* and Fu's *Fs* (with probability of significance), and the population growth rate, *g*, for each mtDNA clade or subclade. Significantly negative values of *D* and *Fs* are indicated in bold.

Clade / Subclade	Pairwise difference[1]	Tajima's *D*	Fu's *Fs*	*g*[1,2]
Clade 1	21.3 ± 9.42	-0.889 (p = 0.219)	**-23.44** (p = 0.015)	
subclade 1A	10.3 ± 4.74	**-1.865** (p = 0.006)	**-24.35** (p = 0.000)	851.8 ± 32.68
subclade 1B	8.1 ± 3.79	**-1.688** (p = 0.017)	**-24.49** (p = 0.000)	1046.3 ± 49.63
subclade 1C	4.2 ± 2.11	**-1.516** (p = 0.038)	**-25.52** (p = 0.000)	1541.2 ± 104.24
Clade 2	13.4 ± 6.02	**-1.477** (p = 0.028)	**-23.52** (p = 0.009)	
subclade 2A	6.4 ± 3.04	**-2.011** (p = 0.001)	**-24.33** (p = 0.001)	1027.8 ± 17.87
subclade 2B	4.9 ± 2.44	**-2.033** (p = 0.001)	**-25.57** (p = 0.000)	2236.7 ± 67.07
subclade 2C	3.6 ± 1.86	0.388 (p = o.713)	-0.49 (p = 0.445)	820.9 ± 301.2
subclade 2D	5.8 ± 2.86	-0.108 (p = 0.514)	-1.49 (p = 0.270)	217.4 ± 75.92
subclade 2E	0.9 ± 0.66	-1.270 (p = 0.094)	0.82 (p = 0.639)	106.4 ± 40.90

[1] mean ± one standard deviation

[2] Fluctuate, version 1.4 (Kuhner et al., 1998); TS:TV ratio = 5.0, 10 short chains with 10 sampling increments and 1000 steps per chain, and 10 long chains with 20 sampling increment and 20000 steps per chain.

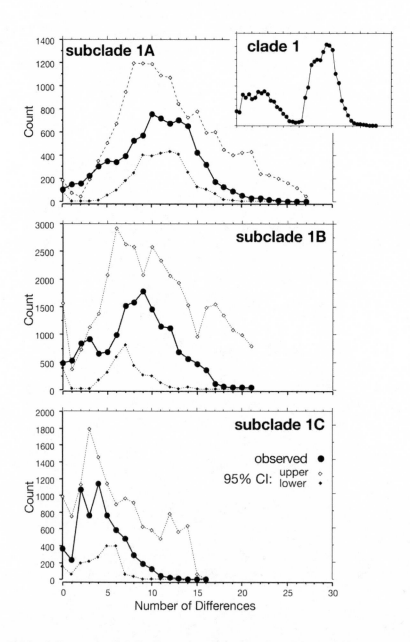

Figure 138. Mismatch distributions for the coastal mtDNA clade 1 (upper right inset) and each subclade. The observed distribution is indicated by the solid line connecting black circles; the upper and lower 95% confidence intervals based on a sudden expansion model are indicated by open and closed diamonds, respectively.

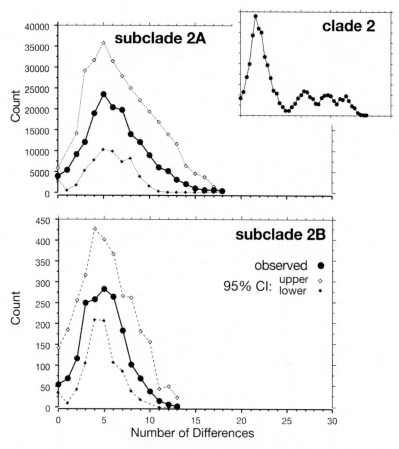

Figure 139. Mismatch distributions for the desert mtDNA clade 2 (upper right inset) and subclades 2A and 2B. The observed distribution is indicated by the solid line connecting black circles; the upper and lower 95% confidence intervals based on a spatial expansion model are indicated by open and closed diamonds, respectively.

The qualitative patterns in the mismatch distributions supporting a history of population expansion are also consistent with the general star-like haplotype phylogenies for each subclade depicted in Figs. 7 through 12 (Slatkin and Hudson, 1991) as well as with high estimates of the coalescent growth rate parameter, g, computed using the Metropolis-Hastings Markov Chain (MHMC) algorithm implemented in the program Fluctuate 1.4 (Kuhner et al., 1998). This method assesses the goodness-of-fit of a model of exponential growth (or decline), and generates Bayesian estimates of the growth parameter (g) and its standard deviation. Following the arguments presented in Lessa et al. (2003), we use a conservative measure of population growth where $g > 3$ times its standard deviation. We provide estimates of g and its standard deviation for each subclade in Table 50, using the genealogical relationships among haplotypes within each clade/subclade, empirical base frequencies, a transition/transversion ratio of 5:1 (the empirical ratio for each clade varied between 4.81 and 4.93 to 1), and historically fluctuating population sizes. A generation time of one year was assumed.

The estimate of g derived from the analysis is positive and large, considerably greater than our conservative lower boundary, for each subclade for which both Tajima's D and Fu's Fs values, as well as the mismatch distributions, suggest a history of expansion (Table 51). On the other hand, g is not greater than 0, by our conservative baseline, in the three Arizona subclades of Clade 2 for which these other approaches support population stability. In summary, therefore, we have four different, although not independent, indicators of the population history of each mtDNA subclade that are fully concordant in either indicating a relative recency of population expansion or stability over their respective coalescent histories: star-like phylogenies, significantly negative Tajima's D and Fu's Fs values, unimodal mismatch distributions with goodness-of-fit to demographic and/or spatial expansion models, and a significantly positive growth parameter, g.

The time (in generations), t, of a possible population expansion can be estimated through $\tau = 2ut$, where τ is the mode of the mismatch distribution and u is the mutation rate per nucleotide of the sequence considering that $u = 2\mu k$, with μ the mutation rate per nucleotide and k the number of nucleotides (Rogers and Harpending, 1992). For the *cyt-b* sequence of this study, k is 801, μ is 0.028 per million years (Arbogast and Slowinski, 1998; Zheng et al., 2003), or 2.8×10^{-8} per generation, if we assume a generation time of one year, and thus our estimate of u is 2.24×10^{-5}. Arlequin3 uses a nonlinear least-squares approach to estimate these parameters and provides approximate 95% confidence intervals by a parametric bootstrap approach; our analyses are based on 500 replicates.

We provide empirical estimates of τ for each clade and subclade and estimates of the absolute time of expansion assuming a generation time of 1 year in Table 52. Consistent with the differences in pairwise divergence values, the estimated age of Clade 1 is nearly an order of magnitude greater then that of Clade 2. Each subclade of these two clades differs in age as well, although there is considerable overlap in the respective 95% confidence intervals. Importantly, in virtually all cases, the initiation of population expansion of each clade/subclade is relatively old, certainly predating the last glacial maximum and subsequent habitat shifts of the late Wisconsin and Holocene that are so well documented by the packrat midden record of western North America (summarized in Betancourt et al., 1990). Even if one were to assume a longer generation time of two years, nearly all estimates of the temporal depth of each clade or subclade would still be substantial (i.e., the time of expansion of the desert subclades 2A or 2B, which exhibit the strongest unimodal mismatch distributions [Fig. 138] and sharpest signal of expansion [Table 51] would decrease from 98-104,000 years to 49-52,000 years ago).

There is a general concordance between the dates for the nodes at the base of each mtDNA clade or subclade (Table 50 and Fig. 137) and the coalescent-based expansion times (Table 52) for the included sets of haplotypes within each. The coalescent estimates are typically older for each subclade, as might be expected because the clock-based estimates in particular are subject to limited numbers of 3rd position transversions in the more recently diverged clades. For example, the clock divergence of 0.61 Ma for the division between subclade 2A and 2B is based on an average of 0.5 3rd position transversions (Fig. 137) and thus must be viewed with some skepticism. However, both sets of estimates are well within the errors of either set, and use of a generation time greater than 1 year would reduce the coalescent-based estimates to bring their means below the mean dates derived from a molecular clock (as above). Interestingly, diversification of subclades within both clades apparently occurred substantially after their respective origins, regardless of method of age estimation. However, both estimates suggest that not only did the subclades in Clade 1 diversify earlier than those of Clade 2, but that the demographic expansion that apparently characterized subclades 1A and 1B occurred substantially earlier than the spatial expansion that clearly characterized the Clade 2 subclades 2A and 2B. Both sets of analyses support the timing of the expansion of these desert subclades in the Late Pleistocene.

Table 52. Calculated measures of τ, the time to expansion, and absolute time based on a generation time of 1 year for a model of sudden demographic expansion (with approximate 95% confidence interval) and one for spatial expansion for each mtDNA clade and subclade of the desert woodrat complex that exhibited a consistent pattern of expansion. The p-value for the goodness-of-fit for both models is given. Best fit models are indicated in bold.

Clade / Subclade	Demographic expansion model			Spatial expansion model		
	τ	time (years)	p	τ	time (years)	p
Clade 1	**29.019** (20.640, 34.284)	**647,745** (460,714 – 765,268)	**0.322**	27.096 (23.054, 58.123)	-----	0.026
subclade 1A	**11.656** (6.938, 15.357)	**260,179** (154,866 – 342,790)	**0.906**	8.904 (6.155, 12.661)	-----	0.021
subclade 1B	**9.776** (5.795, 13.735)	**218,214** (129,353 – 306,585)	**0.890**	6.679 (4.540, 11.294)	-----	0.035
subclade 1C	3.369 (1.998, 9.185)	75,201 (44,598 – 205,022)	0.116	**3.050** (1.589, 5.890)	**68,080** (35,469 – 122,708)	**0.396**
Clade 2	3.860 (2.329, 13.457)	86,161 (51,987 – 300,379)	0.318	**3.634** (1.938, 18.082)	**81,116** (43,259 – 403,616)	**0.538**
subclade 2A	4.823 (3.329, 11.362)	107,656 (74,308 – 253,616)	0.704	**4.400** (2.771, 7.705)	**98,214** (61,853 – 171,987)	**0.918**
subclade 2B	4.939 (2,827, 6.971)	110,246 (63,103 – 155,603)	0.592	**4.690** (2.741, 6.501)	**104,688** (61,183 – 145,112)	**0.894**

Nested Clade Analysis

We use Templeton's (1998) method of the spatial distribution of genetic variation, or Nested Clade Analysis (NCA), to further examine the history of mtDNA *cyt-b* subclades. This approach has the advantage of discrimination between phylogeographic associations due to recurrent but restricted gene flow from historical events such as past population fragmentation or range expansion events. For each subclade, we constructed a haplotype network using the parsimony-based algorithm developed by Templeton et al. (1992), as implemented in the program TCS 1.21 (Clement et al., 2000). We then nested the inferred haplotype relationships based on published rules (e.g., Templeton et al., 1987; Crandall, 1996) and calculated clade distances (the geographic range of a particular n-step clade), nested-clade distances (the dispersion of an n-step clade relative to its evolutionary sister clade(s) nested within the same or higher n + 1-step clade), and the difference between interior and tip clades using the program GeoDis 2.4 (Posada et al., 2000). Interior clades are those having connections to more than one other clade and tip clades lie peripherally in the network and can only be connected to an interior clade. We used 10,000 random permutations to test the null hypothesis of no geographic association separately for each clade at each nested level. For interpretation, we followed the inference key provided by Templeton (2004, p. 807-809).

We illustrate the haplotype network of subclade 1A generated by the TCS analysis in Fig. 140. Only four clades in the NCA are significantly associated with geography. Two are interior clades nested within a larger tip clade, all positioned in the southern part of Baja California Sur (Fig. 140, map; from La Purisima [locality BCS-41] in the north to La Laguna [BCS-120]). The fourth is a tip clade that contains the more localized insular samples from San José and San Francisco and the immediately adjacent mainland sample San Evaristo (BCS-74). Together, these two inclusive clades largely correspond to the southern phyletic cluster of haplotypes identified in the Bayesian analysis (Fig. 7). That these two clade groups have apparently had different histories is suggested by Templeton's (2004) inference key. Although the insular and adjacent mainland samples exhibit contiguous range expansion, it is not possible to distinguish between fragmentation and isolation by distance for the clade occupying the southern peninsula. On the other hand, although many northern samples form a reasonable cluster within the network (including the basal haplotype), there is no apparent internal geographic association of their included haplotypes (X^2 = 114.44, p = 0.4537). The combination of a southern groups of localities with an overall history of expansion and fragmentation and a northern group that has apparently been temporally stable is generally inconsistent with the coalescent analyses that support expansion, based on both significantly negative Tajima's D and Fu's Fs values and a high g-statistic, and significant support for demographic expansion (Table 51) contrasting with a high average pairwise difference of 10.3 steps among haplotypes (Table 51).

There are seven haplotype clades within subclade 1B that exhibit significant associations between their position in the parsimony network and geography (Fig. 141). Three of these are nested within larger clades, of which one is an interior and the other a tip clade. The other two statistically supported clades are tip clades. Of the four higher-order significant clades, three exhibit contiguous range expansion with high support (p = 0.0005 to 0.0001), and all three are geographically positioned at the northern margins of subclade 1B (Fig. 141, map, largely through San Gorgonio Pass and Morongo Valley and in the vicinity of Tejon Pass). These haplotypes comprise two of the phyletic clusters identified in the Bayesian analysis (Fig. 8, clusters b and c). The fourth geographically significant clade includes all population samples in the southern distribution of the subclade in Baja California (Fig. 141, map). Collectively, this group has apparently undergone a history of allopatric fragmentation. As with subclade 1A, the combination of expansion on one distributional border with stability, or fragmentation, on the other has yielded the overall pattern suggested by the coalescent analyses of demographic expansion (Table 52) contrasting with relatively high pairwise divergence (Table 51) and multimodal mismatch distribution (Fig. 138).

The few individual haplotypes within subclade 1C (Fig. 9) encompasses only two significant geographically associated groups (Fig. 142). One of these, including all haplotypes at localities from the Elkhorn Plain in San Luis Obispo Co. north to Stanislaus Co. in coastal California (localities CA-40 to CA-5), exhibits a significant pattern of past fragmentation. The second, including most localities from the southern margins of the subclade distribution, exhibits a pattern of contiguous range expansion. The northern group with the apparently fragmented past is coincidental with the two phyletic clusters with high posterior support delineated in the Bayesian analysis (Fig. 9). Thus, the overall historical pattern of subclade 1C, with fragmentation at one geographic margin and expansion at the other, is the same, but latitudinally reversed, as that exhibited by the geographically adjacent subclade 1B to the immediate south. This suggests that the overlap of these two subclades in the vicinity of Tejon Pass, near where Kern, Ventura, and Los Angeles counties converge, is relatively recent.

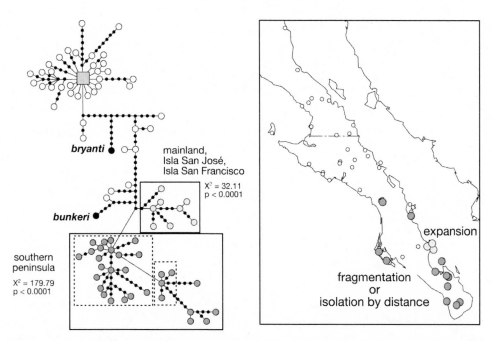

Figure 140. Left – Parsimony network of subclade 1A (large circles) and unsampled (small black circles) haplotypes. The large gray square is the basal haplotype. Boxes (dashed, if nested within a larger clade) enclose clades that exhibit a significant geographical association, with haplotypes included within each coded by a shade of gray. The statistical support for the geographic association for each inclusive clade is provided. Haplotypes of insular taxa from Cedros (*bryanti*) and Coronados (*bunkeri*) are indicated. Right – Map of subclade localities, with those associated statistically indicated by the same gray tone as in the network; unfilled circles are those localities and haplotypes that lack a geographic association.

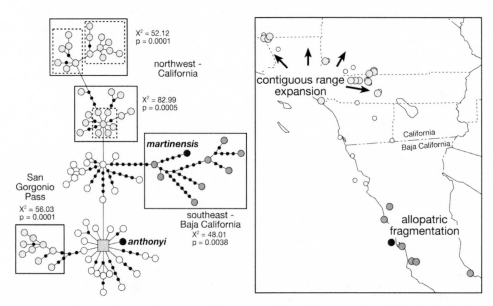

Figure 141.　Left – Parsimony network of subclade 1B (large circles) and unsampled (small black circles) haplotypes.　The large gray square is the basal haplotype.　Boxes (dashed, if nested within a larger clade) enclose haplotype clades that exhibit a significant geographical association, with haplotypes included within each coded by a shade of gray.　The statistical support for this association for each inclusive clade is provided.　Haplotypes of the insular taxa from Todos Santos (*anthonyi*) and San Martín (*martinensis*) are indicated.　Right – Map of subclade localities, with those associated statistically indicated by the same gray tone as in the network; unfilled circles are those localities and haplotypes that lack a geographic association.　Arrows indicate the directions of hypothesized range expansion.

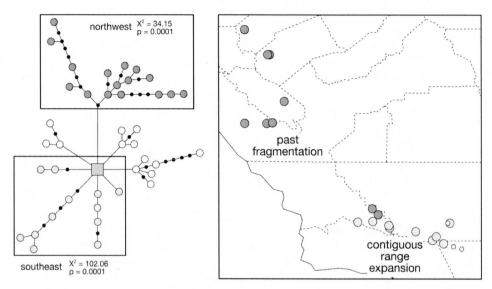

Figure 142. Left – Parsimony network of subclade 1C (large circles) and unsampled (small black circles) haplotypes. The large gray square is the basal haplotype. Boxes (dashed, if nested within a larger clade) enclose haplotype clades that exhibit a significant geographical association, with haplotypes included within each coded by a shade of gray. The statistical support for this association for each inclusive clade is provided. Right – Map of subclade localities, with those associated statistically indicated by the same gray tone as in the network; unfilled circles are those localities and haplotypes that lack a geographic association.

The NCA for the desert subclade 2D shows an overall significant association between haplotype position in the network and geographic position, but our geographic sampling is inadequate to discriminate between fragmentation and isolation-by-distance according to Templeton's inference key. For subclade 2E, it is not possible to even reject the hypothesis of no geographic association. However, seemingly clear historical population-geographic signals are present for the three remaining subclades of Clade 2. These signals are, however, different from those described above for subclades in Clade 1, as might be expected given the clade-specific difference in the coalescence patterns summarized in Tables 51 and 52, above.

Clade 2A has the largest geographic range and greatest overall sampling, both for localities and numbers of individuals per locality (Table 4 and Fig. 10). Moreover, several haplotypes are exceedingly abundant and very broadly distributed, unlike the pattern found in most other subclades, particularly those of

Clade 1.   There are two major clusters of haplotypes in the parsimony network (Fig. 143), one made up of six clusters of largely one-step haplotypes diverging from a single common one, and a second cluster separated from the first by a long branch of 6 unsampled haplotypes.   Among the larger cluster, four of the significant clades are tip clades and one is an interior clade. For each of these the centrally placed haplotype is one that is both numerically common and widespread geographically. This overall assemblage of haplotypes is concordant with phyletic cluster "a" identified in the Bayesian analysis, which occurs throughout the desert regions of eastern California, Nevada, and western Utah (Fig. 10). Each of the clades exhibits a signature of contiguous range expansion but the overall clade has a pattern of either long-distance colonization possibly coupled with subsequent fragmentation or past fragmentation followed by range expansion, based on Templeton's inference key.   Given the low probability of long-distance colonization in woodrats, the combination of fragmentation followed by expansion is the most likely inference.   Importantly, all desert localities, from extreme southeastern California (Tumco Mine, CA-205) to northeastern California (Cedarville, CA-424) across northern Nevada to northwestern Utah (Carrington Island, UT-5) contain haplotypes within this portion of the network, and most of these are separated from phyletic sisters by single steps. The hypothesis of range expansion suggested by NCA is completely consistent with each of the components of the coalescent analyses, including the strongly unimodal mismatch distribution (Fig. 139), significantly negative Tajima's D and Fu's Fs, large population growth statistic g (Table 51), and a significant fit to a model of spatial expansion (Table 52).   The Holocene temperature record perhaps provides an explanation for a continuing expansion of members of this subclade at the northern terminus of its range in Idaho and northwestern Utah (Smith and Betancourt, 2003).

The second major haplotype cluster in the network (Fig. 143) forms a single statistically significant association between haplotype position and geography.   This is the same group of haplotypes identified in the Bayesian analysis as phyletic cluster "b" (Fig. 10), which occurs throughout the Tehachapi Mts. and western parts of the Kern River Plateau in Kern and Tulare counties in south-central California.   Contrary to the majority of haplotypes and localities that exhibit range expansion in subclade 2A, this set is nested in such a way as to suggest either isolation-by-distance or restricted gene flow. Importantly, this is the set of subclade 2A haplotypes that are uniformly present in individuals of the "coastal" morphological group (Fig. 47) that also have a coastal nuclear genetic background based on microsatellite loci (Fig. 51). Thus, the combination of the phyletic separation of this haplotype group (Fig. 10) and its internal signal of isolation-by-distance or restricted gene flow suggest both a separate and an older history than the expansion of phyletic cluster "a" as it spread to occupy its current desert region. These data also suggest that the hybridization event that positioned this group of haplotypes within these otherwise morphological and nuclear "coastal" animals was an earlier episode than that which is occurring presently in Kelso Valley, eastern Kern Co. (Fig. 52).

Contrary to subclade 2A, subclade 2B is restricted geographically and lacks any internal phyletic structure (Fig. 11). The NCA identifies two significant associations between haplotype position within the parsimony network and geography, a tip clade and the total cladogram (Fig. 144). Templeton's inference key suggests that range expansion underlies the distribution of the total cladogram but that either isolation-by-distance or fragmentation has been responsible for the inclusive tip clade. The latter seems more probably, since the haplotypes included in this tip clade are known only from both the southwestern (Virgin Mts., NV-138) and northeastern (Rock Canyon Corral, UT-34) margins of the subclade range, but not those in the geographic middle. The pattern of expansion of the total clade is in accord with all measures stemming from the coalescence analyses, including strongly unimodal mismatch distribution with a low average pairwise difference (Fig. 137), significantly negative Tajima's D and Fu's Fs and very large growth estimate (Table 50), and strong fit to a spatial expansion model (Table 52). However, the nested signature of isolation-by-distance or fragmentation suggests that populations are beginning to differentiate, although not to the degree that the substantial signature of historical expansion has been overridden.

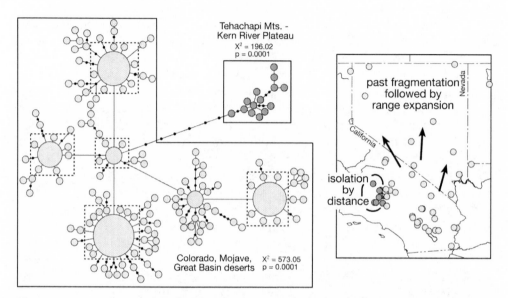

Figure 143. Left – Parsimony network of subclade 2A (large circles) and unsampled (small black circles) haplotypes. Boxes (dashed, if nested within a larger clade) enclose haplotype clades that exhibit a significant geographical association, with haplotypes included within each coded by a shade of gray. The statistical support for this association for each inclusive clade is provided. The central haplotypes of each cluster are drawn proportional to their numerical representation, with the largest circle comprised of 57 individuals from 17 localities. Right – Map of subclade localities, with those associated statistically indicated by the same gray tone as in the network; unfilled circles are those localities and haplotypes that lack a geographic association. Arrows indicate the direction of hypothesized range expansion.

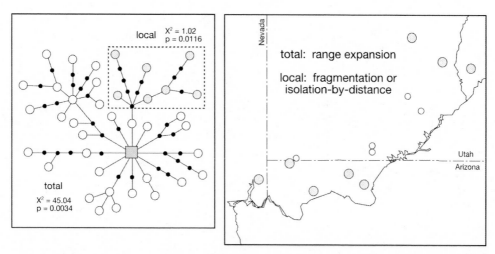

Figure 144. Left – Parsimony network of subclade 2B (large circles) and unsampled (small black circles) haplotypes. The large gray square is the basal haplotype. Boxes (dashed, if nested within a larger clade) enclose haplotype clades that exhibit a significant geographical association, with haplotypes included within each coded by a shade of gray. The statistical support for this association for each inclusive clade is provided. Right – Map of subclade localities, with those associated statistically indicated by the same gray tone as in the network; unfilled circles are those localities and haplotypes that lack a geographic association.

Finally, for subclade 2C the apparent history has been even more different. Although few localities were sampled and few haplotypes identified (Table 4), there is significant geographic signal for a tip clade within the parsimony network as well as for the total cladogram (Fig. 145). The inclusive tip clade contains all haplotypes and localities from north of Flagstaff, in the western Painted Desert of Coconino Co., a group that forms a strong phyletic cluster in the Bayesian analysis (Fig. 12). These localities are linearly aligned from north to south, and the clade exhibits an isolation-by-distance pattern. In contrast, the total cladogram apparently has had a history of allopatric fragmentation, with those samples from along the lower Colorado River well separated from those north of Flagstaff. These results are fully in accord with the coalescent analyses where there is no signature of population expansion (both non-significant Tajima's D or Fu's Fs and lack of fit to either demographic or spatial expansion models; Tables 51 and 52).

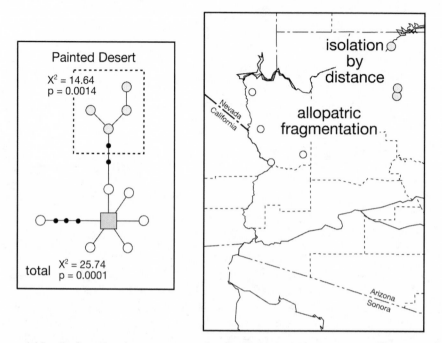

Figure 145. Left – Parsimony network of subclade 2C haplotypes (large circles) and unsampled haplotypes (small black circles). The large gray square is the basal haplotype. Boxes (dashed, if nested within a larger clade) enclose haplotype clades that exhibit a significant geographical association, with haplotypes included within each coded by a shade of gray. The statistical support for this association for each inclusive clade is provided. Right – Map of subclade localities, with those associated statistically indicated by the same gray tone as in the network; unfilled circles are those localities and haplotypes that lack a geographic association.

VICARIANCE BIOGEOGRAPHY AND DESERT WOODRAT mtDNA

"In its extreme form, vicariance is an example of 'biology is passive, physical factors drive evolution' " (Penny and Phillips, 2004, p. 521).

There is a rich literature directly tying mtDNA clade structure to historical biogeography, stemming from initial studies of this molecule using restriction enzyme analyses (e.g., Avise et al., 1987; Riddle, 1996; see summary in Avise, 2000). The vast majority of these studies have interpreted geographic structure of reciprocally monophyletic molecular clades as primary evidence for vicariance as the process linking clade structure to geographic position. Although there is most likely a vicariant connection between these two features of evolutionary history, the connection must be established from empirical evidence and not assumed. Mechanisms that restrict gene flow, whether based on a population attribute (such as small effective size) or an ecological one (such as a dispersal sink) can also generate the geographic position of reciprocally monophyletic clades. Irwin (2002), for example, has presented models to illustrate how population structure can arise in continuously distributed species in the absence of past physical barriers.

Several groups of authors (Hafner and Riddle, 1997; Murphy and Aguirre-Léon, 2002; Lawlor et al., 2002; Lindell et al., 2005; Riddle, 1995; Riddle and Hafner, 2004, 2006a, b; Riddle and Honeycutt, 1990; Riddle et al., 2000a, b, c; Upton and Murphy, 1997; Zink et al., 2001) have presented analyses of vertebrate distribution patterns and vicariant history for the arid lands of North America, an area including nearly the entire range of members of the desert woodrat complex examined herein. These and other authors have amassed the limited geological data that document the presence of some historical barriers coincidental with phylogeographic boundaries suggested by mtDNA clades, and have hypothesized others where sharp clade boundaries exist but where geological data are absent. Some of these authors have also thoroughly discussed the origin of both the islands and their associated flora and fauna on the two sides of Baja California. In this section, we interpret the clade structure and historical inferences stemming from both coalescent and nested clade analyses of the *Neotoma lepida* group in the context of these historical hypotheses.

Our hypotheses of the timing of lineage origin and subsequent diversification of desert woodrats, within a geographic context, are presented in serial pictorial form in the maps, Fig. 146 and 147. This depiction is based on the data presented in Tables 50 and 52 regarding the time frame and on the phylogeny of mtDNA clades illustrated in Fig. 6 and summarized in Fig. 137. The sequential events in the history of the complex are as follows: (1) initial split to form the coastal Clade 1 and desert Clade 2; (2) origin of the insular population on Ángel de la Guarda in the Gulf of California; (3) geographic expansion of Clade 1 in coastal California and Baja California and the split of the desert Clade 2 by the Colorado River into ancestral subclades 2AB and 2CDE; (4) subdivision of Clade 1 and subclade 2CDE, each into three subclades, and initial expansion of Clade 2AB into the Mojave Desert to contact the coastal Clade 1 members in the vicinity of the Tehachapi Mountains; and (5) extensive expansion of subclade 2AB with the split of this subclade into two geographic components.

The initial split within the *Neotoma lepida* group is between the two major mtDNA clades (Fig. 146, map A), the coastal Clade 1 and desert Clade 2, which we date in the Early Pleistocene (about 1.6 Ma; Table 50). We position the ancestral population in southern California and northern Baja California because the next two events in the history of the complex involve occupation of areas adjacent to this general region (Fig. 146, maps B and C). A phylogeographic division between northwestern Baja California and eastern California has been interpreted to result from a vicariant event involving flooding of the Salton Trough and a northern extension of the Gulf of California into the lower Colorado River basin as far north as today's Lake Mojave on the Nevada-Arizona border (the Bouse Embayment; see Riddle and Hafner, 2006a, b, as examples). While there is more recent geological evidence to suggest that the Bouse "embayment" was not a northward marine incursion but rather resulted from lake formation (Spencer and Pearthree, 2005), the origin of the water barrier through the region in question is unimportant to our historical scenario. However, the timing of this event, whatever its origin, is of central concern. Both the flooding of the Salton Trough and the Bouse lake/embayment are of Miocene-Pliocene age (about 5.5-5.3 Ma; e.g., Carreño and Helenes, 2002) and the period of maximal flooding was apparently over by 3.3 Ma. These events thus predate the divergence of Clades 1 and 2 by a considerable degree, even given the rather large error estimate (Table 50). Thus, neither flooding of the Salton Trough or the development of lakes forming the Bouse "embayment" can be argued to be a primary vicariant event underlying the initial clade diversification in the *Neotoma lepida* group. Dispersal either across a flooded area or across the floodplain habitat, with either serving as a substantial dispersal sink because of a general lack of suitable habitat, is more likely to underlie clade formation.

The second event in the history of this woodrat complex was the origin of the population on Isla Ángel de la Guarda in the northcentral Gulf of California (mtDNA Clade 1D; Fig. 146, map B), which is estimated at about 0.8 Ma, in the terminal part of the Early Pleistocene (Table 50; Fig. 137). Ángel de la Guarda is one of the few non-landbridge islands in the Gulf, and its origin has been estimated at 2-3 Ma (Lindell et al., 2005). Under the assumption that this date is correct, derivation of subclade 1D (the insular form *insularis*) must have resulted from overwater dispersal as even the early boundary to the estimated range of its origination is well after the Pliocene origin of the island.

Our data suggest that coastal Clade 1 expanded from its historic range in the mid Pleistocene (at approximately 0.5 Ma), likely both north into central California and south into the southern part of the Baja peninsula (Fig. 146, map C). Simultaneously, the desert Clade 2 became split into two geographic units (ancestral subclades 2AB and 2CDE) separated by the lower Colorado River. The origin of these two ancestral subclades, however, is unlikely to have resulted from a single vicariant event, as both lake/embayment and river channel formation in the lower Colorado basin pre-date the splitting event by a considerable period. Across-river dispersal is the mostly likely origin for the Arizona subclade 2CDE.

The next stage in the history of the *Neotoma lepida* group was apparently the near-simultaneous formation of the mtDNA subclades within Clade 1, subsequent to its expansion to occupy most, or all, of its current distribution along coastal California and Baja California (Fig. 147, map D). The events that generated this breakup are completely unknown; they could have been either strictly vicariant or an ecologically based process. Phylogeographic structure in other taxa has been interpreted as evidence for a mid-peninsular seaway (Riddle et al., 2000a, b; Upton and Murphy, 1997; and others), but the hypothesized position of such a barrier is well south of the position of contact between the coastal woodrat subclades 1A and 1B (Figs. 6 and 147F). Moreover, although different authors provide a wide range of possible dates for this seaway (from 7 Ma [Lindell et al., 2005] to 1 Ma [Riddle and Hafner, 2006a, b]), this range is again well before the dates for the subdivision of Clade 1 (ca. 0.25 Ma; comparisons between subclades 1A, 1B, and 1C in Table 50 and Fig. 137). Because our analyses suggest a complex history for each of these subclades subsequent to their origin, ranging from range expansion at one geographic margin to either stability or fragmentation at the other (NCA results and Figs. 140-142), it may never be possible to determine their actual distribution at the time of subdivision and thus uncover the historical processes underlying their origin. Arizona subclades 2C, 2D, and 2E also divided during this time period. Each of these has either remained stable in geographic position or experienced fragmentation in its recent past (NCA analysis and Fig. 145). Their current boundaries seem to be the Bill Williams River (between subclades 1C and 1D) and the Gila River (subclades 1D and 1E). We are unaware of the times of origin of these rivers, so it is conceivable that subclade structure in western Arizona resulted from the vicariant formation of river systems. However, it seems more likely that both rivers are older than the likely mid Pleistocene origin of the subclades and that the river channels have served as barriers subsequent to a step-wise dispersal of desert woodrats across both. Fragmentation of the habitats across the Coconino Plateau may be responsible for the breakup of subclade 2C into its eastern and western segments in more recent times (NCA, Fig. 145). Finally, we suggest that the mid Pleistocene was also a time for the initial expansion of the desert ancestral subclade 2AB into the western Mojave Desert, as the introgression event that resulted in the origin of the desert subclade 2A haplotype lineage that now characterizes the morphological and nuclear DNA coastal group of populations in the Tehachapi Mts. and Kern River Plateau (see Figs. 47 and 143, and accompanying discussion) is clearly a relatively early event in the history of this subclade.

We hypothesize that the last episode in the history of these woodrats was the rapid and extensive expansion of the desert subclade 2AB north from southern California (Fig. 147, map E). This episode is complex, with at least three separate components, the first of which was the divergence of subclade 2A and 2B (circle 1 in Fig. 147, map E). The second was the contact between subclade 2A and coastal Clade 1 populations in the Tehachapi Mts.-Kern River Plateau (circle 2 in Fig. 147, map E), and the last was the continued northward expansion of subclade 2A through the Great Basin Desert (circle 3 in Fig. 147, map E). The division of subclades 2A and 2B must have preceded the other events enumerated, because these two haplotype groups have reached reciprocal monophyly. The deep canyon of the Virgin River, which currently forms the boundary between these two subclades, is certainly older than subclade divergence in the late Pleistocene. Thus, the formation of the river itself cannot have served as a primary vicariant event in the origin of these subclades. The haplotype cluster now present in the Tehachapi Mts.-Kern River Plateau remains nested within subclade 2A, which suggests that this historical episode post-dated the division of subclades 2A and 2B. The western expansion of subclade 2A into this area also must have represented the initial contact between desert and coastal woodrats, resulting in limited hybridization and either the introgression of the desert subclade 2A mtDNA into the coastal species or the incorporation of that mtDNA as the coastal form replaced the desert taxon in this area. Finally, subclade 2A may be continuing to expand at its current northern terminus in southern Oregon and Idaho (Smith and Betancourt, 2003).

With the information currently available, most of the history of the *Neotoma lepida* group appears governed by population processes, such as spatial and demographic expansion, waif dispersal across pre-existing barriers, and perhaps shifting ranges and consequent competitive interactions of other species of woodrats (such as *N. macrotis* in central California [Cameron, 1971] and *N.*

albigula in western Arizona). The specific vicariant events that have been posited to underlie the phylogeographic structure of many other, largely co-distributed desert taxa, are apparently not directly tied to the diversification of these woodrats. As the quote at the beginning of this section acknowledges, biology may matter.

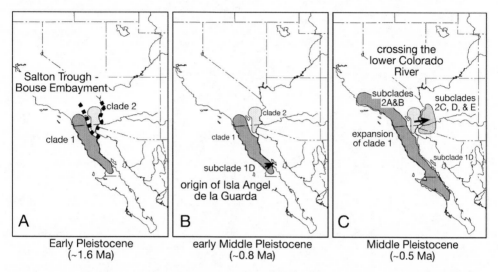

Figure 146. Sequential hypotheses of the temporal divergence of mtDNA clades and subclades (and taxa, see section below) of the *Neotoma lepida* group, beginning with the initial split of an ancestral population into the two clades (1 and 2) in the Early Pleistocene (map A) followed by the origin of the insular population on Ángel de la Guarda (map B) and then separation of the interior desert Clade 2 by what is now the Colorado River and expansion of the coastal Clade 1 (map C). Suggested vicariant or dispersal events are indicated, as discussed in the text and in Table 52. The timing of events is based on data provided in Tables 49 and 50.

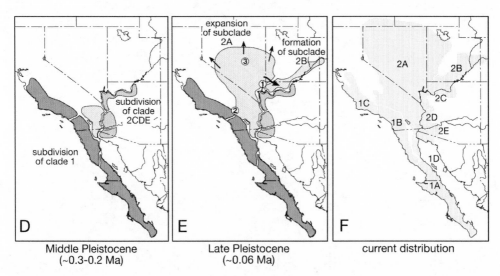

| Middle Pleistocene | Late Pleistocene | current distribution |
| (~0.3-0.2 Ma) | (~0.06 Ma) | |

Figure 147. Continuation of sequential hypotheses of the temporal divergence of mtDNA clades and subclades (from Fig. 139, above) of the *Neotoma lepida* group, beginning with the subdivision of the coastal Clade 1 into three subclades and similar subdivision on the Arizona Clade 2CDE into three subclades in the Middle Pleistocene, as well as the initial expansion of the desert Clade 2AB to contact the coastal Clade 1 (map D) followed by the secondary extensive spatial expansion of the desert Clade 2AB and its subdivision into subclades 2A and 2B (circled 1), contact with the coastal Clade 1 in the Tehachapi Mts. (circled 2), and further northward expansion of subclade 2A (circle 3) in the Late Pleistocene (map E). The final panel (map F) illustrates the current ranges of each clade and subclade (from Fig. 6).

SYSTEMATICS OF THE DESERT WOODRAT COMPLEX

SPECIES AND SPECIES BOUNDARIES

We define the *Neotoma lepida* group as the monophyletic assemblage of taxa strongly supported by both mitochondrial DNA and nuclear DNA sequences (Edwards and Bradley, 2002; Matocq et al., 2007; Figs. 4 and 5). This group is identical to that proposed by Goldman (1932) but excludes *N. stephensi* (which is instead related to *N. mexicana*, *N. picta*, and *N. ishmica*) and *N. goldmani* (which is sister to a clade composed of *N. albigula*, *N. floridana*, and *N. magister*; Matocq et al., 2007). By current taxonomy (Musser and Carleton, 2005), six species are recognized within the complex: the continental *Neotoma lepida* (west and north of the Colorado River, including the length of Baja California and some of the islands in the Gulf of California), *Neotoma devia* (east and south of the Colorado River in Arizona and northwestern Sonora), and four insular taxa on both sides of the Baja peninsula, *Neotoma anthonyi* (Todos Santos), *Neotoma martinensis* (San Martín), *Neotoma bryanti* (Cedros) off the Pacific coast, and *Neotoma bunkeri* (Coronados) in the southern Gulf. The patterns of character variation in both morphological and molecular diversity that we summarize above, however, challenge the validity of each of these six "species," but in a different manner depending upon the taxon in question. What are, then, the fundamental species units within this complex of woodrats?

Sites and Marshall (2003, 2004) summarize both the major species concepts in the current literature and the objective, testable, and operational criteria that have been used to delimit species in nature. We thus make no attempt to provide such a review here, only to use the structure in these synopses to address the question of species boundaries within this group of woodrats. In so doing, we stress several important issues: First, a conceptual, or ontological, definition of species is distinct from the criteria upon which species are delimited. Second, species boundaries are often "fuzzy", because of the retention of ancestral polymorphisms, the failure to complete sorting, and/or reticulation due to hybridization subsequent to initial separation. As a consequence, different operational criteria may either fail to delimit boundaries properly or, more likely, give conflicting results. Third, the emphasis that one places on population criteria, such as the importance of gene flow, versus phylogenetic criteria, or lineage distinctness, may lead to one conclusion, or may support differing sets of boundaries. It should not be surprising, therefore, that the final decisions on species boundaries recognized for any larger taxonomic group may rely on the qualitative "judgment" of the investigators in question, rather than on a particular statistically defensible delineation. Finally, we agree with Sites and Marshall (2004, p. 201) that "regardless, the delimitation of species requires that one have clearly defined operational criteria by which individuals can be tested for species membership, and the criteria must be understood within the context of what kind of entity (interbreeding versus historical lineage) each method is designed to test."

In our conceptual view, species are those entities in nature that have a uniquely defined evolutionary trajectory (past and future) and are diagnosable by morphological and/or molecular, chromosomal, or other kinds of characters. Evidence for an independent evolutionary trajectory comes from both character-based tree topologies that depict hypotheses of lineage distinctness and from evidence of genetic isolation. The characters in question may be derived from multiple gene sequences (both mitochondrial and nuclear) or other types of molecular data (allozyme or microsatellite allele distributions), or may be the results of the analysis of morphological traits, or, preferably, some combination of all. Operationally, therefore, we use the concordant topologies of the mtDNA *cyt-b* gene (Fig. 5) and nucDNA *Fgb-I7* intron sequences (Fig. 13) to define the evolutionary lineages within the *Neotoma lepida* group and use the distribution of microsatellite alleles and morphological characters within and across included population samples to assess the evolutionary independence of these lineages. In those instances where individuals of two distinct lineages co-occur, we are able to examine directly their ability to exchange genes by testing for panmixia across microsatellite loci.

Sites and Marshall (2004, Table 1) summarize 12 operational "programs" designed to define species in nature on objective grounds and discuss strengths and limitations along with the types of data suitable for each. Seven of these are nontree-based methods; five depend on lineage delimitation based on the construction of phylogenetic trees. For illustrative purposes only, we choose three of these operational methods to apply to our diverse datasets (Table 53). One of these is a nontree-based method (Correlated Distance Matrices, or Corr-D; following Puorto et al., 2001); the others are tree-based methods, the Wiens-Penkrot phylogenetic method (Wiens and Penkrot, 2002) and Templeton's tests of cohesion (Templeton, 1989, 2001, 2004). We describe the application of each of these "programs" for all currently recognized species within the *Neotoma lepida* group as well as the molecular and morphological units we have uncovered in the analyses presented herein. For simplicity, we also summarize the conclusions of each application in Table 53.

Correlated distance matrices (Corr-D)

Puorto et al. (2001) used matrix comparisons to test the statistical association of mtDNA clade membership and a multivariate summary of morphological variation. In our case, we simply used the visual correspondence between clades defined by both mtDNA and nucDNA sequences and our analyses of qualitative and quantitative craniodental, glans penis, and color character variation. These comparisons provided tests of species status for the groups that our analyses define, as follows:

Clade 1 versus Clade 2.—There is excellent correspondence between clade structure for both mtDNA and nucDNA at this level and the distribution of morphological characters, be these the multivariate discrimination of craniodental variables (Fig. 23), qualitative craniodental features (M1 anteroloph [Fig. 24, 25], vomerine structure [Fig. 26], and lacrimal-maxillary suture position [Fig. 27, 28]), or qualitative phallic features (Fig. 30). The Corr-D operational criterion would conclude that these units are separate species.

Clade 1D versus Clade 1A+B+C.—There is an exact correspondence between clade structure (mtDNA only) and morphological characters (phallic morphology [Fig. 30 and 31], craniodental "shape" parameters [Fig. 101], and color [Fig. 105]) for this pair of groups. Separate species status of Clade 1D is supported.

Subclade 1B, 1C, and 1A.—Although each subclade is a well supported phylogenetic lineage by mtDNA (Fig. 5), all share the same set of qualitative morphological features (Figs. 25, 28, 30) and overlap extensively in craniodental morphometric space (Fig. 23). Where subclades co-occur geographically, individuals cannot be separated by the very robust discriminant analyses (Fig. 49). Rather, many differences among subclades when all samples are included are clinal, without a major step at the geographic position of clade boundaries (Fig. 82). Moreover, where subclades 1B and 1C are in syntopy (near Ft. Tejon, Kern Co., locality CA-60) all 18 polymorphic microsatellite loci are in Hardy-Weinberg equilibrium, supporting complete panmixia among individuals belonging to both mtDNA subclades. Separate species status for each subclade is thus not supported.

Subclade 2A+B versus Subclade 2C+D+E (*N. lepida* versus *N. devia*).— There is a well supported split separating mtDNA subclades on both sides of the Colorado River (Fig. 5) concordant with microsatellite assignments to similar exclusive groups without evidence of assignment intermediacy (Fig. 131). Karyotypes are also different between subclades (at least, as far as is known;

Mascarello and Hsu, 1976; Mascarello, 1978). Qualitative morphological features are shard by all subclades, but craniodental variables do distinguish clade groups (Figs. 125, 127), although to a much lesser extent that the level between coastal and desert Clade 1 and Clade 2. Phallic characters are shared (contra Mascarello, 1978). Thus, molecular evidence for independent evolutionary lineages is strong, but diagnosability by morphological features is limited. However, the lack of evidence for gene exchange (contra Hoffmeister, 1986) suggests that the two subclade groups are both independent lineages and genetically isolated. Overall, therefore, while demarcation is less clear, separate species status is supported.

Subclade 2A versus Subclade 2B (within *N. lepida*) and Subclade 2C versus Subclade 2D versus Subclade 2E (within *N. devia*).—Each subclade is well-supported in the mtDNA cyt-b tree (Fig. 5), but the *Fbg-I7* tree fails to resolve subclades 2A and 2B, or subclades 2C and 2D, because these pairs respectively contain only one, and the same, haplotype (Fig. 13). There is no consistent pattern of either significant difference, or similarity, in craniodental morphometric (PCA "size" and "shape", CVA; Figs. 127, 128) or colorimetric analyses among subclades (Fig. 129). Thus, the inconsistency and/or non-concordance across character sets falsify separate species status for each subclade.

Wiens-Penkrot phylogenetic method

Wiens and Penkrot (2002) consider species to be sets of populations that are strongly supported, exclusive, and concordant with geography. Their approach is applicable to either molecular or morphological datasets, and presumably when both data types are available, they should be concordant in identifying clade structure, exclusivity, and geographic position.

Five of the six currently recognized species in the *Neotoma lepida* group (Musser and Carleton, 2005) either fail the test of phylogenetic exclusivity (*Neotoma lepida*, which includes all of our Clade 1 and subclades 2A and 2B of Clade 2) or lack molecular phylogenetic support as anything but unique single haplotypes nested within larger phylogenetic clades (the four insular species *N. anthonyi*, *N. martinensis*, *N. bryanti*, and *N. bunkeri*). Each of these "species" is also non-exclusive in morphological characteristics, although we have admittedly not examined the few qualitative morphological characters we identify above, or coded continuous characters, for a formal cladistic analysis. Only *Neotoma devia* is phylogenetically exclusive by molecular characters (both mtDNA and nucDNA). Although its morphological distinctness is marginal, we know of no way to determine if the general character similarities between *N. devia* and subclade 2A+B ("desert" *N. lepida*) are due to the retention of a common ancestral morphology or to similar selection regimes on both sides of the Colorado River. Thus, overall morphological similarity cannot automatically reject the current hypothesis of species status for *N. devia*. Our data falsify the species status for *Neotoma lepida* (sensu Musser and Carleton, 2005, and previous authors), *Neotoma anthonyi*, *Neotoma martinensis*, *Neotoma bryanti*, and *Neotoma bunkeri*. On the other hand, *Neotoma devia* is supported as a distinct species.

Beyond *Neotoma devia*, what, then, are the species units within the *Neotoma lepida* group based on the Wiens-Penkrot approach? The coastal Clade 1A+B+C, insular Clade 1D, and desert Clade 2A+B are each exclusive in molecular and morphological character analyses, well supported in the phylogenetic analyses, and consist of populations occupying an internally common geographic range. Each of these three groups satisfies the Wiens-Penkrot criteria for distinct species.

The remaining question, therefore, is whether or not the subclades within either Clade 1 (subclades 1A, 1B, 1C) or Clade 2 (2A, 2B) warrant species status. Each subclade is exclusive, well supported, and consists of a unique geographic range, thus meeting the molecular tree-based definition of Wiens-Penkrot. In contrast, each fails to meet the Wiens-Penkrot requirement of morphological exclusivity, although for the reasons given above for *Neotoma devia* the sharing of common morphology could be due either to the retention of symplesiomorphic characters or to independent expression resulting from a common selective regime (i.e., convergence). However, given that the geographic ranges of each subclade include quite different physiographic and floristic units, a common selective regime seems unlikely as a basis for their uniform morphology.

Thus, the Wiens-Penkrot criteria support four consistently recognizable and valid species: (1) the "coastal" Clade 1A+B+C (which includes the insular taxa currently recognized as species, *N. bryanti*, *N. anthonyi*, *N. martinensis*, and *N. bunkeri*); (2) *Neotoma insularis* from Isla Ángel de la Guarda; (3) *Neotoma lepida* (including only the "desert" subclades 2A and 2B); and (4) *Neotoma devia* (the "desert" subclades 2C, D, and E). The status of the last species, *N. devia*, is somewhat equivocal. We also question whether or not both *N. lepida* and *N. devia* should be further subdivided at the species level. We believe that current data do not support such action.

Templeton's tests of cohesion

Templeton's (2001) approach also tests hypothesized species boundaries statistically, but through a set of nested null hypotheses evaluating the correlation between genotypes and/or phenotypes and geography (the Nested Clade Analysis originally described by Templeton, 1998). The method addresses two hypotheses: First (H_1), all organisms belong to a single evolutionary lineage; and, second (H_2), populations of separate lineages identified by rejection of the first hypothesis are genetically exchangeable and/or ecologically interchangeable among themselves. Species are then recognized after rejection of both hypotheses.

Templeton's NCA analysis is unnecessary to document that each of the nine mtDNA subclades we identify are separate, and exclusive, evolutionary lineages (Fig. 5; Matocq et al., 2007). Hence, his hypothesis H_1 is rejected for the entire *Neotoma lepida* group, except at the deepest phylogenetic level of the monophyly of this group relative to other woodrats. Alternatively, H_1 is supported for each subclade. The question, therefore, is whether any of the groups of these subclades exhibit genetic and/or ecological exchangeability sufficient to reject hypothesis H_2.

Our molecular Clades 1 and 2 satisfy the criterion for rejection of hypothesis H_2, because, despite the limited hybridization that has taken, and still does take, place at two limited areas of contact, the parental populations retain the unique genetic and morphological profiles indicative of non-exchangeability. These genetic and morphological entities replace one another ecologically and are, therefore, not ecologically interchangeable. By this reasoning, Clade 1 and Clade 2 are different cohesion species. We would further argue that the insular subclade 1D is an independent phylogenetic lineage (hypothesis H_1 rejected) and, by nature of its allopatric status as well as marked morphological distinctness, is genetically and ecologically non-exchangeable (hypothesis H_2 rejected). Therefore, *N. insularis* from Isla Ángel de la Guarda is a valid cohesion species. Within the three "coastal" groups of subclades (1A, 1B, and 1C), subclades 1B and 1C exhibit genetic exchangeability, although each is a separate mtDNA lineage, because individuals of these two subclades co-occur with apparent panmixia. Thus, H_2 is accepted, and members of subclades 1B and 1C form a single cohesion species. The circumstances for subclade 1A relative to these other two is less clear, as we have insufficient data to directly test the criteria of hypothesis H_2.

Hypotheses H_1 (all organisms belong to the same evolutionary lineage) and H_2 (there is both genetic and ecologic exchangeability) are also rejected when applied to the two "desert" clade clusters separated by the Colorado River (subclades 2A + 2B versus subclades 2C + 2D + 2E). Both groups clearly form independent clades defined by mtDNA and nucDNA, and the exclusivity of microsatellite assignments of individuals into unique groups separated by the river supports the lack of genetic exchangeability, even if the degree of morphological separation of these groups varies geographically. Rejection of H_2 is required, regardless of whether or not the lack of genetic exchangeability results solely because samples are allopatric and separated by the apparently impermeable barrier of the Colorado River. Thus, both *N. lepida* Thomas (subclades 2A + 2B) and *N. devia* (subclades 2C + 2D + 2E) meet Templeton's test of cohesion species. For the separate mtDNA lineages within either the western desert *N. lepida* or Arizona *N. devia*, although H_1 is rejected for each, the second level hypothesis of genetic and/or ecological exchangeability (H_2) cannot be rejected for subclades within either. These sets of mtDNA subclades, therefore, fail to meet Templeton's criteria of cohesion species.

The conclusions regarding species status, or lack thereof, based on each of these three different sets of objective, operational, and testable "programs" are consistent and uniform (Table 53). Given our conceptual species framework and this uniformity of operational criteria, we can reject species status for each of the four insular taxa currently recognized: *bryanti, anthonyi, martinensis,* and *bunkeri.* None of these taxa are exclusive relative to mainland populations of *Neotoma "lepida,"* including those on the adjacent peninsula, either phylogenetically or in morphological characters. Rather, the mtDNA haplotypes of each insular taxon are nested within broadly distributed mainland haplotype clades (subclade 1A in the case of *bryanti* and *bunkeri*; subclade 1B in the case of *anthonyi* and *martinensis* [Figs. 7, 8, 140, and 141]). Moreover, each shares the morphological characters of the coastal Clade 1 and is only marginally different from adjacent mainland samples in pairwise comparisons using multivariate analyses of craniodental or color variables (Figs. 99, 100, and 103). However, our molecular Clades 1 and 2 clearly conform to separate species, as do some of the subclades within each. Specifically, the coastal California and Baja California group of subclades (1A, 1B, and 1C), the insular subclade 1D (from Isla Ángel de la Guarda), the desert subclades (2A and 2B) west and north of the Colorado River, and the Arizona/Sonora group of desert subclades (2C, 2D, and 2E) are each uniformly accepted as valid species, although the decision regarding the last species is more equivocal (Table 53). We thus recognize four species within the *Neotoma lepida* group. In the following section, we provide a synopsis of the nomenclatural history of each species, discuss the subspecies concept and describe those that we recognize, list primary synonyms, describe and map distributions, and provide remarks regarding areas of uncertainty and future research.

Table 53. Summary conclusions of species boundary "tests" based on the application of the inclusive criteria of three different sets of operational "programs" (as summarized by Sites and Marshall, 2004).

Operational "program"	Taxon / Group comparison	Data summary	Species status conclusion
Correlated Distance Matrices (Corr-D)	currently recognized species *N. lepida, N. anthonyi, N. bryanti, N. bunkeri,* and *N. martinensis*	no correspondence of molecular clade membership and distribution of morphological character sets	test fails for each "species"
	"coastal" Clade 1 versus "desert" Clade 2	exact correspondence of molecular clade membership and distribution of all morphological character sets	separate species
	Clade 1D (*insularis* Townsend) versus Clade 1A+B+C	exact correspondence of molecular clade membership and distribution of all morphological character sets	separate species
	subclades 1A, 1B, and 1C	well supported, separate lineages , but all share the same set of qualitative and quantitative morphological characters; panmixia between subclades 1B and 1C	same species
	Clades 2A+B (desert *lepida* Thomas) versus Clades 2C+D+E (*devia* Goldman)	well supported separate lineages without evidence of gene exchange ; distinct karyotypes; limited distinctness in craniodental and phallic characters	separate species, but equivocal
	subclade 2A versus 2B, or subclades 2C versus 2D versus 2E	well supported separate lineages; morphological distinctness limited and inconsistent across characters; overall non-concordance of datasets	2A and 2B same species 2C, 2D, and 2E same species

Table 53 (continued)

Wiens-Penkrot phylogenetic method	currently recognized species N. lepida, N. anthonyi, N. bryanti, N. bunkeri, and N. martinensis	none are exclusive clades by either molecular or morphological character sets	test fails for each "species"
	"coastal" Clade 1 versus "desert" Clade 2	strongly supported, exclusive clades by molecular and morphological character sets concordant with geography	separate species
	Clade 1D (insularis Townsend) versus Clade 1A+B+C	strongly supported, exclusive clades by molecular and morphological character sets concordant with geography	separate species
	subclades 1A, 1B, 1C	mtDNA molecular, but not nucDNA molecular or morphological, well supported, exclusive molecular clades concordant with geography; inconsistent exclusivity across character sets	same species
	Clades 2A+B (desert lepida Thomas) versus Clades 2C+D+E (devia Goldman)	chromosomal and molecular exclusivity (both mtDNA, nucDNA, microsatellite assignments); partial exclusivity in morphological character sets	separate species, but equivocal
	subclade 2A versus 2B and subclade 2C versus 2D versus 2E	mtDNA clade exclusivity along; lack of exclusivity in morphological character sets	2A and 2B same species 2C, 2D, and 2E same species

Table 53 (continued)

Templeton's cohesion test			
	currently recognized species *N. lepida, N. anthonyi, N. bryanti, N. bunkeri,* and *N. martinensis*	H_1 – not rejected H_2 – not rejected	test fails for each "species"
	"coastal" Clade 1 versus "desert" Clade 2	H_1 – rejected H_2 – rejected	separate species
	Clade 1D (*insularis* Townsend) versus Clade 1A+B+C	H_1 – rejected H_2 – rejected	separate species
	subclades 1A, 1B, 1C	H_1 – rejected H_2 – not rejected (at least for 1B and 1C, the only two subclades for which the hypothesis can be directly tested)	same species (but status of Subclade 1A equivocal)
	Clades 2A+B (desert *lepida* Thomas) versus Clades 2C+D+E (*devia* Goldman)	H_1 – rejected H_2 – rejected	separate species
	subclade 2A versus 2B and subclade 2C versus 2D versus 2E	H_1 – rejected H_2 – not rejected	2A and 2B same species 2C, 2D, and 2E same species

SPECIES ACCOUNTS

In the accounts below, we provide synonymies for all available names that can be assigned to each of these species, delineate the geographic ranges of each, and provide an abbreviated morphological diagnosis. We also provide our view on subspecies and delineate those that we recognize. The four species that we recognize within the *Neotoma lepida* group, based on our application of the objective operational criteria summarized above (Table 53) are:

    Neotoma bryanti: comprising the "coastal" subclades 1A, 1B, and 1C distributed from Alameda Co., California south to the southern tip of Baja California and including the presently recognized insular "species" *N. anthonyi*, *N. martinensis*, and *N. bunkeri*, and all other insular named taxa (excluding *insularis* from Ángel de la Guarda), and also those samples from the Tehachapi Mts. and Kern River Plateau that are "coastal" in their morphology and microsatellite loci but possess the "desert" subclade 2A mtDNA.

    Neotoma insularis: the insular taxon that occurs on Isla Ángel de la Guarda, Baja California, Mexico, that comprises subclade 1D.

    Neotoma lepida: comprising the "desert" subclades 2A and 2B, distributed throughout the interior deserts west and north of the Colorado River in northeastern Baja California, eastern California, Nevada, southeastern Oregon, southwestern Idaho, Utah, extreme western Colorado, and northwestern Arizona.

    Neotoma devia: comprising the "desert" subclades 2C, 2D, and 2E east and south of the Colorado River in Arizona and northwestern Sonora, Mexico.

Key to species in the desert woodrat group

1.   Maxillofrontal suture intersecting lacrimal bone anterior to midpoint; small vomerine portion to the incisive foramen; auditory bullae small relative to size of skull; glans penis with or without greatly elongate hood but without strongly reflected distal tip; occurs in coastal California and Baja California.. 2

1'.   Maxillofrontal suture intersecting lacrimal bone posterior to midpoint; large vomerine portion to the incisive foramen; auditory bullae inflated relative to size of skull; glans penis with greatly elongate hood with strongly reflected distal tip; occurs east of coastal California in the deserts of western USA, northeastern Baja California, and northwestern Sonora in Mexico 3

2.   Anteroloph of M1 with deep anteromedial notch; glans penis thin with elongate baculum and hood, the latter with straight, tapered, and bifurcated fleshy tip .... *Neotoma bryanti*

2'.   Anteroloph of M1 with shallow anteromedial notch; glans penis stout with short baculum and hood, the latter with straight but blunt fleshy tip .... *Neotoma insularis*

3.   Occurs in deserts of eastern California, Nevada, Oregon, Idaho, Utah and Arizona west and north of the Colorado River .... *Neotoma lepida*

3'.   Occurs in deserts of northern and western Arizona south and east of the Colorado River .... *Neotoma devia*

Subspecies

The formal recognition of subspecies has been a dominant component of mammalian systematics over the past century or longer, and thus not surprisingly a large number of formal infraspecific taxa are currently recognized within the *Neotoma lepida* group.   Hall (1981), for example, lists 31 subspecies in his synoptic concept of *N. lepida*. In the systematic accounts that follow, we list those subspecies that we consider valid, but first we provide our views on subspecies, with special reference to the criteria we use in our recognition of these taxa.

The subspecies is an ill-defined and often illusory concept (Wilson and Brown, 1953) with a contentious history as a useful paradigm to recognize formally geographic units within species (Grinnell, 1935; Wilson and Brown, 1953; Brown and Wilson, 1954; Lidicker, 1962; Fjeldså, 1985).   We agree, however, with the position held by Patton and Smith (1990, pp. 105-110) in their review of pocket gophers of California, an extension of Grinnell's (1935, p. 403-404) four criteria: (1) "centers of differentiation," those areas of geographic uniformity of characters delimited by sharp clines at the boundaries; (2) these "centers" should represent evolutionary responses to history, not just ontogenetic or direct environmental influence; (3) subspecific units are thus to have an underlying phylogenetic basis; and (4) the use of the trinomial indicates incompleteness of differentiation (or, in Grinnell's view, the retention of genetic compatibility).   As Patton and Smith (1990, p. 108) state, however, "emphasis should be placed less on 'centers of differentiation' and more on the identification of geographic plateaus of character uniformity." Thus, in the accounts that follow, we base those subspecies we recognize on the combination of evolutionary uniqueness (largely, but not exclusively derived from our phylogenetic analyses of molecular characters) coupled with relative character uniformity in morphological traits over geography, with boundaries between adjacent taxa delimited by areas of sharp character transition. We note, however, that others might draw boundaries in different geographic places or recognize subspecies as valid that we do not, or not recognize one or more of our subspecies at all.

Neotoma bryanti Merriam
Bryant's Woodrat

Neotoma bryanti Merriam, 1887, American Naturalist, 21: 191. Type locality: "Cerros Island, off Lower California, in lat. 28° 12' N" [Isla Cedros, Baja California, Mexico].

Synonyms.—Listed under subspecies, below.

Diagnosis.—A large bodied woodrat within the *Neotoma lepida* group distributed along coastal California from east of San Francisco Bay south to the cape region of Baja California, including islands on both the Pacific and Gulf sides of the peninsula. Tail proportionally and absolutely long, averaging 85% of head-and-body length with 30 vertebral elements (range 29-34). Pelage relatively stiff and coarse (Grinnell and Swarth, 1913); overall color tones of body (dorsal, dorsal tail stripe, flank, and venter) dark (Fig. 19), although considerable geographic variation exists (Figs. 45, 84, 85, 99). Skull with absolutely and proportionately small auditory bullae (averaging 6.7 x 7.1 mm in length and width dimensions; Fig. 23); septum of incisive foramen comprised of small vomerine portion and elongated vacuity (Fig. 26); and contact of lacrimal with frontal equal to or greater than contact with maxilla (Figs. 27, 28). Anteromedial flexus of anteroloph of M1 deeply notched in all age classes except very old individuals (Figs. 24, 25). Glans penis relatively thin with elongate baculum and hood, the latter with straight, tapered, and bifurcated tip (Fig. 29).

We recognize five subspecies within *N. bryanti*, two primarily continental and three strictly insular (Fig. 148).

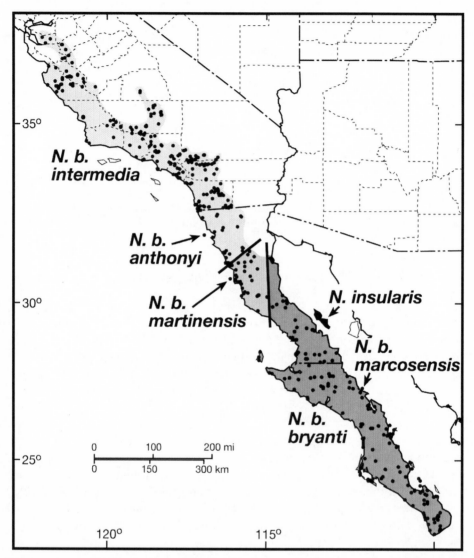

Figure 148. Geographic distribution of *Neotoma bryanti* Merriam and its five
subspecies along the west coast of California and throughout mainland Baja
California and associated islands, and Neotoma insularis from Isla Ángel de la
Guarda in the Gulf of California. Points are those localities from which we have
examined specimens assignable to this species. Thick lines and an intermediate
gray tone indicate the transitional area between the mainland subspecies *N. b.*
bryanti and *N. b. intermedia* in northern Baja California. Arrows identify each
insular subspecies or species.

Neotoma bryanti bryanti Merriam

Synonyms:

1887. *Neotoma bryanti* Merriam, see above.

1898. *Neotoma arenacea* J. A. Allen, Bulletin of the American Museum of Natural History, 10 (8): 150. Type locality: San José del Cabo, Lower California" [Baja California Sur, Mexico].

1903. *Neotoma bella felipensis* Elliot, Field Columbia Museum, publication 79, Zoological series, 3: 217. Type locality: "San Felipe, Gulf of California, Lower California" [Baja California, Mexico].

1905. *Neotoma nudicauda* Goldman, Proceedings of the Biological Society of Washington, 18: 28. Type locality: "Carmen Island, Lower California, Mexico" [Isla del Carmen, Baja California Sur].

1909. *Neotoma intermedia pretiosa* Goldman, Proceedings of the Biological Society of Washington, 22: 139. Type locality: "Matancita (called also Soledad), 50 miles north of Magdalena Bay, Lower California, Mexico (altitude 50 feet)" [Baja California Sur].

1909. *Neotoma intermedia perpallida* Goldman, Proceedings of the Biological Society of Washington, 22: 139. Type locality: "San Jose Island, off east coast of Lower California, Mexico" [Isla San José, Baja California Sur].

1909. *Neotoma intermedia vicina* Goldman, Proceedings of the Biological Society of Washington, 22: 140. Type locality: "Espiritu Santo Island, off east coast of southern Lower California, Mexico" [Isla Espíritu Santo, Baja California].

1909. *Neotoma abbreviata* Goldman, Proceedings of the Biological Society of Washington, 22: 140. Type locality: "San Francisco Island (near San Jose Island), off east coast of southern Lower California, Mexico" [Isla San Francisco, Baja California Sur].

1931. *Neotoma intermedia ravida* Nelson and Goldman, Proceedings of the Biological Society of Washington, 44: 107. Type locality: "Comondú, southern Lower California, Mexico (altitude 700 feet)" [Baja California Sur].

1931. *Neotoma intermedia notia* Nelson and Goldman, Proceedings of the Biological Society of Washington, 44: 108. Type locality: "La Laguna, Sierra de la Victoria, southern Lower California, Mexico (altitude 5500 feet)" [= Sierra La Laguna, Baja California Sur].

1932. *Neotoma lepida felipensis*: Goldman, Journal of Mammalogy, 13: 64 (name combination).

1932. *Neotoma lepida pretiosa*: Goldman, Journal of Mammalogy, 13: 64 (name combination).

1932. *Neotoma lepida ravida*: Goldman, Journal of Mammalogy, 13: 64 (name combination).

1932. *Neotoma lepida arenacea*: Goldman, Journal of Mammalogy, 13: 65 (name combination).

1932. *Neotoma lepida notia*: Goldman, Journal of Mammalogy, 13: 65 (name combination).

1932. *Neotoma lepida perpallida*: Goldman, Journal of Mammalogy, 13: 65 (name combination).

1932. *Neotoma lepida vicina*: Goldman, Journal of Mammalogy, 13: 65 (name combination).

1932. *Neotoma lepida latirostra* Burt, Transactions of the San Diego Society of Natural History, 7 (16): 180. Type locality: Danzante Island (latitude 25° 47' N., longitude, 111° 11' W.), Gulf of California, Lower California, Mexico" [Isla Danzante, Baja California Sur].

1932. *Neotoma bunkeri* Burt, Transactions of the San Diego Society of Natural History, 7 (16): 181. Type locality: "Coronados Island (latitude 26° 06' N., longitude, 111° 18' W.), Gulf of California, Lower California, Mexico" [Isla Coronados, Baja California Sur].

1932. *Neotoma lepida nudicauda*: Burt, Transactions of the San Diego Society of Natural History, 7 (16): 182 (name combination).

1932. *Neotoma lepida abbreviata*: Burt, Transactions of the San Diego Society of Natural History, 7 (16): 182 (name combination).

1945. *Neotoma lepida molagrandis* Huey, Transactions of the San Diego Society of Natural History, 10 (16): 307. Type locality: "Santo Domingo Landing [lat. 28° 15'N.], Baja California, Mexico (more precisely, at the site of the old well near the edge of a mesa-like shelf, some 3 miles inland from the landing beach, elevation about 50')."

1957. *Neotoma lepida aridicola* Huey, Transactions of the San Diego Society of Natural History, 12(15): 287. Type locality: "El Barril (near 28° 20' N), Gulf of California, Baja California, Mexico."

Distribution (Fig. 148).—Baja California from approximately San Felipe on the gulf side and the vicinity of Punta Prieta on the Pacific side of the state of Baja California in the north to the Cape region in Baja California Sur in the south, including Cedros (type locality), Magdalena, and Margarita islands on the Pacific coast and the islands of Coronados, Carmen, Danzante, San José, San Francisco, Partida, and Espíritu Santo in the Gulf of California. There is a broad transition zone between this subspecies and *N. b. intermedia* across the region between Punta Prieta and El Rosario.

Remarks.—The dorsal and ventral aspects of the skull of the holotype (USNM 186481) are illustrated in Fig. 149. We arbitrarily define the transition zone between *N. b. bryanti* and *N. b. intermedia* as that region of discordance between the mtDNA clade (Fig. 6) and colorimetric boundary (Fig. 105) and the sharp step cline in size-free craniodental shape (Figs. 97, 99). Additional analyses of this area should focus on the integration of molecular genetic attributes with morphological characterization, as we have done in both the Tehachapi and San Gorgonio Pass Transects (above) to determine both the steepness of the morphological character cline and evidence of panmixia, or lack thereof, between co-occurring individuals of mtDNA subclade 1A and 1B haplotypes. There are areas of relatively sharp craniodental "size-free" and colorimetric transitions at various points along the length of the Baja peninsula, but these are localized (between one combination of samples but not for another in the same general geographic region), and we choose not to recognize any of these as boundaries between formal taxa. Similarly, although each of the insular taxa from the six landbridge islands off the southern Gulf coast differ in one degree or another, most do so simply in overall size, a singular character upon which we also choose not to base formal taxon decisions. Each of these taxa, which include *bunkeri*, uniformly considered as a distinct species by every author since its original description in 1932 (but see comment by Musser and Carleton, 2005, p. 1054), we thus list as synonyms. We acknowledge, however, that the degrees of difference between some of these insular forms and samples from the mainland (particularly for *abbreviata* Goldman) are only marginally less than those used to justify some insular taxa that we do recognize.

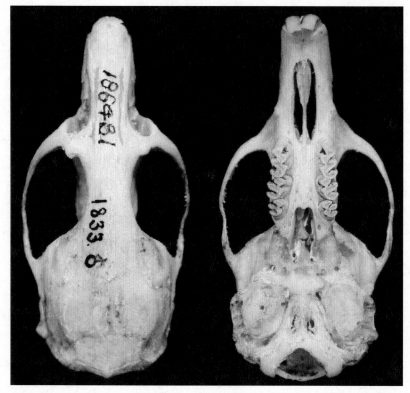

Figure 149. Dorsal and ventral views of the skull of the holotype (USNM 186481) of *Neotoma bryanti*.

Neotoma bryanti intermedia Rhoads

Synonyms:

1894. *Neotoma intermedia* Rhoads, American Naturalist, 28: 68. Type locality: "Dulzura, San Diego Co., Cal." [California].

1894. *Neotoma intermedia gilva* Rhoads, American Naturalist 28: 70. Type locality: "Banning, San Bernardino County, Cal." [California].

1894. *Neotoma californica* Price, Proceedings of the California Academy of Sciences, 2nd series, 3: 154. Type locality: "Bear Valley, San Benito County, California."

1894. *Neotoma desertorum sola* Merriam, Proceedings of the Biological Society of Washington, 9: 126. Type locality: "San Emigdio, Kern County, California."

1932. *Neotoma lepida gilva*: Goldman, Journal of Mammalogy, 13: 63 (name combination).

1932. *Neotoma lepida intermedia*: Goldman, Journal of Mammalogy, 13: 64 (name combination).

1934. *Neotoma lepida egressa* Orr, Proceedings of the Biological Society of Washington, 47: 109 (type locality: one mile east of El Rosario, 200 feet altitude, Lower California, Mexico" [Baja California].

1938. *Neotoma lepida petricola* von Bloeker, Proceedings of the Biological Society of Washington, 51: 203. Type locality: "Abbott's Ranch, 670 feet altitude, Arroyo Seco, Monterey Co., California."

1938. *Neotoma lepida californica*: von Bloeker, Proceedings of the Biological Society of Washington, 51: 201 (name combination).

Distribution (map, Fig. 148).—Alameda Co. east of the San Francisco Bay in central California south along both inner and outer coast ranges, the western foothills of the southern Sierra Nevada, Transverse, and Peninsular ranges, as well as coastal southern California into northwestern Baja California, at least as far as El Rosario where it begins to grade into *N. b. bryanti* (see that account, above).

Remarks.—In the Coastal California Transect, above, we document that character sets (craniodental, colorimetric, and nuclear microsatellite loci) either exhibit no pattern of between-sample differentiation or one that is clinal. Smooth character clines are apparent from north to south or west to east along all regions of this subspecies range. Importantly, although this subspecies includes two mtDNA clades (subclade 1B and 1C), individuals of both interbreed in a panmictic fashion where sympatric and cannot be distinguished morphologically (see discussion in the "Taxonomic considerations" subsection). There seems no justification for continuing to recognize *gilva*, *californica*, or *petricola* as valid infraspecific taxa (contra Hall, 1981). *Neotoma desertorum sola* has long been considered a synonym of *gilva* (e.g., Goldman, 1910, p. 44). We assign *egressa* to this subspecies although it occupies the transitional area between *N. b. bryanti* and *N. b. intermedia* (Fig. 148). This decision is completely arbitrary, but does place more weight on the fact that our samples of *egressa* belong to mtDNA subclade 1B, which otherwise is contained completely within the range of *N. b. intermedia*. However, since mtDNA can clearly introgress across species boundaries (i.e., between *N. bryanti* and *N. lepida* in the Tehachapi Mts. and Kern River Plateau;

see Tehachapi Transect, above), basing subspecies on mtDNA boundaries may be less defensible then doing so on morphological characters.

Neotoma bryanti anthonyi J. A. Allen

Synonyms:
1898. *Neotoma anthonyi* J. A. Allen, Bulletin of the American Museum of Natural History, 10 (8): 151. Type locality: "Todos Santos Island, Lower California" [Isla Todos Santos, Baja California, Mexico].

Distribution (map, Fig. 148).—Known only from Isla Todos Santos.

Remarks.—We recognize this taxon by virtue of its sharp distinction from the adjacent mainland samples in both craniodental size and "size-free" shape parameters (Figs. 97 and 99) and color (Fig. 105). It is diagnosable from other infraspecific taxa of *N. bryanti* except *N. b. martinensis* by its conspicuous blackish outer sides of the hind legs and inner sides of the ankles. However, the mtDNA *cyt-b* haplotype recovered is nested well within the mainland subclade 1B (Fig. 141), which suggests a recent origin for this insular population. This taxon is now apparently extinct (Mellink, 1992b).

Neotoma bryanti martinensis Goldman

Synonyms:
1905. *Neotoma martinensis* Goldman, Proceedings of the Biological Society of Washington, 18: 28. Type locality: "San Martin Island, Lower California, Mexico" [Isla San Martín, Baja California].

Distribution (map, Fig. 148).—Known only from Isla San Martín.

Remarks.—As with *N. b. anthonyi*, we recognize this taxon because of its sharp distinction from adjacent mainland samples in craniodental size and "size-free" characters (Figs. 97 and 99), although it does not differ in color (Fig. 105). It is diagnosable from the other infraspecific taxa of *N. bryanti* except *anthonyi* by its conspicuous blackish outer sides of the hind legs and inner sides of the ankles. The mtDNA haplotype we have obtained from two individuals, however, is quite different from other members of its geographic "clade" (Fig. 141), a group that has apparently undergone a past coalescent history of allopatric fragmentation. This taxon is also likely extinct (Mellink, 1992a).

Neotoma bryanti marcosensis Burt

Synonyms:

1932. *Neotoma lepida marcosensis* Burt, Transactions of the San Diego Society of Natural History, 7 (16): 179. Type locality: San Marcos Island (latitude 27°13'N., longitude, 112°05'W.), Gulf of California, Lower California, Mexico" [Isla San Marcos, Baja California Sur].

Distribution (map, Fig. 148).—Known only from Isla San Marcos, a mid-rift landbridge island in the Gulf of California.

Remarks.—This is the only taxon from the landbridge Gulf of California islands that is differentiated from the adjacent mainland samples in both size and "size-free" craniodental characters (Figs. 95, 97, 101), although it does not differ in color (Fig. 105). Molecularly, our sample of *N. b. marcosensis* is well differentiated from others from both the mainland and other islands in microsatellite allelic divergence (Fig. 106), and the two *cyt-b* haplotypes recovered are within that portion of the parsimony network for which there is no correspondence between phylogenetic position and geographic location (Fig. 140).

Neotoma insularis Townsend
Ángel de la Guarda Woodrat

Neotoma insularis Townsend, 1912, Bulletin of the American Museum of Natural History, 31 (13): 125. Type locality: "Angel del la Guardia Island" [Isla Ángel de la Guarda, Baja California, Mexico].

Synonyms:

1932. *Neotoma lepida insularis*: Burt, Transactions of the San Diego Society of Natural History, 7 (16): 182 (name combination).

Diagnosis.—An insular species characterized by moderate body and cranial size (Fig. 95) and proportionately short tail (approximately 71% of head-and-body length). Overall color tones pale. Skull short and stocky, with noticeably short and broad rostrum and squared zygomatic arches (Fig. 150), long but narrow auditory bullae, septum of incisive foramen with short vomerine portion and elongated vacuity, and frontal contact with the lacrimal much greater than maxillary contact (Fig. 28). Anteromedial flexus of anteroloph of M1 shallow except in young individuals (Fig. 25). Glans penis stout with short baculum and hood, the latter with straight but blunt fleshy tip (Fig. 31).

Distribution.—Known only from Ángel de la Guarda, with the few localities on both the southern and northern ends of this island.

Remarks.—Dorsal and ventral views of skull of holotype (USNM 198405) are illustrated in Fig. 150. Although originally described as a distinct species, *insularis* has been listed as a subspecies by all authors since Burt (1932) placed it in synonymy of *N. lepida* by simple proclamation rather than by apparent examination of any specimens and certainly without analysis of any kind. As delineated above, however, this taxon is clearly and strongly defined as a unique and well-supported molecular clade (mtDNA phylogeny, Fig. 5) and can be diagnosed by a number of morphological attributes for the few specimens that are known. This species is currently considered by the Mexican government to be under threat of extinction, and recent attempts to secure specimens have failed, suggesting that it might already be extinct (Álvarez-Castañeda and Cortés-Calva, 1999). A thorough trapping program is recommended to determine the true status of this species, as remnant populations are at least possible given the overall size and topographic diversity of the island.

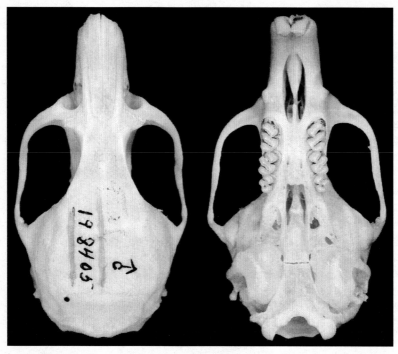

Figure 150. Dorsal and ventral views of holotype (USNM 198405) of *Neotoma insularis*.

Neotoma lepida Thomas
Desert Woodrat

Neotoma lepida Thomas, 1893, Annals and Magazine of Natural History, Ser. 6, Vol. 7: 235. Type locality: "Utah" (refined to "somewhere on 'Simpson's route' between Camp Floyd [a few miles west of Utah Lake], Utah, and Carson City, Nevada" by Goldman, 1932, p. 61).

Synonyms.—Listed under subspecies, below.

Diagnosis.—A relatively small-bodied woodrat limited to the drier deserts of western North America, with an absolutely and proportionately short tail in comparison to *N. bryanti* (Fig. 20; approximately 80% of head-and-body length [Grinnell and Swarth, 1913], with a mean of 25.3 caudal vertebrae). Same number of caudal vertebrae as *N. devia*, but latter has a proportionately longer tail (see below). Pelage typically long and soft; overall color tones pale (Fig. 19), although melanism is present in many populations. Skull with large auditory bullae (averaging 7.2 x 7.5 mm in length and width); elongated vomerine portion to incisive foramen septum, with corresponding short vacuity (Fig. 26); and frontal contact with lacrimal less than half the length of the maxillary contact (Fig. 27, 28). Anteromedial flexus of anteroloph of M1 shallow in all by youngest aged individuals (Fig. 24, 25). A glans penis with a greatly elongated hood with strongly reflected distal tip is shared with *N. devia* but distinct from those of *N. bryanti* and *N. insularis* (Figs. 29 and 31).

 We recognize three subspecies within this widespread taxon (Fig. 151). Two of these (*N. l. lepida* and *N. l. monstrabilis*) correspond to mtDNA subclades 2A and 2B that are bounded by the Virgin River in southwestern Nevada, northwestern Arizona, and southwestern Utah (Fig. 5). While most specimens of *N. l. monstrabilis* Goldman that we have examined share the same glans penis morphology that Mascarello (1978) and we document for *Neotoma devia* (Fig. 30), posterior assignments based on canonical analysis of craniodental variables sort specimens from both sides of this river into separate groups with high individual probabilities (Figs. 128, 132). Moreover, a phylogenetic linkage between *lepida* Thomas and *monstrabilis* Goldman with respect to *N. devia* is quite strong, in mtDNA (Fig. 5) and nuclear gene sequences (Matocq et al., 2007) as well as in individual assignments based on microsatellite loci (Figs. 130 and 131). Thus, the transition across the Virgin River appears sharp, marking a strong step in otherwise clinal variation in craniodental characters yet discrete in phallic and molecular characters. The third subspecies we recognize are the insular samples in the Great Salt Lake of northern Utah (*N. l. marshalli*) that differ in craniodental characteristics from nearby samples of *N. l. lepida* (Fig. 116) although it shares mtDNA haplotypes within the same subclade 2A.

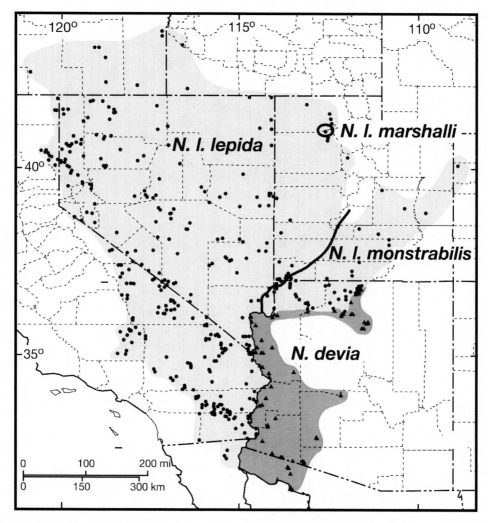

Figure 151. Distribution of *Neotoma lepida* Thomas throughout the Colorado Desert, Mojave Desert, and Great Basin Desert north and west of the Colorado River, and of *Neotoma devia* Goldman from east and south of the Colorado River in Arizona. The solid lines indicate the approximate boundary between the subspecies *N. l. lepida* and *N. l. monstrabilis* in southern Nevada and southwestern Utah and *N. l. marshalli* and *N. l. lepida* in northern Utah. No subspecies are recognized within *N. devia*.

Neotoma lepida lepida Thomas

Synonyms:

1893. *Neotoma lepida* Thomas, see above.
1894. *Neotoma desertorum* Merriam, Proceedings of the Biological Society of Washington, 9: 125. Type locality: "Furnace Creek, Death Valley, California" [Inyo Co.].
1899. *Neotoma bella* Bangs, Proceedings of the New England Zoological Club, 1: 66. Type locality: "Palm Springs, Riverside Co., California."
1910. *Neotoma nevadensis* Taylor, University of California Publications in Zoology, 5 (6): 289. Type locality: "Virgin Valley, Humboldt Co., Nevada, altitude 4800 ft."
1942. *Neotoma lepida grinnelli* Hall, University of California Publications in Zoology, 46 (5): 369. Type locality: "Colorado River, 20 miles above (by river, but about 12 ½ miles north by air-line) Picacho, Imperial Co., California."

Distribution (map, Fig. 151).—This subspecies occurs widely throughout the Colorado Desert of southeastern California and adjacent northeastern Baja California and north through the Mojave Desert and Great Basin Desert of northeastern California, southeastern Oregon, southern Idaho, Nevada, and western Utah.

Remarks.—We have not examined the holotype of *N. lepida* Thomas, which is housed in the Natural History Museum (London), but illustrate the skull of the holotype of *Neotoma desertorum* Merriam in Fig. 152. Nearly all samples examined of this subspecies possess the western desert tip type of the glans penis, as described above and mapped in Fig. 30. Hence, this subspecies differs from *N. l. monstrabilis* east and south of the Virgin River in Nevada, Arizona, and Utah, of which most samples exhibit the eastern desert tip type.

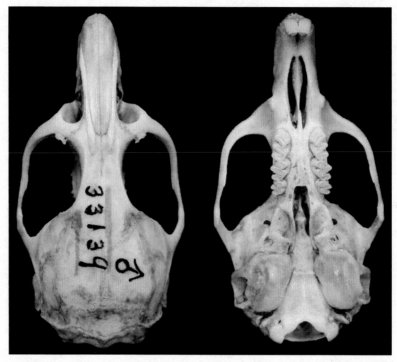

Figure 152. Dorsal and ventral views of the holotype (USNM 33139) of *Neotoma desertorum.*

Neotoma lepida monstrabilis Goldman

Synonyms:

1932. *Neotoma lepida monstrabilis* Goldman, Journal of Mammalogy, 13: 62. Type locality: "Ryan, Kaibab National Forest, Coconino County, Arizona (altitude 6,000 feet)."

1949. *Neotoma lepida sanrafaeli* Kelson, Journal of the Washington Academy of Sciences, 38: 418. Type locality: "Rock Canyon Corral, 5 miles southeast of Valley City, 4,500 feet, Grand County, Utah."

Distribution.—Southern Nevada east of the Virgin River, Arizona north of the Grand Canyon, and southern and eastern Utah and extreme western Colorado throughout the upper Colorado River basin (Fig. 151).

Remarks.—Tip of the glans penis largely of the eastern desert type (Fig. 30), a morphology shared with *N. devia* not *N. l. lepida*.

Neotoma lepida marshalli Goldman

Synonyms:
1939.   *Neotoma lepida marshalli* Goldman, Journal of Mammalogy, 20: 357. Type locality: "Carrington Island, Great Salt Lake, Utah (altitude about 4,250 feet)" [Tooele Co.].

Distribution.—Known only from Carrington and Stansbury islands, Great Salt Lake, Tooele Co., Utah.

Remarks.—We examined no glans penis for this taxon, and thus the overall structure of the glans as well as the tip type is unknown.

Neotoma devia Goldman
Arizona Woodrat

Neotoma intermedia devia Goldman, 1927, Proceedings of the Biological Society of Washington, 40: 205. Type locality: "Tanner Tank (altitude 5,200 feet), Painted Desert, Arizona" [Coconino Co.].

Synonyms:
1927.   *Neotoma intermedia devia* Goldman, as above.
1932.   *Neotoma lepida devia*: Goldman, Journal of Mammalogy, 13: 62 (name combination).
1933.   *Neotoma auripila* Blossom, Occasional Papers of the Museum of Zoology, University of Michigan, No. 273, p. 1. Type locality: Agua Dulce Mountains, 9 miles east of Papago Well, Pima County, Arizona."
1935.   *Neotoma lepida bensoni* Blossom, Occasional Papers of the Museum of Zoology, University of Michigan, No. 315, p. 1. Type locality: "Papago Tanks, Pinacate Mountains, Sonora, Mexico."
1935.   *Neotoma lepida auripila*: Blossom, Occasional Papers of the Museum of Zoology, University of Michigan, No. 315, p. 3 (name combination).
1935.   *Neotoma lepida flava* Benson, Occasional Papers of the Museum of Zoology, University of Michigan, No. 317, p. 7. Type locality: "Tinajas Altas, 1150 feet, Yuma County, Arizona."

1937.   *Neotoma lepida aureotunicata* Huey, Transactions of the San Diego Society of Natural History, 8 (25): 349. Type locality: "Punta Peñascosa, Sonora, Mexico."

1937.   *Neotoma lepida harteri* Huey, Transactions of the San Diego Society of Natural History, 8 (25): 351. Type locality: "10 miles south of Gila Bend (or, exactly, from the summits of a group of lava hills on the east side of the Ajo railroad, about 2 miles north of Black Gap), Maricopa County, Arizona."

Diagnosis.—A small-bodied woodrat restricted to the deserts of western Arizona and northwestern Sonora. Tail proportionately long (86% of head-and-body length) but with number of caudal vertebrae averaging 25.5 and not significantly different from that found in *N. lepida* ($F_{(1,55)}$ = 0.076, p = 0.784). Pelage soft and silky; overall color tones intermediate with melanic individuals and/or populations known at various localities throughout range. Skull with inflated bullae (average 7.1 x 7.4 mm length by width); septum of incisive foramen with short vomer and enlarged vacuity; frontal contact with lacrimal much less than maxilla contact. Anteromedial flexus of anteroloph of M1 shallow to obsolete, except in very young individuals. Morphological features of the glans penis shared with most samples of *N. l. monstrabilis* but not with *N. l. lepida* (see above).

Distribution (map, Fig. 151).—Western Arizona south and east of the Colorado River (see also Hoffmeister, 1986); northwestern Sonora, Mexico.

Remarks.—Dorsal and ventral views of the skull of the holotype (USNM 226376) are illustrated in Fig. 153. As noted above, the species status of *Neotoma devia* is somewhat equivocal as the boundary between it and *N. lepida* is the apparently impermeable barrier of the Colorado River. Hence, it is not possible to test whether the exclusivity of molecular characters (karyotype, mtDNA, nuclear microsatellites, and allozymes) results solely from lack of gene flow due to physical separation or if the two forms are, in fact, genetically incompatible. It is not surprising, therefore, that different authors have reached opposite conclusions regarding the status of *N. devia* (e.g., Mascarello, 1978, and Musser and Carleton, 2005, versus Hoffmeister, 1986). Indeed, as we noted in our description of the glans penis and in our remarks about *Neotoma lepida*, and as discussed by Hoffmeister (1986), the tip of the glans penis is shared between *N. devia* and *N. lepida monstrabilis*, on both sides of the Grand Canyon (Fig. 30) but not by *N. devia* and *N. lepida lepida*, along the lower Colorado River. Thus, details of the tip of the glans penis are discordant with other characters and, contrary to Mascarello (1978), cannot be used to diagnose either species.

Hoffmeister (1986) did recognize 2 subspecies in western Arizona, separating a northern *devia* from a southern *auripila*, and placed the boundary between them at about the Bill Williams River. Other than for the concordant position of mtDNA subclades 2C and 2D, our analyses do not support this boundary. For example, craniodental PC-1 scores ("size") exhibit only clinal variation among our sample groups (Figs. 124 and 132), indicate no difference among samples (craniodental PC-1 [= "shape"] parameters, Fig. 127), or exhibit greater difference between samples of *devia* from north of Flagstaff and south of Lake Mead than between Hoffmeister's two subspecies (canonical analysis [Fig. 128], color [FC-1 scores, Fig. 129], or microsatellite divergence [Fig. 130]). If infraspecific units are to be recognized within *N. devia*, a more defensible boundary would be placed between Painted Desert samples (those from north of Flagstaff that include the type locality of *devia*) and those along the lower Colorado River (which would include *auripila*, *benson*, *flava*, *aureotunicata*, and *harteri*). We choose not to do so.

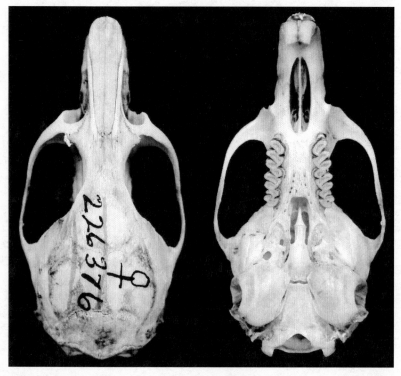

Figure 153. Dorsal and ventral views of the skull of the holotype (USNM 226376) of *Neotoma devia*.

Literature Cited

Allen, J. A.
  1898.   Descriptions of new mammals from western Mexico and Lower California. Bulletin of the American Museum of Natural History, 10: 143-158.

Álvarez-Castañeda, S. T., and P. Cortés-Calva
  1999.   Familia Muridae. Pp. 445-570, in Mamíferos del Noroeste de México (S. T. Álvarez-Castañeda and J. L. Patton, eds.). Centro de Investigaciones Biologícas Noroeste, S. C. La Paz, Baja California, México.

Álvarez-Castañeda, S. T., and A. Ortega-Rubio.
  2003.   Current status of rodents on islands in the Gulf of California. Biological Conservation, 109: 157-163.

Anderson, E. C., and E. A. Thompson.
  2002.   A model-based method for identifying species hybrids using multilocus genetic data. Genetics, 160: 1217-1229.

Angers, B., and L. Bernatchez.
  1998.   Combined use of SMM and non-SMM methods to infer fine structure an evolutionary history of closely related brook char populations from microsatellites. Molecular Biology and Evolution, 15: 143-159.

Arbogast, B. S., S. V. Edwards, J. Wakeley, P. Beerli, and J. B. Slowinski
  2002.   Estimating divergence times from molecular data on phylogenetic and population genetic timescales. Annual Review of Ecology and Systematics, 33: 707-740.

Arbogast, B. S., and J. B. Slowinski
  1998.   Pleistocene speciation and the mitochondrial DNA clock. Science, 282: 1955.

Aris-Brosou, S., and L. Excoffier
    1996.   The impact of population expansion and mutation rate heterogeneity on DNA sequence polymorphism. Molecular Biology and Evolution, 13: 494-504.

Avise, J. C.
    2000.   Phylogeography. The History and Formation of Species. Harvard University Press, Cambridge, Massachusetts.

Avise, J. C., J. Arnold, R. M. Ball, E. Bermingham, T. Lamb, J. E. Neigel, C. A. Reeb, and N. C. Saunders
    1987.   Intraspecific phylogeography: the mitochondrial bridge between population genetics and systematics. Annual Review of Ecology and Systematics, 18: 489-522.

Baker, R. J., and J. T. Mascarello
    1969.   Karyotypic analyses of the genus *Neotoma* (Cricetidae, Rodentia). Cytogenetics, 8: 187-198.

Bangs, O.
    1899.   Descriptions of some new mammals from western North America. Proceedings of the New England Zoological Club, 1: 65-72.

Benson, S. B.
    1935.   Geographic variation in *Neotoma lepida* in Arizona. Occasional Papers of the Museum of Zoology, University of Michigan, 317: 1-9.

Betancourt, J. L., T. R. Van Devender, and P. S. Martin
    1990.   Packrat Middens. The Last 40,000 Years of Biotic Change. The University of Arizona Press, Tucson, Arizona.

Blossom, P. M.
    1933.   A new woodrat from southwestern Arizona and a new rock pocket-mouse from northwestern Arizona. Occasional Papers of the Museum of Zoology, University of Michigan 273: 1-5.
    1935.   Description of a race of desert woodrat (*Neotoma lepida*) from Sonora. Occasional Papers of the Museum of Zoology, University of Michigan, 315: 1-3.

Bohonak, A. J.
    2002.   IBD (Isolation By Distance): a program for analyses of isolation by distance. Journal of Heredity, 93: 153-154.

Brown, W. L., Jr., and E. O. Wilson
    1954.   The case against the trinomen. Systematic Zoology, 5: 157-160.

Burt, W. H.
    1932.   Descriptions of heretofore unknown mammals from islands in the Gulf of California, Mexico. Transactions of the San Diego Society of Natural History, 7: 161-182.
    1960.   Bacula of North American mammals. Miscellaneous Publications, Museum of Zoology, University of Michigan, 113: 1-76, + 25 plates.

Burt, W. H., and F. S. Barkalow, Jr.
    1942.   A comparative study of the bacula of woodrats (Subfamily Neotominae). Journal of Mammalogy, 23: 287-297.

Cameron, G. N.
    1971.   Niche overlap and competition in woodrats. Journal of Mammalogy, 52: 288-296.

Carleton, M. D.
    1980.   Phylogenetic relationships in neotomine-peromyscine rodents (Muroidea) and a reappraisal of the dichotomy within New World Cricetinae. Miscellaneous Publications, Museum of Zoology, University of Michigan, 157: 1-146.

Carreño, A. L., and J. Helenes
    2002.   Geology and ages of the islands. Pp. 14-40, in A New Island Biogeography of the Sea of Cortés (T. J. Case, M. L. Cody, and E. Ezcurra, eds.). Oxford University Press, New York, New York.

Clement, M., D. Posada, and K. A. Crandall
    2000.   TCS: a computer program to estimate gene genealogies. Molecular Ecology, 9: 1657-1660.

Crandall, K. A.
  1996. Multiple interspecies transmissions of human and simian T-cell leukemia/lymphoma virus type I sequences. Molecular Biology and Evolution, 13: 115-131.

Dial, K. P., and N. J. Czaplewski
  1990. Do woodrat middens accurately reflect the animals' environment or diets? The Woodhouse Mesa study. Pp. 43-58, in Packrat Middens: The Last 40,000 Years of Biotic Change (J. L. Betancourt, T. R. Van Devender, and P. S. Martin, eds.). The University of Arizona Press, Tucson, Arizona.

dos Reis, S. F., L. M. Pessôa, and R. E. Strauss
  1990. Application of size-free canonical discriminant analysis to studies of geographic differentiation. Revista Brasiliera de Genética, 3: 509-520.

Edwards, C. W., and R. D. Bradley
  2001. Molecular phylogenetics of the *Neotoma floridana* species group. Journal of Mammalogy, 82: 791-798.
  2002. Molecular systematics of the genus *Neotoma*. Molecular Phylogenetics and Evolution, 25: 489-500.

Edwards, C. W., C. F. Fulhorst, and R. D. Bradley
  2001. Molecular phylogenetics of the *Neotoma albigula* species group: further evidence of a paraphyletic assemblage. Journal of Mammalogy, 82: 267-279.

Edwards, S. V.
  1997. Relevance of microevolutionary processes to higher level molecular systematics. Pp. 251-278 in Avian Molecular Evolution and Systematics (D. P. Mindell, ed.). Academic Press, San Diego, California.

Edwards, S. V., and P. Beerli
  2000. Gene divergence, population divergence, and the vicariance in coalescence time in phylogeographic studies. Evolution, 54: 1839-1854.

Elliot, D. G.
    1903.   A list of mammals collected by Edmund Heller, in the San Pedro Martir and Hanson Laguna Mountains and the accompanying coast regions of lower California with descriptions of apparently new species. Field Columbian Museum Publication No. 79, Zoological Series, 3(12): 199-232.
    1904.   Catalogue of mammals collected by E. Heller in southern California. Field Columbian Museum Publication No. 91, Zoological Series, 3(16): 271-321.

Estoup, A., L. Garnery, M. Solignac, and J.-M. Cournuet
    1995.   Microsatellite variation in honeybee (*Apis mellifera* L.) populations: hierarchical genetic structure and test of the infinite allele and stepwise mutation models. Genetics, 140: 679-695.

Excoffier, L.
    2004.   Patterns of DNA sequence diversity and genetic structure after a range expansion: lessons from the infinite-island model. Molecular Ecology, 13: 853-864.

Excoffier, L., G. Laval, and S. Schneider
    2005.   Arlequin ver 3.0. An integrated software package for population genetics data analysis. Evolutionary Bioinformatics Online 1: 47-50 (available from http://cmpg.unibe.ch/software/arlequin3).

Fjeldså, J.
    1985.   Subspecies recognition in ornithology: history and the current rationale. Fauna Norvegica, Series C, Cinclus, 8: 57-63.

Fu, X.-Y.
    1997.   Statistical tests of neutrality of mutations against population growth, hitchhiking and background selection. Genetics, 147: 915-925.

Gilbert M. T. P., H.-J. Brandelt, M. Hofreiter, and I Barnes
    2005.   Assessing ancient DNA studies. Trends in Ecology and Evolution, 20:541-544.

Goldman, E. A.
  1905.   Twelve new woodrats of the genus *Neotoma*. Proceedings of the Biological Society of Washington, 18: 27-34.

  1909.   Five new woodrats of the genus *Neotoma* from Mexico. Proceedings of the Biological Society of Washington, 22: 139-142.

  1910.   Revision of the woodrats of the genus *Neotoma*. North American Fauna, 31:5-124.

  1927.   A new woodrat from Arizona. Proceedings of the Biological Society of Washington, 40: 205-206.

  1932.   Review of woodrats of the *Neotoma lepida* group. Journal of Mammalogy, 13: 59-67.

  1939.   Nine new mammals from islands in Great Salt Lake Utah. Journal of Mammalogy, 20: 351-357.

Goudet, J.
  2001.   FSTAT, Version 2.9.3. A program to estimate and test gene diversities and fixation indices. Computer program available from http://www.unil.ch/izea/softwares/fstat.html (updated from Goudet, 1995, Journal of Heredity, 86: 485-486).

Grayson, D. K.
  1993.   The Desert's Past. A Natural Prehistory of the Great Basin. Smithsonian Institution Press, Washington, D.C.

Grinnell, J.
  1933.   A review of the recent mammal fauna of California. University of California Publications in Zoology, 40: 71-234.

  1935.   Differentiation in pocket gophers of the *Thomomys bottae* group in California and southern Oregon. University of California Publications in Zoology, 40: 403-416.

Grinnell, J., and H. S. Swarth
  1913.   An account of the birds and mammals of the San Jacinto area of southern California. University of California Publications in Zoology, 10: 197-406.

Grismer, L. L.
  1994.   Ecogeography of the peninsular herpetofauna of Baja California, Mexico, and its utility in historical biogeography. Pp. 89-125 in

Proceedings of a Conference on Herpetology of North American Deserts (J. W. Wright and P. Brown, eds.), Southwestern Herpetological Society, Special Publication No. 5.

Hafner, D. J., and B. R. Riddle
    1997.   Biogeography of Baja California Peninsular Desert mammals. Pp. 39-68 in Life Among the Muses: Papers in Honor of James S. Findley (T. L. Yates, W. L. Gannon, and D. E. Wilson, eds.). The Museum of Southwestern Biology, Albuquerque, New Mexico.

Hall, E. R.
    1942.   A new race of woodrat (*Neotoma lepida*). University of California Publications in Zoology 46(5): 369-370.
    1946.   Mammals of Nevada. University of California Press, Berkeley and Los Angeles, CA.
    1981.   The mammals of North America, vol. 2, 2nd edition. John Wiley & Sons, New York, New York.

Hall, E. R., and K. R. Kelson
    1959.   The Mammals of North America, vol. 2, 1st edition. Ronald Press, New York, New York.

Harpending, H.
    1994.   Signature of ancient population growth in a low-resolution mitochondrial mismatch distribution. Human Biology, 66: 591-600.

Harpending, H., and A. Rogers
    2000.   Genetic perspective on human origins and differentiation. Annual Review of Human Genetics, 1: 361-385.

Hein, J., M. H. Schierup, and C. Wiuf
    2005.   Gene Genealogies, Variation, and Evolution. A Primer in Coalescent Theory. Oxford University Press, New York, New York.

Hill, G. E.
    1998.   An easy, inexpensive means to quantify plumage coloration. Journal of Field Ornithology, 69: 353-363.

Hoekstra, H. E., and M. W. Nachman
    2003.   Different genes underlie adaptive melanism in different populations of the rock pocket mouse. Molecular Ecology, 12: 1185-1194.

Hoffmeister, D. E.
    1986.   Mammals of Arizona. The University of Arizona Press, Tucson, Arizona.

Hoffmeister, D. E., and L. De la Torre
    1959.   The baculum in the woodrat *Neotoma stephensi*. Proceedings of the Biological Society of Washington, 72: 171-172.

Hooper, E. T.
    1960.   The glans penis in *Neotoma* (Rodentia) and allied genera. Occasional Papers, Museum of Zoology, University of Michigan, 613: 1-10.

Huelsenbeck, J. P., F. Ronquist, R. Nielsen, and J. P. Bollback
    2001.   Bayesian inference of phylogeny and its impact on evolutionary biology. Science, 294: 2310-2314.

Huelsenbeck, J. P., and F. Ronquist
    2001.   MrBayes: Bayesian inference of phylogenetic trees. Bioinformatics, 17: 754-755.

Huey, L. M.
    1937.   Descriptions of new mammals from Arizona and Sonora, Mexico. Transactions of the San Diego Society of Natural History, 8: 349-360.
    1945.   A new woodrat, genus *Neotoma*, from the Viscaino Desert region of Baja California, Mexico. Transactions of the San Diego Society of Natural History, 10: 307-310.
    1957.   A new race of woodrat (*Neotoma*) from the Gulf side of central Baja California, Mexico. Transactions of the San Diego Society of Natural History, 12: 287-288.

Irwin, D. E.
    2002.   Phylogeographic breaks without geographic barriers to gene flow. Evolution, 56: 2383-2394.

Irwin, D. M., T. D. Kocher, and A. C. Wilson
  1991. Evolution of the cytochrome *b* gene of mammals. Journal of Molecular Evolution, 32: 128-144.

Jones, C., and R. D. Fisher.
  1973. Comments on the type-specimen of *Neotoma desertorum sola* Merriam 1894 (Mammalia: Rodentia). Proceedings of the Biological Society of Washington, 86: 435-438.

Kelson, K. R.
  1949. Two new woodrats from eastern Utah. Journal of the Washington Academy of Sciences, 39: 417-419.

Knowles, L. L.
  2004. The burgeoning field of statistical phylogeography. Journal of Evolutionary Biology, 17: 1-10.

Knowles, L. L., and W. P. Maddison
  2002. Statistical phylogeography. Molecular Ecology, 11: 2623-2635.

Koop, B. F., R. J. Baker, and J. T. Mascarello.
  1985. Cladistical analysis of chromosomal evolution within the genus Neotoma. Occasional Papers, The Museum, Texas Tech University, 86:1-9.

Kuhner, M. K., J. Yamato, and J. Felsenstein
  1998. Maximum likelihood estimation of population growth rates based on the coalescent. Genetics, 149: 429-434.

Kumar, S., K. Tamura, and M. Nei
  2004. MEGA3: Integrated software for molecular evolutionary genetics and analysis and sequence alignment. Briefings in Bioinformatics, 5: 150-163.

Lawlor, T. E.
  1982. The evolution of body size in mammals: evidence from insular populations in Mexico. The American Naturalist, 119: 54-72.

Lawlor, T. E., D. J. Hafner, P. Stapp, B. R. Riddle, and S. T. Álvarez-Castañeda
 2002. The mammals. Pp. 326-361 in A New Island Biogeography of the Sea of Cortés (T. J. Case, M. L. Cody, and E. Ezcurra, eds.). Oxford University Press, New York, New York.

Leite, Y. L. R.
 2003. Evolution and systematics of the Atlantic tree rats, genus *Phyllomys* (Rodentia, Echimyidae), with description of two new species. University of California Publications in Zoology, 132: 1-118.

Lessa, E. P., J. A. Cook, and J. L. Patton
 2003. Genetic footprints of demographic expansion in North America, but not Amazonia, during the Late Quaternary. Proceedings of the National Academy Sciences, USA, 100: 10331-10334.

Lewis, P. O.
 2001. A likelihood approach to estimating phylogeny from discrete morphological character data. Systematic Biology, 50: 913-925.

Lewis, P. O., and D. Zaykin.
 2002. GDA (version 1.1). Computer software available from http://hydrodictyon.eeb.uconn.edu/people/plewis/software.ph.

Lidicker, W. Z., Jr.
 1962. The nature of subspecies boundaries in a desert rodent and its implications for subspecies taxonomy. Systematic Zoology, 11: 160-171.

Lieberman, M., and D. Lieberman
 1970. The evolutionary dynamics of the desert woodrat, *Neotoma lepida*. Evolution, 24: 560-570.

Lewis, P. O., and D. Zaykin.
 2002. GDA (version 1.1). Computer software available from http://hydrodictyon.eeb.uconn.edu/people/plewis/software.ph.

Lindell, J., F. R. Méndez-de la Cruz, and R. W. Murphy
 2005. Deep genealogical history without population differentiation: discordance between mtDNA and allozyme divergence in the

zebra-tailed lizard (*Callisauruss draconoides*). Molecular Phylogenetics and Evolution, 36: 682-694.

Marshall, T. C., J. Slate, L. E. B. Kruuk, and J. M Pemberton
1998. Statistical confidence for likelihood-based paternity inference in natural populations. Molecular Ecology, 7: 639-655.

Matocq, M. D.
2002. Morphological and molecular analysis of a contact zone in the *Neotoma fuscipes* species complex. Journal of Mammalogy, 83: 866-883.

Matocq, M. D., and E. A. Lacey.
2004. Philopatry, kin clusters, and genetic relatedness in a population of woodrats (*Neotoma macrotis*). Behavioral Ecology, 15:647-653.

Matocq, M. D., Q. R. Shurtliff, and C. R. Feldman
2007. Phylogenetics of the woodrat genus *Neotoma* (Rodentia: Muridae): Implications for the evolution of phenotypic variation in male external genitalia. Molecular Phylogenetics and Evolution, 42: 637-652.

Mascarello, J. T.
1978. Chromosomal, biochemical, mensural, penile, and cranial variation in desert woodrats (*Neotoma lepida*). Journal of Mammalogy, 59: 477-495.

Mascarello, J. T., and T. C. Hsu
1976. Chromosome evolution in woodrats, genus *Neotoma* (Rodentia: Cricetidae). Evolution, 30: 152-169.

Mellink, E.
1992a. Status de los Heteromyidos y Cricetidos endémicos del Estado del Baja California. Informe Técnico. Comunicaciones Académicos, serie Ecología, Centro de Investigaciones Científicos y Educación Superior de Enemata, 1-10.
1992b. The status of *Neotoma anthonyi* (Rodentia, Muridae, Cricetinae) of Todos Santos Islands, Baja California, Mexico. Bulletin of the Southern California Academy of Sciences, 91: 137-140.

Merriam, C. H.
1887.   Description of a new species of wood-rat from Cerros Island, off Lower California. The American Naturalist, 21: 191-193.
1894a.   Abstract of a study of the American woodrats, with descriptions of fourteen new species and subspecies of the genus *Neotoma*. Proceedings of the Biological Society of Washington, 9: 117-128.
1894b.   A new subfamily of murine rodents – the Neotominae – with descriptions of a new genus and species and a synopsis of the known forms. Proceedings of the Academy of Natural Sciences, 46: 225-252.

Munsell (Color Company)
1976.   Munsell Book of Color: Glossy Finish Collection. Munsell/MacBeth/Kollmorgen Corp., Baltimore, Maryland.

Murphy, R. W., and G. Aguirre-Léon
2002.   Nonavian reptiles: origins and evolution. Pp. 181-220 in A New Island Biogeography of the Sea of Cortés (T. J. Case, M. L. Cody, and E. Ezcurra, eds.). Oxford University Press, New York, New York.

Murphy, R. W., F. Sanchez-Piñero, G. A. Polis, and R. L. Aalbu
2002.   Appendix 1.1, New measurements of area and distance for islands in the Sea of Cortés. Pp. 447-464 in A New Island Biogeography of the Sea of Cortés (T. J. Case, M. L. Cody, and E. Ezcurra, eds.). Oxford University Press, New York, New York.

Musser, G. G., and M. D. Carleton
1993   Family Muridae. Pp. 501-755 in Mammal Species of the World. A Taxonomic and Geographic Reference, 2nd Edition (D. E. Wilson and D. M. Reeder, eds.). Smithsonian Institution Press, Washington, DC.
2005.   Superfamily Muroidea. Pp. 894-1531 in Mammal Species of the World. A Taxonomic and Geographic Reference, 3rd Edition, volume 2 (D. E. Wilson and D. M. Reeder, eds.). Johns Hopkins University Press, Baltimore, Maryland.

Myers, P, and M. D. Carleton
    1981.   The species of *Oryzomys* (*Oligoryzomys*) in Paraguay and the identity of Azara's "Rat sixieme ou rat a tarse noir." Miscellaneous Publications Museum of Zoology, University of Michigan 161: 1-41.

Nachman, M. W., H. E. Hoekstra, and S. L. D'Agostino
    2003.   The genetic basis of adaptive melanism in pocket mice. Proceedings of the National Academy of Sciences, USA, 100: 5268-5273.

Nelson, E. W., and E. A. Goldman
    1931.   Two new woodrats from Lower California. Proceedings of the Biological Society of Washington, 44: 107-110.

Nylander, J. A. A.
    2004.   MrModelTest, version 2.2. Program distributed by author. Evolutionary Biology Centre, Uppsala University, Sweden.

Orr, R. T.
    1934.   Two new woodrats from Lower California, Mexico. Proceedings of the Biological Society of Washington, 47: 109-112.

Paetkau, D., W. Calvert, I. Stirling, and C. Strobeck
    1995.   Microsatellite analysis of population structure in Canadian polar bears. Molecular Ecology, 4: 347-354.

Page, R. D. M.
    1996.   TREEVIEW: An application to display phylogenetic trees on personal computers. Computer Applications in the Biosciences, 12: 357-358.

Patton, J. L., and S. T. Álvarez-Castañeda
    2005.   Phylogeography of the desert woodrat, *Neotoma lepida*, with comments on systematics and biogeographic history. Pp. 375-388, in Contribuciones Mastozoológicas en Homenaje a Bernardo Villa (V. Sánchez-Cordero and R. A. Medellín, eds.). Instituto de Biología e Instituto de Ecología, UNAM, México.

Patton, J. L., and P. V. Brylski
    1987.   Pocket gophers in alfalfa fields: causes and consequences of habitat-related body size variation. The American Naturalist, 130: 493-506.

Patton, J. L., and M. F. Smith
    1990.   The evolutionary dynamics of the pocket gopher *Thomomys bottae*, with emphasis on California population. University of California Publications in Zoology, 123: 1-161.

Penny, D., and M. J. Phillips
    2004.   The rise of birds and mammals: are microevolutionary processes sufficient for macroevolution? Trends in Ecology and Evolution, 19: 516-522.

Planz, J. V.
    1992.   Molecular phylogeny and evolution of the American woodrats, genus *Neotoma* (Muridae). PhD dissertation, University of North Texas, Denton, Texas.
    1999.   Arizona woodrat / *Neotoma devia*. Pp. 600-601 in The Smithsonian Book of North American Mammals (D. E. Wilson and S. Ruff, eds.). Smithsonian Institution Press, Washington, D.C.

Posada, D.
    2005.   Collapse, version 1.2: Describing haplotypes from sequence alignments. Distributed free from http://darwin.uvigo.es/.

Posada, D., K. A. Crandall, and A. R. Templeton
    2000.   GEODIS: A program for the cladistic nested analysis of the geographical distribution of genetic haplotypes. Molecular Ecology, 9: 487-488.

Price, W. W.
    1894.   Description of a new wood-rat from the Coast Range of central California. Proceedings of the California Academy of Sciences, 2nd series, 3: 154-156.

Pritchard, J. K., M. Stephens, and P. Donnelly.
2000.   Inference of population structure using multilocus genotype data. Genetics, 155: 945-959.

Pritchard, J. K., and W. Wen.
2003.   STRUCTURE (version 2).   Available from http://pritch.bsd. uchicago.edu.

Puorto, G., M. D. Salamão, R. D. G. Theakston, R. S. Thorpe, D. A. Warrell, and W. Würster
2001.   Combining mitochondrial DNA sequences and morphological data to infer species boundaries: phylogeography of lancehead pitvipers in the Brazilian Atlantic forest, and the status of *Bothrops pradoi* (Squamata: Serpentes: Viperidae). Journal of Evolutionary Biology, 14: 527-538.

Queney, G., N. Ferrand, S. Weiss, F. Mougel, and M. Monnerot.
2001.   Stationary distributions of microsatellite loci between divergent populations groups of the European rabbit (*Oryctolagus cuniculus*). Molecular Biology and Evolution, 18: 2169-2178.

Rannala, B., and Z. H. Yang
1996.   Probability distribution of molecular evolutionary trees: a new method of phylogenetic inference.   Journal of Molecular Evolution, 43: 304-311.

Ray, N., N. Currat, and L. Excoffier
2003.   Intra-deme molecular diversity in spatially expanding populations. Molecular Biology and Evolution, 20(1): 76-86.

Raymond, M., and F. Rousset.
1995.   GENEPOP (version 1.2): population genetics software for exact tests and ecumenicism. Journal of Heredity, 86: 248-249.

Rebman, J. P., J. L. Leon de la Luz, and R. V. Moran
2002.   Appendix 4.1. Vascular plants of the Gulf islands. Pp. 465-511, in A New Island Biogeography of the Sea of Cortés (T. J. Case, M. L. Cody, and E. Ezcurra, eds.).   Oxford University Press, New York, New York.

Rhoads, S. N.
    1894.   Descriptions of three new rodents from California and Oregon.
             The American Naturalist, 28: 67-68.

Rice, W. R.
    1989.   Analyzing tables of statistical tests. Evolution 43: 223-225.

Riddle, B. R.
    1995.   Molecular biogeography in the pocket mice (*Perognathus* and
             Chaetodipus) and grasshopper mice (*Onychomys*)—the Late
             Cenozoic development of a North American aridlands rodent
             guild. Journal of Mammalogy, 76: 283-301.
    1996.   The molecular phylogeographic bridge between deep and
             shallow history in continental biotas. Trends in Ecology and
             Evolution, 11: 207-211.

Riddle, B. J., and D. J. Hafner
    2004.   The past and future roles of phylogeography in historical
             biogeography. Pp. 93-110 in Frontiers of Biogeography: New
             Directions in the Geography of Nature (M. V. Lomolino, and L.
             R. Heaney, eds.).   Sinauer Associates, Inc., Sunderland,
             Massachussets.
    2006a.   Biogeografía histórica de los desiertos cálidos de Norteamérica.
             Pp. 57-65 in Genética y mamíferos Mexicanos: presente y futuro
             (E. Veasquez-Domínguez and D. J. Hafner, eds.). New Mexico
             Museum of Natural History and Science, Bulletin 32,
             Albuquerque, New Mexico.
    2006b.   A step-wise approach to integrating phylogeographic and
             phylogenetic biogeographic perspectives on the history of a core
             North American warm deserts biota.   Journal of Arid
             Environments, 66: 435-461.

Riddle, B. R., D. J. Hafner, and L. F. Alexander
    2000b.   Comparative phylogeography of Bailey's pocket mouse
             (*Chaetodipus baileyi*) and the *Peromyscus eremicus* species
             group: historical vicariance of the Baja California Peninsular
             desert. Molecular Phylogenetics and Evolution, 17: 161-172.
    2000c.   Phylogeography and systematics of the *Peromyscus eremicus*
             species group and the historical biogeography of North

American warm desert. Molecular Phylogenetics and Evolution, 17: 145-160.

Riddle, B. R., D. J. Hafner, L. F. Alexander, and J. F. Jaeger
2000a. Cryptic vicariance in the historical assembly of a Baja California Peninsular Desert biota. Proceedings of the National Academy of Sciences, USA, 97: 14438-14443.

Riddle, B. R., and R. L. Honeycutt
1990. Historical biogeography in North American arid regions: an approach using mitochondrial –DNA phylogeny in grasshopper mice (genus *Onychomys*). Evolution, 44: 1-15.

Robinson-Rechavi, M., and D. Huchon
2000. RRTree: Relative-rate tests between groups of sequences on a phylogenetic tree. Bioinformatics, 16: 296-297.

Rogers, A. R., and H. Harpending
1992. Population growth model makes waves in the distribution of pairwise genetic differences. Molecular Biology and Evolution, 9: 552-569.

Ronquist, F., and J. P. Huelsenbeck
2003. MRBAYES 3: Bayesian phylogenetic inference under mixed models. Bioinformatics, 19: 1572-1574.

Schneider, S., and L. Excoffier
1999. Estimation of demographic parameters from the distribution of pairwise differences when the mutation rates vary among sites: application to human mitochondrial DNA. Genetics, 152: 1079-1089.

Sites, J. W., Jr., and J. C. Marshall
2003. Delimiting species: A Renaissance issue in systematic biology. Trends in Ecology and Evolution, 18: 462-470.
2004. Operational criteria for delimiting species. Annual Review of Ecology, Evolution, and Systematics, 35: 199-227.

Slatkin, M. and R. R. Hudson
    1991.   Pairwise comparisons of mitochondrial DNA sequences in stable and exponentially growing populations. Genetics, 129: 555-562.

Smith, F. A.
    1992.   Evolution of body size among woodrats from Baja California, Mexico. Functional Ecology, 6: 265-273.

Smith, F. A., and J. L. Betancourt
    2003.   The effect of Holocene temperature fluctuations on the evolution and ecology of *Neotoma* (woodrats) in Idaho and northwestern Utah. Quaternary Research, 59: 160-171.

Smith, M. F., and J. L. Patton
    1988.   Subspecies of pocket gophers: causal bases for geographic differentiation in *Thomomys bottae*. Systematic Zoology, 37: 163-178.
    1991.   Variation in mitochondrial cytochrome *b* sequence in natural populations of South American akodontine rodents (Muridae: Sigmodontinae). Molecular Biology and Evolution, 8: 85-103.
    1993.   Diversification of South American muroid rodents: evidence from mitochondrial DNA sequence data for the Akodontine tribe. Biological Journal of the Linnean Society, 50: 149-177.
    1999.   Phylogenetic relationships and the radiation of sigmodontine rodents in South America: evidence from cytochrome b. Journal of Mammalian Evolution, 6: 89-128.

Sousa, B., L. M. E. Svensson, and J. L. Patton
    2007.   Characterization of 18 microsatellite loci for woodrats of the *Neotoma lepida* group (Rodentia, Cricetidae, Neotominae). Molecular Ecology Notes, 7: 868-870

Spencer, J. E., and P. A. Pearthree
    2005.   Abrupt initiation of the Colorado River and initial incision of the Grand Canyon. Arizona Geology, 35: 1-4.

Steppan, S. J., R. M. Adkins, and J. Anderson
    2004.   Phylogeny and divergence-date estimates of rapid radiations in muroid rodents based on multiple nuclear genes. Systematic Biology, 53: 533-553.

Strauss, R. E.
    1985.   Evolutionary allometry and variation in body form in the South American catfish genus *Corydoras* (Callichthyidae). Systematic Zoology, 34: 381-396.

Swofford, D. L.
    2002.   PAUP*: Phylogenetic Analysis Using Parsimony. Version 4.0b10. Sinauer Associates, Inc., Publishers, Sunderland, Massachussets.

Tajima, F.
    1989.   Statistical method for testing the neutral mutation hypothesis by DNA polymorphism. Genetics, 123: 585-595.

Takezaki, N., and M. Nei.
    1996.   Genetic distances and reconstruction of phylogenetic trees from microsatellite DNA. Genetics, 144: 389-399.

Taylor, W. P.
    1910.   Two new rodents from Nevada. University of California Publications in Zoology, 5: 283-302.

Templeton, A. R.
    1989.   The meaning of species and speciation. Pp. 3-27 in Speciation and Its Consequences (D. Otte and J. A. Endler, eds.). Sinauer Associates, Sunderland, Massachussets.
    1998.   Nested clade analysis of phylogeographic data: testing hypotheses about gene flow and population history. Molecular Ecology, 7: 381-397.
    2001.   Using phylogeographic analyses of gene trees to test species status and boundaries. Molecular Ecology, 10: 779-791.
    2004.   Statistical phylogeography: methods of evaluating and minimizing inference errors. Molecular Ecology, 13: 789-809.

Templeton, A. R., E. Boerwinkle, and C. F. Sing
    1987.   A cladistic analysis of phenotypic associations with haplotypes inferred from restriction endonuclease mapping. I. Basic theory and an analysis of alcohol dehydrogenase activity in *Drosophila*. Genetics, 117: 343-351.

Templeton, A. R., K. A. Crandall, and C. F. Sing
    1992.   A cladistic analysis of phenotypic associations with haplotypes inferred from restriction endonuclease mapping and DNA sequence data. III. Cladogram estimation. Genetics, 619-633.

Thomas, O.
    1893.   On two new members of the genus *Heteromys* and two of *Neotoma*. Annals and Magazine of Natural History, ser. 6, 12: 233-235.

Townsend, C. H.
    1912.   Mammals collected by the 'Albatross' expedition in Lower California in 1911, with descriptions of new species. Bulletin of the American Museum of Natural History, 31: 117-130.

Upton, D. E., and R. W. Murphy
    1997.   Phylogeny of the side-blotched lizards (Phrynosomatidae: *Uta*) based on mtDNA sequences: support for a midpeninsular seaway in Baja California. Molecular Phylogenetics and Evolution, 8: 104-113.

Vähä, J.-P., and C. R. Primmer
    2006   Efficiency of model-based Bayesian methods for detecting hybrid individuals under different hybridization scenarios and with different numbers of loci. Molecular Ecology, 15: 63-72.

Verts, B. J., and L. N. Carraway
    2002.   *Neotoma lepida*. Mammalian Species, 699: 1-12.

von Bloeker, J. C., Jr.
    1938.   Geographic variation in *Neotoma lepida* in west-central California. Proceedings of the Biological Society of Washington, 51: 201-204.

Walsh, P. S., D. A Metzger, and R. Higughi
    1991.   Chelex 100 a medium for simple extraction of DNA for PCR-based typing from forensic material. Biotechniques, 10: 506-513.

Waser, P. M., and C. Strobeck
    1998.   Genetic signatures of interpopulation dispersal. Trends in Ecology and Evolution, 13: 43-44.

Weir, B. S., and C. C. Cockerham
    1984.   Estimating F-statistics for the analysis of population structure. Evolution, 38: 1358-1370.

Wiens, J. J., and T. A. Penkrot
    2002.   Delimiting species using DNA and morphological variation and discordant species limits in spiny lizards (*Sceloporus*). Systematic Biology, 51: 69-91.

Wilson, E. O., and W. L. Brown, Jr.
    1953.   The subspecies concept and its taxonomic application. Systematic Zoology, 2:97-111.

Wickliffe, J. W., F. G. Hoffmann, D. S. Carrol, Y. V. Dunina-Barkovskaya, R. D. Bradley, and R. J. Baker
    2003.   Intron 7 (*FGB-I7*) of the fibrinogen B beta polypeptide (*FGB*): a nuclear DNA phylogenetic marker for mammals. Occasional Papers, The Museum, Texas Tech University, 219: 1-6.

Wiggins, I. L.
    1980.   Flora of Baja California. Stanford University Press, Palo Alto, California.

Zheng, X., B. S. Arbogast, and G. J. Kenagy
    2003.   Historical demography and genetic structure of sister species: deermice (*Peromyscus*) in the North American temperate rain forest. Molecular Ecology, 12: 711-724.

Zink, R. M., A. E. Keeson, T. V. Line, and R. C. Blackwell-Rago
    2001.   Comparative phylogeography of some aridland bird species. Condor, 103: 1-10.

Appendix

Lists of localities from which we examined specimens of the *Neotoma lepida* group. Locality numbers are those referred to in the text. Georeferenced coordinates provided and used to map localities are those obtained from each museum collection via the Mammal Networked Information Systems (MaNIS; http://manisnet.org), from the curators of non-MaNIS collection participants, or by use of TopoZone maps (http://www.topozone.com/). Data are given in decimal degrees using the WGS84 datum. Because MaNIS data are dynamic, and thus subject to change as locality coordinates are refined, all data from non-MVZ specimens are from 1 January 2005; those from MVZ are from 1 January 2006; and those from CIB, which is not a participant of the MaNIS network, are from 10 December 2005. Museum acronyms are: ANSP = Academy of Natural Sciences, Philadelphia; BYU = Monte L. Beam Museum, Brigham Young University; CIB = Centro de Investigaciones Biológicas del Noroeste; CSULB = California State University Long Beach; LACM = Los Angeles Country Museum of Natural History; KU = Museum of Natural History, University of Kansas; MCZ = Museum of Comparative Zoology, Harvard University; MVZ = Museum of Vertebrate Zoology, University of California, Berkeley; SDNHM = San Diego Natural History Museum; UCLA = Dickey Collection, University of California Los Angeles; UIMNH = University of Illinois Museum of Natural History (now at the Museum of Southwestern Biology, University of New Mexico); UNT – University of North Texas; USNM = United States National Museum, Washington, D.C.

Locality number	County	Specific locality	Latitude	Longitude	Museum
United States: Arizona					
AZ-1	Mohave Co.	Beaver Dam Mts.	36.98311	-113.82020	MVZ
AZ-2	Mohave Co.	near mouth Beaverdam Creek, above Littlefield	36.90080	-113.93480	USNM
AZ-3	Mohave Co.	Littlefield, 1500 ft	36.88720	-113.92970	USNM
AZ-4	Mohave Co.	Elbow Canyon, W slope Virgin Mts.	36.79720	-113.93410	USNM
AZ-5	Mohave Co.	Limekiln Canyon, w slope Virgin Mts., 4500 ft	36.68750	-114.01800	USNM
AZ-6	Mohave Co.	Grand Wash, 8 mi S. Pakoon Spring 1800 ft	36.23700	-114.00720	USNM

ID	County	Location	Latitude	Longitude	Collections
AZ-6a	Mohave Co.	Pakoon Spring	34.41560	-113.95780	UIMNH
AZ-7	Mohave Co.	Mokaac Wash	36.90334	-113.56356	MVZ
AZ-8	Mohave Co.	10 mi N Wolf Hole	36.86350	-113.56930	MVZ, USNM
AZ-9	Mohave Co.	6 mi. N Wolfhole	36.81560	-113.56140	MVZ
AZ-10	Mohave Co.	4 mi N Wolf Hole	36.80240	-113.55640	SDNHM
AZ-11	Mohave Co.	3 mi NW Diamond Butte	36.60090	-113.41030	USNM
AZ-12	Mohave Co.	1 mi W Diamond Butte, 5000 ft	36.57230	-113.39690	USNM
AZ-13	Mohave Co.	Hurricane Ledge, 6 mi N Mt. Trumbull, 6000 ft	36.48940	-113.32270	USNM
AZ-14	Mohave Co.	3 mi S Trumbull Spring, Mt. Trumbull	36.40200	-113.14600	USNM
AZ-14	Mohave Co.	Mt. Trumbull, Nixon Spring	36.40220	-113.14660	MVZ, SDNHM
AZ-15	Mohave Co.	head of Toroweap Valley	36.39810	-113.06130	MVZ
AZ-16	Mohave Co.	Toroweap Valley	36.34078	-113.05669	MVZ
AZ-17	Mohave Co.	lower end Toroweap Valley	36.23440	-113.07790	MVZ, USNM
AZ-18	Mohave Co.	Fern Glen, right bank Colorado River, river mile 168	36.25970	-112.91880	USNM
AZ-19	Mohave Co.	Kanab Wash, Kaibab Indian Reservation	36.86970	-112.58610	MVZ
AZ-20	Coconino Co.	6 mi SE Fredonia	36.89120	-112.47530	USNM
AZ-20a	Coconino Co.	2.5 mi N Fredonia	36.99060	-112.53050	UIMNH
AZ-21	Coconino Co.	Ryan	36.69022	-112.33710	MVZ, USNM
AZ-22	Coconino Co.	Thunder River, 3 mi above mouth, Kaibab National Forest	36.39460	-112.45670	USNM
AZ-23	Coconino Co.	Lees Ferry, n side	36.86690	-111.58430	USNM
AZ-23a	Coconino Co.	1.5 mi W Glen Canyon Dam	36.94900	-111.49700	UINHM
AZ-24	Coconino Co.	Soap Creek, 15 mi SW Lees Ferry	36.73800	-111.69400	USNM
AZ-25	Coconino Co.	Jacobs Pools	36.72200	-111.90400	USNM
AZ-26	Coconino Co.	Houserock Valley, 15 mi E Jacobs Pools	36.73280	-112.04820	USNM
AZ-26	Coconino Co.	Houserock Valley, 15 mi W Colorado Bridge	36.73280	-112.04820	USNM
AZ-27	Coconino Co.	North Canyon, Houserock Valley	36.55800	-112.01850	USNM
AZ-28	Coconino Co.	east slope Kaibab Plaetau, ca. 3 mi W House Rock Buffalo Ranch	36.47997	-112.00745	MVZ
AZ-29	Coconino Co.	Buck Farm Canyon, right bank Colorado River, river mile 41	36.40500	-111.88070	USNM

ID	County	Locality	Latitude	Longitude	Collection
AZ-30	Coconino Co.	Shinumo Creek, Grant Canyon, 3000 ft	36.23700	-112.34900	USNM
AZ-31	Coconino Co.	Phantom Ranch, Grand Canyon	36.10500	-112.09520	USNM
AZ-32	Coconino Co.	left bank Colorado River, Cardenas Creek, river mile 71	36.08700	-111.86300	USNM
AZ-33	Coconino Co.	left bank Colorado River, above Tanner Rapid, river mile 68	36.10240	-111.83150	USNM
AZ-34	Coconino Co.	2 mi W Lees Ferry	36.82650	-111.64290	MVZ
AZ-35	Coconino Co.	S side Grand Canyon Bridge	36.81630	-111.62870	MVZ
AZ-36	Coconino Co.	6 mi S Grand Canyon Bridge, Marble Canyon	36.78780	-111.68980	USNM
AZ-36a	Coconino Co.	6.5 mi W & 6 mi S Marble Canyon	36.75280	-111.75410	UIMNH
AZ-37	Coconino Co.	3.7 mi S Navajo Spring	36.71858	-111.64772	MVZ
AZ-38	Coconino Co.	left bank Colorado River, river mile 19	36.66560	-111.74040	USNM
AZ-39	Coconino Co.	mouth, Bright Angel Creek	36.10050	-112.09600	USNM
AZ-40	Coconino Co.	Pipe Creek, 2500 ft, Grand Canyon	36.09900	-112.11100	USNM
AZ-41	Coconino Co.	Indian Gardens, 3800 ft, Grand Canyon	36.09460	-112.11240	USNM
AZ-42	Coconino Co.	Grand Canyon	36.05400	-112.13900	USNM
AZ-43	Coconino Co.	Grand Canyon, S side, 6 mi N Bass Camp	36.22860	-112.33990	USNM
AZ-43a	Coconino Co.	1 mi NNW Supai, Navajo Falls	36.24880	-112.69680	UIMNH
AZ-44	Coconino Co.	lower end Prospect Valley, Hualpai Indian Reservation	36.17900	-113.08000	USNM
AZ-45	Coconino Co.	Cameron	35.87580	-111.41290	LACM
AZ-46	Coconino Co.	2 mi S Cameron	35.85680	-111.42410	USNM
AZ-47	Coconino Co.	5 mi SW Cameron	35.85860	-111.44590	USNM
AZ-47a	Coconino Co.	4 mi W & 1.5 mi S Cameron	35.86200	-111.48170	UIMNH
AZ-48	Coconino Co.	5 mi W Cameron	35.86636	-111.50952	MVZ
AZ-49	Coconino Co.	ca. 1 mi NE Tanner Tank	35.71277	-111.53078	MVZ
AZ-50	Coconino Co.	Tanner Tank, 5200 ft	35.69360	-111.55100	USNM
AZ-50a	Coconino Co.	Slide Reservoir, Kaibab Forest	36.65580	-112.46780	MVZ
AZ-51	Mohave Co.	Rampart Cave, Colorado River, Lake Mead NRA	36.09920	-113.93320	USNM
AZ-52	Mohave Co.	Pierce Ferry, Colorado River	36.11670	-114.00080	USNM
AZ-53	Mohave Co.	4 mi SW Pierce Ferry	36.07970	-114.06250	USNM
AZ-54	Mohave Co.	Hoover Dam Ferry	36.01610	-114.73370	MVZ

ID	County	Location	Latitude	Longitude	Museum
AZ-55	Mohave Co.	Colorado River, Willow Beach	35.87030	-114.65940	MVZ, SDNHM
AZ-56	Mohave Co.	22 mi SE Hoover Dam (on Highway 93)	35.78174	-114.51934	MVZ
AZ-57	Mohave Co.	Mud Spring, 12 mi WSW Chloride	35.38720	-114.35660	MVZ
AZ-58	Mohave Co.	Kingman, 3300 ft	35.18900	-114.05200	USNM
AZ-59	Mohave Co.	7 mi E Davis Dam [by Hwy. 68], Black Mts.	35.18700	-114.46690	MVZ
AZ-60	Mohave Co.	31 mi N & 2.5 mi W Camp Mohave	35.09460	-114.60250	MVZ
AZ-61	Mohave Co.	1 mi E (by road) Sitgreaves Pass, Black Mountains	35.03758	-114.34830	MVZ
AZ-62	Mohave Co.	9.3 mi NE Oatman, by Goldroad, Black Mts.	35.03850	-114.33090	MVZ
AZ-63	Mohave Co.	foot of the Needles, Colorado River	34.67610	-114.43770	MVZ
AZ-64	Mohave Co.	9.0 mi S I-40 by Hwy. 95, Mohave Mts.	34.64260	-114.33140	MVZ
AZ-65	Mohave Co.	Burro Creek Campground	34.53590	-113.45390	BYU
AZ-66	Mohave Co.	Chemehuevis Mts., Gold Spring	34.47720	-114.16440	SDNHM
AZ-66	Mohave Co.	Chemehuevis Mts., Lucky Star Mine	34.47721	-114.16441	SDNHM
AZ-67	Mohave Co.	Aubrey Hills	34.35385	-114.15496	MVZ
AZ-68	Maricopa Co.	New River Valley, 30 mi N Phoenix	33.91590	-112.13600	MVZ
AZ-68a	Maricopa Co.	10 mi N Sun City	33.72670	-112.27250	UIMNH
AZ-69	Maricopa Co.	0.5 mi E Black Gap	32.74729	-112.81994	MVZ
AZ-69	Maricopa Co.	1.2 mi E Black Gap	32.73877	-112.80641	MVZ
AZ-69	Maricopa Co.	10 mi. S Gila Bend	32.75270	-112.82460	SDNHM
AZ-70	La Paz Co.	Parker	34.15000	-114.28910	USNM
AZ-71	La Paz Co.	Osborne Well, Buckskin Mts.	34.12903	-114.07076	MVZ
AZ-72	La Paz Co.	10 mi E Quartzite	33.80160	-114.22060	MVZ
AZ-73	La Paz Co.	5.5 mi W Quartzite, 1.0 mi S I-10	33.64570	-114.31980	MVZ
AZ-74	La Paz Co.	west slope Dome Rock Mts., La Paz Wash	33.61275	-114.39852	MVZ
AZ-74a	La Paz Co.	7.5 mi N & 0.5 mi W Wendon	33.88340	-113.65090	UIMNH
AZ-74b	La Paz Co.	7 mi N Wendon	33.92250	-113.54540	UIMNH
AZ-75	Yuma Co.	Kofa Queen Canyon, Kofa Game Range	33.27610	-113.96190	USNM
AZ-76	Yuma Co.	base, Castle Dome	33.08530	-114.17920	MVZ
AZ-77	Yuma Co.	Castle Dome Mine, west slope Castle Dome Mts.	33.05637	-114.17498	MVZ
AZ-78	Yuma Co.	Telegraph Pass, 22 mi E Yuma	32.66100	-114.32600	USNM
AZ-79	Yuma Co.	Tinajas Altas	32.31170	-114.05160	MVZ, SDNHM
AZ-80	Yuma Co.	0.75 mi W Tule Tank	32.22590	-113.81080	MVZ

AZ-80	Yuma Co.	Tule Tank	32.22650	-113.79600	MVZ
AZ-81	Yuma Co.	Pinacate Lava, Tule Desert, 8 mi NW U.S.-Mexico Mont 179	32.10780	-113.49840	MVZ
AZ-82	Pima Co.	7 mi. E Papago Wells	32.10510	-113.22540	SDNHM
AZ-82	Pima Co.	Agua Dulce Mts., 7 mi. E Papago Wells	32.10510	-113.22540	MVZ
AZ-83	Pima Co.	O'Neill Hills, 8 mi E Papago Well	32.08810	-113.19960	USNM
AZ-84	Pima Co.	9 mi E Papago Well	32.10410	-113.17080	MVZ

United States: California

CA-1	Alameda Co.	Arroyo Mocho, 7 mi E Livermore	37.67000	-121.91000	MVZ
CA-2	Alameda Co.	7 mi SE Livermore	37.62000	-121.67000	MVZ
CA-3	Alameda Co.	Calaveras Dam	37.49000	-121.81000	MVZ
CA-4	Alameda Ca.	Del Puerto Canyon	37.43830	-121.33480	MVZ
CA-5	Stanislaus Co.	Del Puerto Canyon	37.44480	-121.29702	MVZ
CA-6	Stanislaus Co.	2 mi S Orestimba Peak	37.29000	-121.21000	MVZ
CA-7	Merced Co.	Romero Creek, Romero Ranch, near Santa Nella	37.11999	-121.09999	LACM, MVZ
CA-8	Merced Co.	Sweeney's Ranch, 22 mi S Los Baños, Herrero Canyon	36.90190	-121.00800	MVZ
CA-9	Santa Clara Co.	Pacheco Pass summit	37.06600	-121.21900	USNM
CA-9	Santa Clara Co.	summit, Pacheco Pass	37.06600	-121.21900	MVZ
CA-10	San Benito Co.	Bell's Station	37.03660	-121.31080	USNM
CA-11	San Benito Co.	12 mi SE [by Road] Paicines	36.67000	-121.18000	MVZ
CA-12	San Benito Co.	Griswold Canyon	36.55173	-120.83970	MVZ
CA-13	San Benito Co.	Griswold Canyon	36.53673	-120.83455	MVZ
CA-14	San Benito Co.	San Benito Store, San Benito Valley	36.50970	-121.08190	USNM
CA-15	San Benito Co.	1 mi N Cook P.O.	36.50000	-121.10000	USNM
CA-16	San Benito Co.	Bear Valley	36.48200	-121.15500	USNM
CA-17	San Benito Co.	6 mi ESE San Benito	36.48000	-121.03000	MVZ
CA-18	San Benito Co.	2 mi NNE New Idria	36.44000	-120.67000	MVZ
CA-19	San Benito Co.	1 mi SE summit San Benito Mt.	36.35000	-120.63000	MVZ
CA-20	San Benito Co.	9.1 mi NE King City (Monterey Co.)	36.28277	-120.98768	MVZ
CA-21	San Benito Co.	Fremont Peak, Gavilan Range	36.75800	-121.50300	USNM

CA-22	Fresno Co.	Stanley, 8 mi W Huron	36.19820	-120.26120	USNM
CA-23	Monterey Co.	Strawberry Canyon	36.82000	-121.73000	MVZ
CA-24	Monterey Co.	Stonewall Creek, 6.3 mi NE Soledad	36.46000	-121.30000	MVZ
CA-25	Monterey Co.	mouth of Stonewall Creek	36.40999	-121.30630	MVZ
CA-26	Monterey Co.	Bear Valley, head of Carmel River	36.37120	-121.53670	USNM
CA-27	Monterey Co.	Posts	36.22830	-121.76440	USNM
CA-28	Monterey Co.	Tassajara Creek, 6 mi below Tassajara Springs	36.21940	-121.50220	USNM
CA-29	Monterey Co.	Abbott Ranch, Arroyo Seco	36.22999	-121.45999	MVZ
CA-30	Monterey Co.	Arroyo Seco	36.27000	-121.34000	MVZ
CA-30	Monterey Co.	Arroyo Seco, 7 mi SW Greenfield	36.27186	-121.34763	MVZ
CA-31	Monterey Co.	Santa Lucia Peak	36.14550	-121.41940	USNM
CA-32	Monterey Co.	Cone Peak	36.05110	-121.46910	USNM
CA-33	Monterey Co.	Milpitas Ranch, S base Santa Lucia Peak	36.07140	-121.30380	USNM
CA-34	Monterey Co.	5.2 mi NE King City	36.27004	-121.06419	MVZ
CA-35	Monterey Co.	Lewis Creek	36.23000	-120.98000	MVZ
CA-36	Monterey Co.	Priest Valley, 2400 ft	36.18700	-120.69900	MVZ
CA-37	San Luis Obispo Co.	San Luis Obispo, 500 ft	35.28280	-120.65960	USNM
CA-38	San Luis Obispo Co.	13.3 mi NW (by road) New Cuyama	35.04427	-119.89468	MVZ
CA-39	San Luis Obispo Co.	Carrizo Plains	35.19140	-119.79290	USNM
CA-40	San Luis Obispo Co.	Crocker Grade, Temblor Range	35.20715	-119.70368	MVZ
CA-41	San Luis Obispo Co.	Elkhorn Plain Ecological Reserve	35.13513	-119.62623	MVZ
CA-42	San Luis Obispo Co.	0.4 mi S Wells Ranch, Caliente Range	35.05458	-119.69585	MVZ
CA-43	San Luis Obispo Co.	Beam Flat, Elkhorn Hills	35.01912	-119.49248	MVZ
CA-44	San Luis Obispo Co.	Elkhorn Hills	35.00560	-119.49133	MVZ
CA-45	San Luis Obispo Co.	MU Ranch, south end Carrizo Plains	34.98590	-119.48180	MVZ
CA-46	Santa Barbara Co.	Peach Tree Ranch, San Rafael Mts.	34.70300	-119.70490	USNM
CA-47	Santa Barbara Co.	Pt. Arguello	34.57720	-120.65070	LACM
CA-48	Santa Barbara Co.	Jalama Beach	34.51080	-120.50040	LACM
CA-49	Ventura Co.	Mt. Pinos	34.81000	-119.14000	MVZ
CA-50	Ventura Co.	Cuddy Canyon, Frazier Mt.	34.81000	-118.88000	MVZ
CA-51	Ventura Co.	Whale Rock, Ojai Valley, 1250 ft	34.44860	-119.22340	UCLA
CA-52	Ventura Co.	4 mi E Fillmore, Santa Clara River bottom	34.38990	-118.91690	LACM

CA-53	Ventura Co.	Point Mugu	34.08560	-119.06090	LACM
CA-53a	Ventura Co.	Little Sycamore Canyon	34.05530	-118.96340	CSULB
CA-54	Tulare Co.	8 mi E Porterville	36.07010	-118.89960	USNM
CA-55	Tulare Co.	13.2 mi SSE Porterville	36.02888	-118.89027	MVZ
CA-55a	Tulare Co.	Kennedy Meadows, South Fork Kern River	36.00880	-118.12300	LACM
CA-56	Kern Co.	San Emigdio	34.98530	-119.18570	USNM
CA-57	Kern Co.	2 mi NNW Eagle Rest Peak, San Emigdio Mts.	34.92801	-119.13620	MVZ
CA-58	Kern Co.	Wheeler Ridge	35.00000	-119.01000	MVZ
CA-59	Kern Co.	6 mi. W Lebec	34.82220	-118.92370	MVZ
CA-59	Kern Co.	Cuddy Canyon, 1 mi E Frazier Park	34.82138	-118.92392	MVZ
CA-59	Kern Co.	Cuddy Canyon, Frazier Mt.	not found		SDNHM
CA-60	Kern Co.	1.5 mi SE Fort Tejon	34.86662	-118.87703	MVZ
CA-61	Kern Co.	National Cement Plant, Tehachapi Mts.	34.82526	-118.74452	MVZ
CA-62	Kern Co.	Pescadero Creek, Tehachapi Mts.	34.87613	-118.65842	MVZ
CA-63	Kern Co.	El Paso Creek, Tehachapi Mts.	35.01939	-118.71935	MVZ
CA-64	Kern Co.	Joaquin Flat, Tehachapi Mts.	35.02651	-118.69642	MVZ
CA-65	Kern Co.	south side Tejon Creek, Tehachapi Mts.	35.03869	-118.70396	MVZ
CA-65	Kern Co.	Tejon Creek, Tehachapi Mts.	35.05180	-118.69965	MVZ
CA-66	Kern Co.	Kern River, 12 mi below Bodfish	35.53000	-118.62000	MVZ
CA-67	Kern Co.	Rankin Ranch, Walker Basin	35.38000	-118.55000	MVZ
CA-68	Kern Co.	Caliente Creek	35.31512	-118.54950	MVZ
CA-69	Kern Co.	Kern River at Bodfish	35.58000	-118.49000	MVZ
CA-70	Kern Co.	1.3 mi NW Lake Isabella	35.63685	-118.48968	MVZ
CA-71	Kern Co.	2.8 mi NE Lake Isabella	35.65563	-118.43937	MVZ
CA-72	Kern Co.	Rancheria Creek, east end Walker Basin	35.39925	-118.45877	MVZ
CA-73	Kern Co.	Fay Creek, 6 mi W Weldon	35.69000	-118.29000	MVZ
CA-74	Kern Co.	0.7 mi E Weldon	35.67937	-118.25115	MVZ
CA-75	Kern Co.	0.5 mi E Onyx	35.69540	-118.20993	MVZ
CA-75	Kern Co.	Onyx	35.69000	-118.21000	MVZ
CA-76	Kern Co.	Brown Spring, 3.5 mi S Weldon	35.61617	-118.25268	MVZ
CA-77	Kern Co.	Harris Grade, 1.5 mi SW Sageland	35.46656	-118.23478	MVZ
CA-78	Kern Co.	NW part of Kelso Valley	35.46000	-118.21000	MVZ

ID	County	Location	Latitude	Longitude	Museum
CA-79	Kern Co.	St. John Mine, north end Kelso Valley	35.45100	-118.22049	MVZ
CA-80	Kern Co.	Kelso Valley, ca. 1 mi SW Whitney Well	35.43073	-118.25121	MVZ
CA-80	Kern Co.	Kelso Valley, ca. 1 mi SW Whitney Well	35.42600	-118.25107	MVZ
CA-81	Kern Co.	1.3 mi SW Schoolhouse Well, Kelso Valley	35.36111	-118.23492	MVZ
CA-82	Kern Co.	2 mi N Sorrell's Ranch, Kelso Valley	35.37000	-118.21000	MVZ
CA-83	Kern Co.	Hoffman Summit on Jawbone Canyon Rd.	35.37369	-118.16339	MVZ
CA-84	Kern Co.	Cameron	35.09660	-118.29760	USNM
CA-85	Kern Co.	3 mi SE (by rd) Oak Creek Pass, Tehachapi Mts.	35.01415	-118.34268	MVZ
CA-86	Kern Co.	3 mi S Mohave	34.99000	-118.12000	MVZ
CA-86a	Kern Co.	near Rosamond	34.86420	-118.16340	CSULB
CA-87	Kern Co.	0.5 mi E Warren Station	35.10000	-118.08000	MVZ
CA-87a	Kern Co.	Pine Tree Canyon, 5 mi S & 7.5 mi W Cantil	35.23080	-118.10140	LACM
CA-88	Kern Co.	Red Rock Canyon	35.32000	-117.94000	MVZ
CA-89	Kern Co.	Onyx, 11.75 mi S, 7.75 mi E; Bird Spring Canyon	35.49680	-118.05280	LACM
CA-90	Kern Co.	W slope Walker Pass	35.66000	-118.03000	MVZ
CA-91	Kern Co.	Freeman Canyon	35.65100	-118.00627	MVZ
CA-92	Kern Co.	2.5 mi S and 4.5 mi E Walker Pass	35.63000	-117.98000	MVZ
CA-92	Kern Co.	2.6 mi SE Walker Pass, Freeman Canyon	35.63000	-117.97000	MVZ
CA-92	Kern Co.	6 mi WNW Freeman Junction	35.64920	-117.99840	LACM
CA-92a	Kern Co.	3 mi WNW Freeman Junction	35.62340	-117.95480	LACM
CA-93	Kern Co.,	4 mi S Inyo County line, Hwy. 395	35.72000	-117.85000	MVZ
CA-94	Kern Co.	8.2 mi SE Inyokern	35.55540	-117.72568	MVZ
CA-95	Kern Co.	3 mi SSW Inyokern	35.55540	-117.72568	MVZ
CA-96	Los Angeles Co.	0.4 mi W Gorman	34.79703	-118.86111	MVZ
CA-97	Los Angeles Co.	4.5 mi E (by road) Gorman	34.77825	-118.77710	MVZ
CA-98	Los Angeles Co.	Gorman, 10 mi S, 4 mi E, Bainbridge Flats	34.65940	-118.82380	LACM
CA-99	Los Angeles Co.	4 mi W Three Points	34.73872	-118.65551	MVZ
CA-100	Los Angeles Co.	3.8 mi SW (by road) Three Points	34.70951	-118.54447	MVZ
CA-101	Los Angeles Co.	Fairmont, Antelope Valley	34.73497	-118.42416	MVZ
CA-102	Los Angeles Co.	2.1 mi SW Red Mountain	34.56155	-118.55845	MVZ
CA-103	Los Angeles Co.	San Francisquito Canyon	34.48148	-118.54078	LACM, MVZ
CA-104	Los Angeles Co.	San Francisquito Canyon, 1 mi W Power Plant #1	34.42780	-118.59220	LACM

CA-105	Los Angeles Co.	Mint Canyon C G	34.43590	-118.43540	LACM
CA-105	Los Angeles Co.	Mint Canyon, North Oaks, Rowler Flats	34.43590	-118.43540	LACM
CA-105	Los Angeles Co.	Mint Canyon; Mint Canyon Dr, 100 yd N	34.43590	-118.43540	LACM
CA-105	Los Angeles Co.	Mint Canyon; Oaks, 3.5 mi SW	34.43590	-118.43540	LACM
CA-105a	Los Angeles Co.	5 mi W Quartz Hill	34.64550	-118.30550	LACM
CA-106	Los Angeles Co.	Acton, 4 mi N, 2 mi W	34.47910	-118.22870	LACM
CA-107	Los Angeles Co.	Acton, 2 mi SW; Kashmere Canyon	34.44820	-118.21190	LACM
CA-108	Los Angeles Co.	San Fernando1 mi NE; Lopez Canyon	34.29980	-118.39920	LACM
CA-109	Los Angeles Co.	San Fernando	34.28190	-118.43900	LACM, UANM
CA-110	Los Angeles Co.	Tujunga Wash, near San Fernando	34.25236	-118.38929	MVZ
CA-111	Los Angeles Co.	Tujunga Valley	34.25999	-118.33000	MVZ
CA-112	Los Angeles Co.	Big Tujunga Canyon, 1.5 mi N & 1.4 mi E Sunland	34.27470	-118.31370	LACM
CA-112	Los Angeles Co.	San Gabriel Mts.; Big Tujunga Canyon	34.27470	-118.31370	LACM
CA-112	Los Angeles Co.	Sunland; Tujunga Wash, .03 mi E Foothill Blvd	34.27690	-118.30650	LACM
CA-113	Los Angeles Co.	Santa Monica Mountains, on Mulholland Dr, near Encino	34.13000	-118.49950	LACM
CA-113a	Loa Angeles Co.	Sepulveda Pass	34.09650	-118.47720	CSULB
CA-114	Los Angeles Co.	Santa Monica Mountains; Zuma Beach	34.01720	-118.81710	LACM
CA-115	Los Angeles Co.	Escondido Canyon; .05 mi E Puritan Road	34.03040	-118.76370	LACM
CA-116	Los Angeles Co.	Baldwin Hills	33.99530	-118.36400	MVZ
CA-116	Los Angeles Co.	Baldwin Hills, Stocker St & Fairfax Ave	33.99530	-118.36400	LACM
CA-116	Los Angeles Co.	Baldwin Hills; Kenneth Hahn State Rec Area	33.99530	-118.36400	LACM
CA-116	Los Angeles Co.	La Brea Hills, Army missile site 29	33.99530	-118.36400	LACM
CA-116	Los Angeles Co.	Los Angeles, Baldwin Hills	33.99530	-118.36400	LACM
CA-117	Los Angeles Co.	Palos Verdes Hills, Point Vicente	33.74110	-118.41090	LACM
CA-117	Los Angeles Co.	Rancho Palos Verdes, 1.2 mi NW Royal Palms State Beach	33.74450	-118.38700	SDNHM
CA-117	Los Angeles Co.	Rancho Palos Verdes, Palos Verdes Hills	33.73667	-118.39500	SDNHM
CA-118	Los Angeles Co.	San Jose Hills (=Covina Hills), between Covina and Pomona	34.05920	-117.84260	LACM

ID	County	Location	Latitude	Longitude	Source
CA-119	Los Angeles Co.	Azusa, 2 mi W; San Gabriel Wash	34.13570	-117.94530	MVZ
CA-120	Los Angeles Co.	San Gabriel Wash, near Azusa	34.15190	-117.92133	LACM, MVZ
CA-121	Los Angeles Co.	Duarte; Fish Canyon	34.15720	-117.92400	LACM
CA-121	Los Angeles Co.	Fish Canyon; N Duarte; Gun Range, 1 mi S	34.15720	-117.92400	LACM
CA-122	Los Angeles Co.	San Gabriel Canyon	34.16080	-117.90870	LACM
CA-123	Los Angeles Co.	Azusa	34.13000	-117.90000	MVZ
CA-124	Los Angeles Co.	Claremont; NE of Baseline Rd & Padua Rd	34.14390	-117.70780	LACM
CA-124	Los Angeles Co.	Palmer Canyon, Padua Hills Fire Station	34.14390	-117.70780	LACM
CA-125	Los Angeles Co.	San Antonio Canyon	34.16000	-117.67693	MVZ
CA-125a	Los Angeles Co.	San Dimas Canyon	34.15930	-117.76690	MVZ
CA-126	Los Angeles Co.	4.5 mi NE Shoemaker	34.46090	-117.86663	MVZ
CA-126a	Loa Angeles Co	South Fork Campground, Angeles National Forest	34.39440	-117.82030	CSULB
CA-127	Los Angeles Co.	18 mi E Palmdale [Lovejoy Buttes]	34.60672	-117.83218	MVZ
CA-127a	Los Angeles Co.	Alpine Butte, SW Palmdale	34.63110	-117.98970	LACM
CA-127b	Los Angeles Co.	4.5 mi S Pearblossom	34.45200	-117.90210	LACM
CA-128	Orange Co.	Buena Park, 6 mi N	33.96700	-117.99320	LACM
CA-129	Orange Co.	0.5 mi SW La Habra	33.92550	-117.93980	LACM
CA-130	Orange Co.	Brea, 2 mi W; on Deadera Rd	33.91160	-117.94390	LACM
CA-131	Orange Co.	Carbon Canyon; Brea, 4 mi E	33.92020	-117.82260	LACM
CA-132	Orange Co.	Yorba Linda	33.88860	-117.81310	LACM
CA-132	Orange Co.	Yorba Linda; Yorba Linda JC	33.88860	-117.81310	LACM
CA-133	Orange Co.	Santa Ana Canyon	33.85000	-117.81000	MVZ
CA-133a	Orange Co.	Olive	33.83610	-117.84620	CSULB
CA-134	Orange Co.	Irvine Park	33.79530	-117.74840	LACM
CA-134a	Orange Co.	Irvine	33.66950	-117.82310	CSULB
CA-134b	Orange Co.	Orange	33.78780	-117.85310	CSULB
CA-135	Orange Co.	Silverado Canyon, 2 mi up; Irvine Ranch	33.75640	-117.67760	LACM
CA-136	Orange Co.	Trabuco Canyon	33.66000	-117.56000	MVZ
CA-136	Orange Co.	Trabuco Canyon Campground	33.66000	-117.56000	LACM
CA-136	Orange Co.	Trabuco Canyon Rd, 1 mi E	33.66000	-117.56000	LACM
CA-136	Orange Co.	Trabuco Canyon, Santa Ana Mts.	33.65999	-117.56000	LACM
CA-136	Orange Co.	Trabuco Canyon; O'Neill Park, 3 mi N	33.66000	-117.56000	LACM

CA-136	Orange Co.	Trabuco Canyon; O'Neill Park, 7 mi S	33.66000	-117.56000	LACM
CA-136	Orange Co.	Trabuco Canyon; Rancho Mission Viejo	33.66000	-117.56000	LACM
CA-136a	Orange Co.	1 mi W Lower San Juan Camp	33.55210	-117.47210	LACM
CA-136a	Orange Co.	San Juan Guard Station	33.59170	-117.51380	CSULB
CA-137	Orange Co.	El Toro, 4 mi SE & 4 mi N Rancho Mission Viejo	33.64700	-117.68370	LACM
CA-138	Orange Co.	Corona del Mar; Narcissus & Fifth St, 200 yd N	33.60120	-117.86850	LACM
CA-139	Orange Co.	Laguna Beach, 4.1 mi N; Laguna Canyon Rd	33.58090	-117.76240	LACM
CA-140	Orange Co.	near Niguel Hill	33.51250	-117.73420	LACM
CA-141	Orange Co.	Capistrano	33.50170	-117.66260	USNM
CA-142	Orange Co.	Dana Point	33.46060	-117.71420	MVZ
CA-143	San Diego Co.	1.6 mi. S, 0.4 mi. W San Onofre Mt.	33.34640	-117.48610	SDNHM
CA-144	San Diego Co.	Camp Pendleton, 3 mi N of Las Pulgas Gate	33.34000	-117.47000	MVZ
CA-145	San Diego Co.	2.6 mi. N, 1.1 mi. E Las Flores Mission	33.32510	-117.43480	SDNHM
CA-146	San Diego Co.	1.3 mi. N, 4.1 mi. W Morro Hill	33.32420	-117.34010	SDNHM
CA-146	San Diego Co.	5.9 mi. N, 1.6 mi. W San Luis Rey	33.32680	-117.34100	SDNHM
CA-147	San Diego Co.	4.5 mi. N, 2.8 mi. W Morro Hill	33.35460	-117.30920	SDNHM
CA-147a	San Diego Co.	15 mi N & 1.58 mi W Fontuna Mountain	33.05090	-117.05550	SDNHM
CA-148	San Diego Co.	Point Loma	32.69150	-117.24420	MVZ
CA-149	San Diego Co.	San Diego	32.71000	-117.15000	MVZ
CA-150	San Diego Co.	Alvarado Canyon	32.78120	-117.08810	SDNHM
CA-151	San Diego Co.	Mission Gorge	32.81230	-117.07090	SDNHM
CA-152	San Diego Co.	near mouth of Tiajuana River	32.54999	-117.12000	MVZ
CA-153	San Diego Co.	Chula Vista	32.64000	-117.08000	MVZ
CA-154	San Diego Co.	Bonita	32.65780	-117.03000	SDNHM
CA-155	San Diego Co.	Proctor Valley, 4 mi. E San Miguel Road	32.66280	-116.93090	SDNHM
CA-156	San Diego Co.	Jamul Creek, 2 mi E El Nido PO	32.63090	-116.89720	USNM
CA-156	San Diego Co.	Jamul Creek, near El Nido	32.63090	-116.89720	USNM
CA-157	San Diego Co.	Dulzura	32.64000	-116.78000	ANSP, LACM, MVZ, UCLA, USNM
CA-158	San Diego Co.	Dulzura, 1.2 mi S, 1.6 mi E	32.63040	-116.76500	LACM
CA-159	San Diego Co.	6 mi N Foster	32.99000	-116.92000	MVZ

CA-160	San Diego Co.	El Monte Oaks	32.99000	-116.83000	MVZ
CA-161	San Diego Co.	Witch Creek	33.08000	-116.71000	MVZ
CA-162	San Diego Co.	Santa Ysabel	33.10920	-116.67310	SDNHM
CA-163	San Diego Co.	Echo Valley, 5 mi N of Descanso	32.89000	-116.65000	MVZ
CA-164	San Diego Co.	7 mi N & 0.5 mi E Julian	33.19370	-116.60630	LACM
CA-165	San Diego Co.	Julian	33.07000	-116.60000	MVZ
CA-166	San Diego Co.	Arkansas Canyon	33.11890	-116.54810	SDNHM
CA-166a	San Diego Co.	Sentenac Canyon, 5.5 mi E Banner	33.12950	-116.43200	CSULB
CA-167	San Diego Co.	San Felipe Canyon	33.15000	-116.54000	MVZ
CA-167	San Diego Co.	San Felipe Valley	33.14999	-116.54000	MVZ
CA-168	San Diego Co.	Grapevine Springs	33.17000	-116.52000	MVZ
CA-169	San Diego Co.	San Felipe Valley, 2.3 mi NW Hwy. 78 on S2	33.12000	-116.50000	MVZ
CA-170	San Diego Co.	south end San Felipe Valley	33.10237	-116.48437	MVZ
CA-171	San Diego Co.	0.67 mi N & 6 mi E Julian	33.08790	-116.50690	LACM
CA-171	San Diego Co.	1 km NW Hwy. 78 on Cty. Rd. S2	33.09000	-116.50000	LACM
CA-172	San Diego Co.	Granite Mt.	33.05120	-116.47920	SDNHM
CA-173	San Diego Co.	4.3 mi W (by road) Vallecitos	32.98000	-116.46000	MVZ
CA-174	San Diego Co.	La Puerta Valley	32.97000	-116.44000	MVZ, SDNHM
CA-175	San Diego Co.	3.2 mi W Vallecito Stage Station	32.98000	-116.42000	MVZ
CA-176	San Diego Co.	3.25 mi S & 2.25 mi E Scissors Crossing, Earthquake Valley	33.03000	-116.40000	MVZ
CA-177	San Diego Co.	5 mi NW Borrego P.O.	33.31000	-116.41000	MVZ
CA-178	San Diego Co.	near Borrego Springs	33.25590	-116.37500	SDNHM
CA-178a	San Diego Co.	5.5 mi W & 1 mi E Ocotillo Wells	33.07230	-116.11350	SDNHM
CA-179	San Diego Co.	Vallecito	32.97060	-116.34780	MVZ, SDNHM
CA-180	San Diego Co.	3.25 mi N Manzanita, McCain Valley	32.68999	-116.27000	MVZ
CA-181	San Diego Co.	3 mi NW Jacumba	32.61000	-116.23000	USNM
CA-182	San Diego Co.	1 mi W Jacumba	32.61230	-116.20090	MVZ
CA-183	San Diego Co.	Jacumba	32.60999	-116.18000	USNM
CA-184	San Diego Co.	old Hwy 80, 1.2 mi N & 4.5 mi E Jacumba	32.62200	-116.15870	LACM
CA-185	San Diego Co.	5 mi E Jacumba	32.64228	-116.10304	MVZ
CA-185a	San Diego Co.	3 mi E & 2 mi N Jacumba, near Mica Gem Mine	32.64280	-116.12390	LACM

ID	County	Location	Latitude	Longitude	Institution
CA-186	San Diego Co.	Mt. Springs	32.67000	-116.09999	MVZ
CA-187	San Diego Co.	Carrizo Canyon, 6.9 mi S and 0.2 mi E Mt. Palm Springs	32.84000	-116.19000	MVZ
CA-187a	San Diego Co.	Camp Elliott	not found		SDNHM
CA-188	Imperial Co.	1 mi E Mt. Springs	32.67000	-116.08000	MVZ
CA-188a	Imperial Co.	Smuggler's Cave Basin, Jacumba Range	32.63500	-116.09220	LACM
CA-189	Imperial Co.	1.6 mi S of Imperial-Riverside County line, by Hwy. 86	33.39999	-116.08000	MVZ
CA-190	Imperial Co.	Salton Sea [sw1/4 sec 4, T9S, Rclade1Clade2E]	33.41490	-115.73150	LACM
CA-191	Imperial Co.	Salton Sea [s1/2 sec 10, T9S, Rclade1Clade2E]	33.40220	-115.70800	LACM
CA-192	Imperial Co.	Salton Sea [ne 1/4 sec 12, T9S, Rclade1Clade2E]	33.41090	-115.66730	LACM
CA-193	Imperial Co.	Salton Sea [ne 1/4 sec 13, T9S, Rclade1Clade2E]	33.39720	-115.68230	LACM
CA-194	Imperial Co.	Salton Sea [nw 1/4 sec 20, T9S, R13E]	33.38060	-115.64260	LACM
CA-195	Imperial Co.	Chocolate Mts., 8 km NNE Frink Spring	33.40980	-115.47230	LACM
CA-196	Imperial Co.	Chocolate Mts., 14 km NW Beal Well	33.42840	-115.24630	LACM
CA-197	Imperial Co.	9 mi NE Beal Well	33.39000	-115.24000	MVZ
CA-198	Imperial Co.	Beal Well, Chocolate Mts.	33.33000	-115.33000	MVZ
CA-199	Imperial Co.	Chocolate Mts., 2.8 km S Beal Well	33.31860	-115.34880	LACM
CA-200	Imperial Co.	Chocolate Mts., 3 km SW Pegleg Well	33.23730	-115.25010	LACM
CA-201	Imperial Co.	Chocolate Mts., 12 km E Pegleg Well	33.27800	-115.12250	LACM
CA-202	Imperial Co.	Chocolate Mts., 17 km E Pegleg Well	33.29170	-115.09080	LACM
CA-203	Imperial Co.	2 mi NE Glamis along Hwy 78	32.99750	-115.07190	LACM
CA-204	Imperial Co.	13 mi NE Glamis	33.10450	-114.89430	MVZ
CA-204a	Imperial Co.	Black Mountain, 3 mi E (by rd) CA 78	33.05480	-114.82830	CSULB
CA-205	Imperial Co.	Tumco Mine, Cargo Muchacho Mts.	32.88713	-114.82807	MVZ
CA-206	Imperial Co.	Pilot Knob, Colorado River	32.73000	-114.74000	MVZ
CA-207	Imperial Co.	4 mi S Palo Verde	33.34000	-114.68000	MVZ
CA-208	Imperial Co.	opposite Cibola, Colorado River	33.31000	-114.71000	MVZ
CA-209	Imperial Co.	Milpitas Wash, 27.9 mi NE Glamis	33.27080	-114.72260	LACM
CA-209	Imperial Co.	Milpitas Wash, 10.25 mi S & 3.6 mi W Palo Verde	33.27080	-114.72260	LACM
CA-209	Imperial Co.	Palo Verde, 11 mi SSW; Milpitas Wash	33.27080	-114.72260	LACM
CA-210	Imperial Co.	20 mi above Picacho, Colorado River	33.18000	-114.70000	MVZ

CA-210	Imperial Co.	20 mi by River above Picacho, Colorado River	33.18000	-114.70000	MVZ
CA-211	Imperial Co.	8 mi below Picacho, Colorado River	33.01000	-114.50000	MVZ
CA-212	Imperial Co.	2 mi NW Potholes	32.85000	-114.50000	MVZ
CA-213	Imperial Co.	Potholes, Colorado River	32.82000	-114.50000	MVZ
CA-214	Riverside Co.	Jurupa Mts., 7 mi NW Riverside	34.03000	-117.45000	MVZ
CA-215	Riverside Co.	Riverside	33.95330	-117.39620	USNM
CA-215a	Riverside Co.	Sugarloaf Mountain	33.99340	-117.32230	UNT
CA-216	Riverside Co.	7 mi SE Riverside [Schellenger Ranch]	33.95000	-117.30000	MVZ
CA-217	Riverside Co.	4 mi. SW Perris	33.75130	-117.27630	SDNHM
CA-217a	Riverside Co.	Upper San Joan Campground	33.60640	-117.43310	CSULB
CA-218	Riverside Co.	Wildomar	33.59890	-117.28000	USNM
CA-219	Riverside Co.	S side Mount Russell	33.90030	-117.14920	MVZ
CA-220	Riverside Co.	Eden Hot Springs	33.89000	-117.05000	MVZ
CA-221	Riverside Co.	0.5 mi S Eden Hot Springs	33.88000	-117.05000	MVZ
CA-222	Riverside Co.	Lamb Canyon, 2.5 mi S (by rd) Beaumont	33.90377	-116.98513	MVZ
CA-223	Riverside Co.	Portero Creek, 2 mi E and 5 mi S Beaumont	33.87000	-116.91000	MVZ
CA-224	Riverside Co.	Vallevista, San Jacinto Valley	33.74000	-116.89000	MVZ
CA-225	Riverside Co.	1.5 mi SE Banning	33.88000	-116.78000	MVZ
CA-226	Riverside Co.	1.5 mi S and 1 mi W Banning	33.89000	-116.87000	MVZ
CA-227	Riverside Co.	2 mi W and 1.5 mi N Cabezon	33.93999	-116.83000	MVZ
CA-228	Riverside Co.	2 mi E and 0.5 mi S Banning	33.92000	-116.83000	MVZ
CA-229	Riverside Co.	1.5 mi S by W Cabezon	33.89000	-116.83000	ANSP, MVZ
CA-229	Riverside Co.	Banning	33.89000	-116.83000	MVZ
CA-229	Riverside Co.	Banning, base of San Jacinto Mts.	33.89000	-116.83000	MVZ
CA-230	Riverside Co.	1 mi S & 1 mi E Cabezon	33.90102	-116.80145	MVZ
CA-231	Riverside Co.	0.25 mi E Cabezon	33.91000	-116.78000	MVZ
CA-231	Riverside Co.	Cabezon	33.91750	-116.78720	SDNHM
CA-232	Riverside Co.	1.5 mi S Banning	33.90493	-116.88412	MVZ
CA-232	Riverside Co.	near Cabezon, base of San Jacinto Mts.	33.88000	-116.78000	MVZ
CA-232	Riverside Co.	near Cabezon, San Jacinto Foothills	33.88000	-116.78000	MVZ
CA-233	Riverside Co.	San Gorgonio Pass	33.91670	-116.75080	SDNHM
CA-234	Riverside Co.	1 mi E Cabezon	33.91000	-116.74000	LACM

ID	County	Location	Latitude	Longitude	Museum
CA-235	Riverside Co.	Stubby Canyon, 4.5 mi NE Cabezon	33.94000	-116.71000	MVZ
CA-236	Riverside Co.	3 mi N Whitewater	33.97000	-116.68000	MVZ
CA-237	Riverside Co.	5 mi E and 1 mi N Cabezon	33.93999	-116.69000	MVZ
CA-238	Riverside Co.	1 mi W Whitewater	33.92000	-116.68000	MVZ
CA-238	Riverside Co.	2 mi W Whitewater	33.92000	-116.68000	MVZ
CA-239	Riverside Co.	4 mi E and 0.5 mi N Cabezon	33.90999	-116.69000	MVZ
CA-240	Riverside Co.	Snow Creek Road, 3 mi S Hwy 111	33.89970	-116.67330	LACM
CA-241	Riverside Co.	near Whitewater, Snow Creek	33.89000	-116.68000	MVZ
CA-242	Riverside Co.	Snow Canyon, 7 mi E Cabezon	33.89000	-116.67000	MVZ
CA-243	Riverside Co.	Whitewater Canyon, 3 mi N; Hwy 60	33.95840	-116.64720	LACM
CA-243	Riverside Co.	Whitewater Canyon, 3 mi up; E of Hwy 10	33.95840	-116.64720	LACM
CA-243	Riverside Co.	Whitewater Canyon; Whitewater, 3 mi N	33.95840	-116.64720	LACM
CA-244	Riverside Co.	San Gorgonio Pass, 0.55 mi W & 0.65 mi N Whitewater	33.93300	-116.65060	LACM
CA-246	Riverside Co.	edge San Gorgonio River 0.41 mi W & 0.55 mi N Whitewater	33.92880	-116.65150	LACM
CA-247	Riverside Co.	0.5 mi S & 0.8 mi W Whitewater	33.91660	-116.65252	MVZ
CA-248	Riverside Co.	mouth Whitewater Canyon	33.92758	-116.64200	MVZ
CA-248	Riverside Co.	Whitewater Canyon; Hwy 111	33.92630	-116.64190	MCZ, USNM
CA-249	Riverside Co.	Whitewater	33.92500	-116.63830	MVZ
CA-250	Riverside Co.	Whitewater Station	33.92000	-116.63000	MVZ
CA-251	Riverside Co.	Whitewater, 2 mi NE	33.94510	-116.60730	LACM
CA-252	Riverside Co.	2.6 mi E Whitewater	33.92380	-116.59716	MVZ
CA-253	Riverside Co.	2 mi E Whitewater	33.92000	-116.59000	MVZ
CA-253	Riverside Co.	3 mi E & 0.5 mi S Whitewater	33.92000	-116.59000	LACM
CA-253	Riverside Co.	3 mi E Whitewater	33.92000	-116.59000	MVZ
CA-254	Riverside Co.	2.5 mi E and 2 mi N Whitewater	33.90000	-116.59000	MVZ
CA-255	Riverside Co.	2.5 mi E and 1 mi S Whitewater	33.89000	-116.59000	MVZ
CA-256	Riverside Co.	mouth Blaisdell Canyon	33.87952	-116.61473	MVZ
CA-257	Riverside Co.	Blaisdell Canyon	33.88000	-116.61000	MVZ
CA-257	Riverside Co.	Blaisdell Canyon, 2 mi SE Palm Springs Station	33.88000	-116.61000	MVZ
CA-258	Riverside Co.	4.5 mi SE Palm Springs Station	33.87000	-116.60000	MVZ

CA-259	Riverside Co.	4 mi W Palm Springs	33.88000	-116.58000	MVZ
CA-260	Riverside Co.	3 mi S Palm Springs	33.85000	-116.58000	MVZ
CA-261	Riverside Co.	Chino Canyon	33.84515	-116.58475	MVZ
CA-262	Riverside Co.	Mission Creek, 3.9 mi W & 0.8 mi S Morongo Valley	33.98270	-116.58920	LACM
CA-263	Riverside Co.	0.5 mi N & 4.4 mi W Desert Hot Springs	33.97611	-116.57719	MVZ
CA-264	Riverside Co.	4 mi N & 4.5 mi W Desert Hot Springs	34.00810	-116.57530	UNT
CA-265	Riverside Co.	3.0 mi N & 4.3 mi W Desert Hot Springs	34.00221	-116.57518	MVZ
CA-266	Riverside Co.	2.3 mi N & 4.2 mi W Desert Hot Springs	33.99745	-116.57451	MVZ
CA-267	Riverside Co.	Palm Springs	33.83333	-116.55000	LACM, MCZ
CA-268	Riverside Co.	Palm Springs	33.82999	-116.54000	MVZ, UCLA
CA-269	Riverside Co.	Desert Hot Springs	33.96110	-116.50170	CSULB
CA-270	Riverside Co.	Desert Hot Springs, 5 mi N	33.99460	-116.51830	LACM
CA-271	Riverside Co.	Desert Hot Springs, 7 mi N	34.00100	-116.51740	LACM
CA-271a	Riverside Co.	3 mi S Morongo Valley	34.00530	-116.57970	CSULB
CA-271b	Riverside Co.	3 mi W Morongo Valley	34.04050	-116.62160	CSULB
CA-272	Riverside Co.	1 mi E Palm Canyon	33.79000	-116.51000	MVZ
CA-272	Riverside Co.	Lower Palm Canyon, San Jacinto Mts.	33.79000	-116.51000	MVZ
CA-272	Riverside Co.	Palm Canyon, San Jacinto Mts.	33.78999	-116.51000	MVZ
CA-273	Riverside Co.	Andreas Canyon	33.76000	-116.53000	MVZ
CA-274	Riverside Co.	Dos Palmas Spring, Santa Rosa Mts.	33.72000	-116.55000	MVZ
CA-275	Riverside Co.	Kenworthy, San Jacinto Mts.	33.60999	-116.62000	MVZ
CA-276	Riverside Co.	1.25 mi N, 0.33 mi E Aguanga	33.47000	-116.86000	MVZ
CA-277	Riverside Co.	0.25 mi ENE Aguanga	33.44000	-116.86000	MVZ
CA-278	Riverside Co.	Aguanga	33.43303	-116.84895	MVZ, SDNHM
CA-279	Riverside Co.	Garnet	33.54000	-116.48000	MVZ
CA-279	Riverside Co.	Garnet Queen Mine, Santa Rosa Mts.	33.54000	-116.48000	MVZ
CA-280	Riverside Co.	mouth Tahquitz Canyon	33.50000	-116.30000	MVZ
CA-280	Riverside Co.	N wall Tahquitz Canyon	33.50000	-116.30000	MVZ
CA-281	Riverside Co.	Piñon Crest, Santa Rosa Mts.	33.61085	-116.42115	MVZ
CA-282	Riverside Co.	Cathedral Canyon, 6 mi SE Palm Canyon	33.77000	-116.45000	MVZ
CA-283	Riverside Co.	near Palm Desert; San Bernardino National Forest	33.67290	-116.41150	LACM

		boundary			
CA-284	Riverside Co.	San Jacinto Mts., Carrizo Creek	33.67110	-116.40580	MVZ
CA-285	Riverside Co.	Carrizo Creek, Santa Rosa Mts.	33.67000	-116.40000	LACM, MVZ
CA-286	Riverside Co.	Deep Canyon	33.67510	-116.36990	LACM
CA-286	Riverside Co.	Santa Rosa Mts.; Deep Canyon	33.67350	-116.37170	MVZ
CA-287	Riverside Co.	mouth of Deep Canyon at edge of Deep Canyon Reserve	33.68000	-116.36000	MVZ
CA-288a	Riverside Co.	Lake Matthews	33.84250	-117.46190	CSULB
CA-288	Riverside Co.	7 mi W Indio	33.71000	-116.33600	MVZ
CA-289	Riverside Co.	Thousand Palms, R6E, T4S, S18	33.82000	-116.38000	MVZ
CA-290	Riverside Co.	Indio	33.72060	-116.21560	LACM
CA-290a	Riverside Co.	vic. Ski Valley, Dillon Rd, 5 mi W int. 1000 Palms Canyon Rd.	33.91420	-116.37740	CSULB
CA-290b	Riverside Co.	4.5 mi NE jct Dillon & 1000 Palms Canyon Rd	33.89440	-116.23180	CSULB
CA-291	Riverside Co.	Berdoo Canyon bajada, Little San Bernardino Mts.	33.80568	-116.17952	MVZ
CA-291a	Riverside Co.	White Tank Campground, Joshua Tree Nat. Mon.	33.89420	-116.01610	CSULB
CA-291b	Riverside Co.	Lost Horse Ranger Station	34.01830	-116.19000	CSULB
CA-291c	Riverside Co.	Squaw Tank, Joshua Tree Nat. Mon.	33.92970	-116.07470	CSULB
CA-291d	Riverside Co.	Long Canyon, Joshua Tree Nat. Mon	33.97300	-116.44400	CSULB
CA-291e	San Bernardino Co.	Long Canyon, Joshua Tree Nat. Mon.	34.03470	-116.44500	CSULB
CA-291f	Riverside Co.	29 Palms Hwy, 4 mi N I-10	34.06010	-115.27030	CSULB
CA-292	Riverside Co.	Indio, 7 mi N, 4 mi E; Berdoo Canyon Rd	33.82770	-116.15880	LACM
CA-293	Riverside Co.	Indio, 12 mi NE	33.83660	-116.12420	LACM
CA-294	Riverside Co.	Pinyon Wells	33.86000	-116.09000	MVZ
CA-295	Riverside Co.	Painted Canyon, Mecca Hills	33.61962	-115.99953	MVZ
CA-296	Riverside Co.	ca. 8 mi E (by road) Mecca, Box Canyon, Mecca Hills	33.56000	-115.99000	MVZ
CA-296a	Riverside Co.	Box Canyon, 11 mi E Mecca	33.59180	-115.97980	CSULB
CA-297	Riverside Co.	Shavers Well	33.62000	-115.91000	MVZ
CA-298	Riverside Co.	Cottonwood Mts., 7.5 mi N & 12.1 mi E Mecca	33.71280	-115.90750	LACM
CA-299	Riverside Co.	Mecca, 4 mi N, 10.5 mi E; Shavers Valley	33.64170	-115.86750	LACM
CA-300	Riverside Co.	N end Orocopia Mts.	33.66423	-115.67969	MVZ

ID	County	Location	Latitude	Longitude	Institution
CA-301	Riverside Co.	Indian Rocks, 7.5 mi SE Shaver summit	33.60000	-115.68000	MVZ
CA-302	Riverside Co.	2 mi E Clemens Well	33.52000	-115.65000	MVZ
CA-303	Riverside Co.	Desert Center, 9.9 mi S, 9.8 mi W; Salt Creek Wash	33.58330	-115.55590	LACM
CA-304	Riverside Co.	Red Cloud Wash, west slope Chuckwalla Mts.	33.63226	-115.49904	MVZ
CA-305	Riverside Co.	Desert Center, 5.8 mi S, 3.5 mi W; Lost Pony Mine	33.63450	-115.46710	LACM
CA-306	Riverside Co.	N side Eagle Mt.	33.87000	-115.47000	MVZ
CA-307	Riverside Co.	Desert Center, 14.5 mi S; Bradshaw Rd	33.50700	-115.39830	LACM
CA-308	Riverside Co.	Desert Center, 18.3 mi S, 7.8 mi E; Bradshaw Rd	33.48660	-115.31310	LACM
CA-309	Riverside Co.	Desert Center, 20 mi S, 15 mi E; Bradshaw Rd	33.45140	-115.27170	LACM
CA-310	Riverside Co.	Chocolate Mts., 16 km NW Beal Well	33.43800	-115.22670	LACM
CA-311	Riverside Co.	near Wiley Wells, 7.2 mi off Hwy 60, 500'	33.49340	-114.88910	MVZ
CA-312	Riverside Co.	13 mi N Blythe [by Hwy.95], Maria Mts.	33.74000	-114.50000	MVZ
CA-313	Riverside Co.	west slope Big Maria Mts.	33.80928	-114.69801	MVZ
CA-314	Riverside Co.	E slope Big Maria Mts.	33.79832	-114.56084	MVZ
CA-315	Riverside Co.	Riverside Mts., Colorado River	34.02000	-114.54000	MVZ
CA-316	San Bernardino Co.	Reche Canyon, 4 mi SE Colton	34.02000	-117.27000	MVZ
CA-316	San Bernardino Co.	mouth of Reche Canyon, near Colton	34.04000	-117.28000	MVZ
CA-317	San Bernardino Co.	Reche Canyon	34.04890	-117.28980	USNM
CA-318	San Bernardino Co.	5 mi. SE Colton	34.07390	-117.31370	SDNHM
CA-319	San Bernardino Co.	Cajon Wash, San Bernardino	34.14000	-117.35000	MVZ
CA-319a	San Bernardino Co.	6.4 mi W San Bernardino	34.11670	-117.41670	CSULB
CA-320	San Bernardino Co.	San Timoteo Canyon, Redlands	34.03000	-117.20000	MVZ
CA-321	San Bernardino Co.	Redlands	34.05000	-117.18000	MVZ
CA-322	San Bernardino Co.	Greenspot Pumping Station, ca. 4.5 mi NE Redlands	34.09670	-117.10450	MVZ
CA-323	San Bernardino Co.	San Bernardino Mts.	34.15990	-117.18800	USNM
CA-324	San Bernardino Co.	Lone Pine Canyon	34.31499	-117.54019	MVZ
CA-325	San Bernardino Co.	Mormon Rocks, Cajon Canyon	34.31794	-117.49404	MVZ
CA-326	San Bernardino Co.	Hesperia, 10 mi SE; Deep Creek Campground	34.34610	-117.23230	LACM
CA-327	San Bernardino Co.	Hesperia	34.42640	-117.30090	UCLA

ID	County	Location	Latitude	Longitude	Institution
CA-327	San Bernardino Co.	Hesperia, 6 mi N; Mojave Camp Site	not found		LACM
CA-328	San Bernardino Co.	Victorville	34.53000	-117.29000	MVZ
CA-329	San Bernardino Co.	Oro Grande	34.59000	-117.33000	MVZ
CA-329a	San Bernardino Co.	12 mi N Lucerne Valley	34.80080	-117.00150	LACM
CA-329b	San Bernardino Co.	Deadman's Pt. (between Apple and Lucerne Valley)	34.45480	-117.06610	CSULB
CA-330	San Bernardino Co.	0.7 mi W Seven Oaks Resort, San Bernardino Mts.	34.17845	-116.94088	MVZ
CA-331	San Bernardino Co.	Seven Oaks, San Bernardino Mts.	34.18000	-116.91000	MVZ
CA-332	San Bernardino Co.	Cactus Flat, San Bernardino Mts.	34.31549	-116.81038	MVZ
CA-333	San Bernardino Co.	Cushenbury Springs, San Bernardino Mts.	34.36273	-116.85627	MVZ
CA-334	San Bernardino Co.	Cottonwood Springs, 1 mi W Old Woman Springs	34.40000	-116.72000	MVZ
CA-335	San Bernardino Co.	1.5 mi NE Barstow, Mohave Desert	34.92000	-116.97000	MVZ
CA-336	San Bernardino Co.	12 mi NNE Johannesburg, along Johannesburg Rd.	35.41000	-117.58000	MVZ
CA-337	San Bernardino Co.	2 mi E Searles Station, 9 mi NNE Johannesburg	35.47999	-117.58000	MVZ
CA-338	San Bernardino Co.	west end Morongo Valley	34.04187	-116.58530	MVZ
CA-339	San Bernardino Co.	Morongo Valley	34.04700	-116.58000	LACM
CA-339	San Bernardino Co.	Big Morongo Creek, 3 mi NW Morongo Valley Inn,	34.07090	-116.57330	LACM
CA-340	San Bernardino Co.	mid Morongo Valley, south side	34.06825	-116.54296	MVZ
CA-341	San Bernardino Co.	east end Morongo Valley	34.09810	-116.49143	MVZ
CA-341	San Bernardino Co.	Morongo Pass	34.10260	-116.49240	USNM
CA-341	San Bernardino Co.	Morongo Valley Park, east end Morongo Valley	34.10539	-116.49342	MVZ
CA-342	San Bernardino Co.	1.2 mi E Pioneertown	34.15139	-116.47908	MVZ, UNT
CA-342a	San Bernardino Co.	Giant Rock	34.33310	-116.38860	CSULB
CA-342b	San Bernardino Co.	8 mi N Joshua Tree	34.25100	-116.29830	LACM
CA-343	San Bernardino Co.	Quail Spring, T1S, R7E, Mohave Desert	34.03000	-116.25000	MVZ
CA-344	San Bernardino Co.	6 mi S Lavic	34.64000	-116.31000	MVZ
CA-345	San Bernardino Co.	5 mi S Lavic	34.64000	-116.31000	MVZ
CA-346	San Bernardino Co.	4 mi S Lavic	34.67000	-116.31000	MVZ
CA-347	San Bernardino Co.	3 mi SW Lavic	34.70000	-116.31000	MVZ
CA-348	San Bernardino Co.	E side Pisgah Lava Flow	34.74704	-116.34514	MVZ
CA-349a	San Bernardino Co.	Stewart Ranch, Newberry Springs	34.82720	-116.68840	LACM

CA-349	San Bernardino Co.	Pisgah Lava Flow	34.76470	-116.37733	MVZ
CA-350	San Bernardino Co.	west slope Sheephole Mts.	34.24408	-115.72062	MVZ
CA-351	San Bernardino Co.	Amboy Crater Lava Flow	34.53000	-115.78000	MVZ
CA-352	San Bernardino Co.	1 mi W Amboy	34.57400	-115.79567	MVZ
CA-353	San Bernardino Co.	Pass between Granite and Providence Mts.	34.81000	-115.60000	MVZ
CA-353a	San Bernardino Co.	Providence Mountains	35.03030	-115.52720	CSULB
CA-354	San Bernardino Co.	Cottonwood Basin, Granite Mts.	34.81681	-115.66182	MVZ
CA-355	San Bernardino Co.	Colton Well, Providence Mts.	34.93000	-115.92000	MVZ
CA-356	San Bernardino Co.	Mitchell's Caverns, Providence Mts.	34.93999	-115.52000	MVZ
CA-357	San Bernardino Co.	6 mi S Granite Well, Providence Mts.	35.04000	-115.43000	MVZ
CA-358	San Bernardino Co.	Gold Valley Ranch, Providence Mts.	35.09000	-115.40000	MVZ
CA-359	San Bernardino Co.	5 mi NE Granite Well, Providence Mts.	35.13000	-115.38000	MVZ
CA-360	San Bernardino Co.	Government Holes, Providence Mts.	35.14000	-115.35000	MVZ
CA-361	San Bernardino Co.	2 mi ESE Rock Spring, Lanfair Valley	35.15000	-115.33000	MVZ
CA-362	San Bernardino Co.	Cedar Canyon, Providence Mts.	35.16000	-115.46000	MVZ
CA-363	San Bernardino Co.	2 mi NNE Cima	35.26000	-115.49000	MVZ
CA-364	San Bernardino Co.	3 mi N Cima	35.29000	-115.49000	MVZ
CA-365	San Bernardino Co.	Purdy	35.26000	-115.21000	MVZ
CA-366	San Bernardino Co.	Halloran Spring	35.38420	-115.89480	MVZ
CA-367	San Bernardino Co.	1.4 mi N Halloran Spring	35.40230	-115.89977	MVZ
CA-368	San Bernardino Co.	Yucca Grove Auto Camp	35.40000	-115.79000	MVZ
CA-369	San Bernardino Co.	Mescal Spring, 8.5 mi E Valley Wells	35.49000	-115.54000	MVZ
CA-370	San Bernardino Co.	N side Clark Mt.	35.52000	-115.58000	MVZ
CA-371	San Bernardino Co.	Pachalka Spring, Clark Mt.	35.51000	-115.63000	MVZ
CA-372	San Bernardino Co.	8 mi W Clark Mt.	35.52000	-115.76000	MVZ
CA-372a	San Bernardino Co.	Saratoga Spring	35.68160	-116.42360	CSULB
CA-373	San Bernardino Co.	0.5 mi W Horse Thief Spring, Kingston Mts.	35.80000	-115.88000	MVZ
CA-373	San Bernardino Co.	Horse Spring, Kingston Range	35.77000	-115.88000	MVZ
CA-374	San Bernardino Co.	Beck Spring, Kingston Range	35.78000	-115.93000	MVZ
CA-375	San Bernardino Co.	2 mi N Horse Spring, Kingston Range	35.80000	-115.88000	MVZ
CA-376	San Bernardino Co.	Horn Mine, E base of Turtle Mts., 14 mi NE Blythe jct.	34.20000	-114.78000	MVZ

CA-377	San Bernardino Co.	25 mi S Needles	34.54000	-114.62000	MVZ
CA-377a	San Bernardino Co.	27 mi W Needles, Hwy 66	34.83170	-115.03230	CSULB
CA-378	San Bernardino Co.	opposite the Needles	34.84919	-114.60084	MVZ
CA-378a	San Bernardino Co.	Bannock	34.93670	-114.86390	LACM
CA-378b	San Bernardino Co.	Fenner Wash, 1 mi S & 3 mi W Arrowhead Jct.	34.93150	-114.88250	LACM
CA-379	San Bernardino Co.	3 mi E Mt. Manchester and 12 mi N Needles	35.02000	-114.70000	MVZ
CA-380a	Inyo Co.	Indian Joe Canyon, 0.2 mi S & 3.3 mi W Valley Wells	35.82800	-117.39340	LACM
CA-380	Inyo Co.	Little Lake	35.93660	-117.90670	MVZ
CA-381	Inyo Co.	3 mi N Little Lake	35.97663	-117.92000	MVZ
CA-382	Inyo Co.	Carroll Creek	36.49730	-118.05960	MVZ
CA-282a	Inyo Co.	5.5 mi N Olancha	36.35500	-118.02600	LACM
CA-383	Inyo Co.	Lone Pine	36.60000	-118.06000	MVZ
CA-384	Inyo Co.	Lone Pine Creek	36.59452	-118.20827	MVZ
CA-385	Inyo Co.	Mazourka Canyon, Inyo Mts.	36.89814	-118.08357	MVZ
CA-386	Inyo Co.	Lead Canyon, Saline Valley drainage	36.91926	-117.96028	MVZ
CA-387	Inyo Co.	3.5 mi E Lead Canyon, Lava Flow, Saline Valley	36.92000	-117.90000	MVZ
CA-388	Inyo Co.	Birch Creek, 6.4 mi S Big Pine	37.07160	-118.25580	MVZ
CA-389	Inyo Co.	3.5 mi E Big Pine	37.17127	-118.21183	MVZ
CA-390	Inyo Co.	Deep Springs Lake	37.27863	-118.02261	MVZ
CA-391	Inyo Co.	5 mi ENE Bishop, Silver Canyon	37.39999	-118.31000	MVZ
CA-392	Inyo Co.	7 mi E Laws, Silver Canyon, White Mts.	37.40196	-118.22378	MVZ
CA-393	Inyo Co.	Warm Sulphur Spring, Panamint Valley	36.12245	-117.21420	MVZ
CA-394	Inyo Co.	Wildrose Canyon	36.26544	-117.19363	MVZ
CA-394a	Inyo Co.	Thorndike Camp, Death Valley National Monument	36.23690	-117.07140	CSULB
CA-395	Inyo Co.	Johnson Canyon, Panamint Mts.	36.10050	-117.05181	MVZ
CA-396	Inyo Co.	Panamint City, Johnson Canyon, Panamint Mts.	36.10000	-117.04000	MVZ
CA-397	Inyo Co.	E side Death Valley, 40 mi S Furnace Creek	35.97000	-116.75000	MVZ
CA-398	Inyo Co.	Shoshone	35.97310	-116.27030	MVZ
CA-399	Inyo Co.	Hanaupah Canyon, Panamint Mts.	36.20529	-116.98728	MVZ
CA-400	Inyo Co.	Eagle Borax Works	36.20054	-116.86662	MVZ

CA-401	Inyo Co.	Badwater	36.21999	-116.76000	MVZ
CA-402	Inyo Co.	near Ryan	36.32000	-116.67000	MVZ
CA-403	Inyo Co.	4 mi S mouth Echo Canyon, Funeral Mts.	36.39000	-116.75000	MVZ
CA-404	Inyo Co.	Echo Canyon, Funeral Mts.	36.44770	-116.80120	MVZ
CA-405	Inyo Co.	0.5 mi N Furnace Creek	36.45000	-116.86000	MVZ
CA-405	Inyo Co.	Furnace Creek, Death Valley National Park	36.45720	-116.86530	MVZ, USNM
CA-405	Inyo Co.	Furnace Creek Ranch	36.45720	-116.86530	MVZ
CA-405a	Inyo Co.	Bennett Well, Death Valley National Monument	36.16610	-116.86090	LACM
CA-406	Inyo Co.	Triangle Spring	36.72734	-117.13504	MVZ
CA-407	Inyo Co.	Leadfield Road at Red Pass, Grapevine Mts.	36.82000	-117.03000	MVZ
CA-408	Inyo Co.	Titus Canyon at Entering Titus Canyon Sign, Grapevine Mts.	36.85000	-116.98999	MVZ
CA-409	Inyo Co.	Fall Canyon, Grapevine Mts.	36.83110	-117.17450	MVZ
CA-410	Inyo Co.	Scotty's Ranch	37.00999	-117.38000	MVZ
CA-411	Inyo Co.	1 mi E Scotty's Castle, Grapevine Canyon, Grapevine Mts.	37.03000	-117.31999	MVZ
CA-412	Mono Co.	Fish Lake Valley	37.50000	-117.92000	MVZ
CA-413	Mono Co.	Dutch Pete's Ranch, 4 mi W Benton	37.78999	-118.56000	MVZ
CA-414	Mono Co.	Benton	37.81000	-118.47000	MVZ
CA-415	Lassen Co.	Turtle Mt., Fort Sage Mts.	40.11000	-120.08000	MVZ
CA-416	Lassen Co.	High Rock Ranch, T28N, R17E, S25	40.24000	-120.00000	MVZ
CA-417	Lassen Co.	Honey Lake, 3900 ft	40.23570	-120.30570	USNM
CA-418	Lassen Co.	Amedee	40.32330	-120.18490	USNM
CA-419	Lassen Co.	1 mi N Wendel	40.35999	-120.23000	MVZ
CA-420	Lassen Co.	5 mi N, 2.5 mi W Wendel	40.43000	-120.27000	MVZ
CA-421	Lassen Co.	Shaffer Mt., 3.5 mi NE Litchfield	40.43999	-120.34999	MVZ
CA-422	Lassen Co.	near Secret Valley	40.50020	-120.32300	USNM
CA-422	Lassen Co.	Secret Valley, 5000 ft	40.50020	-120.32300	MVZ
CA-423	Lassen Co.	Pete's Valley	40.50000	-120.45999	MVZ
CA-424	Modoc Co.	8 mi NE Cedarville	41.58266	-120.03254	MVZ
CA-425	Modoc Co.	6 mi E Cedarville	41.57999	-120.06999	MVZ

United States: Colorado					
CO-1	Rio Blanco Co.	5 mi W Rangley	40.10800	-108.89300	USNM
United States: Idaho					
ID-1	Canyon Co.	2 mi S Melba	43.34710	-116.53240	MVZ
ID-2	Owyhee Co.	S Fork Owyhee River, 12 mi N Nevada line	42.15520	-116.41740	MVZ
United States: Oregon					
OR-1	Lake Co.	White Horse Creek	42.59930	-120.76890	USNM
OR-2	Harney Co.	1 mi S Narrows	43.26580	-118.96070	MVZ
OR-3	Harney Co.	Blitzen Valley, Buena Vista	43.27150	-118.85380	MVZ
OR-4	Harney Co.	Diamond, 4300 ft	43.01210	-118.66600	USNM
OR-5	Harney Co.	Borax Spring	42.32810	-118.65780	MVZ
OR-6	Malheur Co.	4 mi. S Adrian	43.70410	-117.09760	MVZ
OR-7	Malheur Co.	16 mi. S Adrian	43.59710	-117.11380	MVZ
OR-8	Malheur Co.	21 mi N McDermitt [Nevada]	42.27050	-117.78910	MVZ
OR-9	Malheur Co.	Watson	43.32770	-117.45400	USNM
United States: Nevada					
NV-1	Washoe Co.	12 Mile Creek, 0.5 mi E California line	41.89270	-119.99680	MVZ
NV-2	Washoe Co.	0.5 mi N Vya	41.59830	-119.86070	MVZ
NV-3	Washoe Co.	Painted Point, 9 mi E Vya	41.59270	-119.69970	MVZ
NV-4	Washoe Co.	10.5 mi S Vya	41.47690	-119.83780	MVZ
NV-5	Washoe Co.	mouth of Little High Rock Canyon	41.28800	-119.29410	MVZ
NV-6	Washoe Co.	15 mi NNW Deephole	40.88430	-119.56660	MVZ
NV-7	Washoe Co.	Rock Creek, Granite Mts.	40.79880	-119.33180	MVZ
NV-8	Washoe Co.	Deep Hole	40.71910	-119.48320	USNM
NV-9	Washoe Co.	1 mi NE Gerlach	40.66200	-119.37010	MVZ
NV-10	Washoe Co.	Horse Canyon, 3 mi NW Pahrum Peak	40.65990	-119.74600	MVZ
NV-11	Washoe Co.	Smoke Creek Desert, 10 mi W Deep Hole	40.67640	-119.61330	USNM
NV-12	Washoe Co.	17.5 mi W Deephole	40.61270	-119.71590	MVZ

ID	County	Locality	Latitude	Longitude	Collection
NV-13	Washoe Co.	17 mi W Deephole	40.61440	-119.70150	MVZ
NV-14	Washoe Co.	13 mi W and 9 mi S Deephole	40.61920	-119.65250	MVZ
NV-15	Washoe Co.	Smoke Creek, 9 mi E California line	40.51100	-119.88720	MVZ
NV-16	Washoe Co.	Smoke Creek	40.44740	-119.66690	USNM
NV-17	Washoe Co.	Sand Pass	40.26160	-119.77940	MVZ
NV-18	Washoe Co.	2 mi NW Flanigan	40.19620	-119.90550	MVZ
NV-19	Washoe Co.	6 mi E California line (10 min Flanigan)	40.16640	-119.91270	MVZ
NV-20	Washoe Co.	0.5 mi NW Flanigan	40.17430	-119.89470	MVZ
NV-21	Washoe Co.	2.5 mi E Flanigan	40.17840	-119.81650	MVZ
NV-22	Washoe Co.	Barrel Spring, 9 mi E and 3 mi N Fort Bidwell	40.07770	-119.72850	MVZ
NV-23	Washoe Co.	11 mi S Pyramid PO, W side Pyramid Lake 4040 ft.	40.04750	-119.67790	USNM
NV-23	Washoe Co.	Pyramid Lake	40.04750	-119.67790	USNM
NV-24	Washoe Co.	2 mi W Sutcliffe Station, Virginia Mts. 4350 ft	39.95180	-119.60400	USNM
NV-24	Washoe Co.	Virginia Mts., 2 mi W Sutcliffe Station	39.95180	-119.60400	USNM
NV-25	Washoe Co.	1 mi SW Pyramid Lake	39.87710	-119.62070	MVZ
NV-26	Washoe Co.	0.5 mi S Pyramid Lake	39.83170	-119.43680	MVZ
NV-27	Washoe Co.	3 mi E Reno	39.52970	-119.69620	MVZ
NV-28	Washoe Co.	1 mi W Wadsworth	39.62730	-119.32030	MVZ
NV-29	Storey Co.	3 mi SW Wadsworth, Truckee Canal	39.61340	-119.31970	MVZ
NV-30	Humboldt Co.	Virgin Valley, Sheldon National Wildlife Refuge	41.81567	-119.09500	MVZ
NV-31	Humboldt Co.	Virgin Valley	41.84540	-119.01630	MVZ
NV-32	Humboldt Co.	mouth Alder Creek, Pine Forest Mts.	41.83210	-118.74430	MVZ
NV-33	Humboldt Co.	Big Creek Ranch, Pine Forest Mts.	41.66880	-118.58790	MVZ
NV-34	Humboldt Co.	1.5 mi N Quinn River Crossing	41.59440	-118.43030	MVZ
NV-35	Humboldt Co.	4.5 mi S Quinn River Crossing	41.53010	-118.41460	MVZ
NV-36	Humboldt Co.	5 mi S Quinn River Crossing	41.51910	-118.41750	MVZ
NV-37	Humboldt Co.	Soldier Meadow, T40N, R25E	41.39180	-119.16350	MVZ
NV-38	Humboldt Co.	Jackson Creek Ranch, 17.5 mi S & 5 mi W Quinn River	41.31880	-118.56130	MVZ
NV-39	Humboldt Co.	10 mi W and 6 mi N Sulphur	40.90230	-118.92610	MVZ
NV-40	Humboldt Co.	2 mi E Antelope	40.88870	-118.49280	MVZ
NV-41	Humboldt Co.	36 mi NE Paradise Valley	41.69170	-117.24930	MVZ

ID	County	Location	Latitude	Longitude	Museum
NV-42	Humboldt Co.	Granite Creek	41.05880	-117.22790	USNM
NV-43	Pershing Co.	3.5 mi SE Sulphur	40.86230	-118.70250	MVZ
NV-44	Pershing Co.	8 mi S Sulphur	40.80710	-118.72400	MVZ
NV-45	Pershing Co.	El Dorado Canyon, Humboldt Range	40.52740	-118.24740	MVZ
NV-46	Pershing Co.	8 mi NE Gerlach (Washoe Co.)	40.35785	-119.28436	MVZ
NV-47	Pershing Co.	30 mi W and 4 mi N Lovelock	40.27330	-118.79100	MVZ
NV-48	Pershing Co.	3.5 mi W Toulon	40.07360	-118.71600	MVZ
NV-49	Pershing Co.	2 mi W Toulon	40.07350	-118.71840	MVZ
NV-50	Pershing Co.	1 mi W Toulon	40.06440	-118.66430	MVZ
NV-51	Pershing Co.	Toulon	40.06720	-118.64490	MVZ
NV-52	Pershing Co.	S side Granite Peak, East Range	40.22650	-117.79880	MVZ
NV-53	Pershing Co.	8.5 mi E and 5 mi N Granite Peak	40.34940	-117.60940	MVZ
NV-54	Pershing Co.	18 mi NE Iron Point	41.16610	-116.96590	MVZ
NV-55	Lyon Co.	4 mi W Hazen	39.58890	-119.12560	MVZ
NV-56	Lyon Co.	12 mi E (by road) Yerington	39.04000	-118.96000	MVZ
NV-57	Lyon Co.	12 mi S Yerington, W Walker River	38.83410	-119.19340	MVZ
NV-58	Lyon Co.	19.8 mi S, 4.2 mi E Yerington	38.80900	-119.04170	MVZ
NV-59	Lyon Co.	East Walker River, 2 mi NW Morgans Ranch	38.56280	-118.97790	MVZ
NV-60	Douglas Co.	Topaz Lake	38.69570	-119.54410	MVZ
NV-61	Douglas Co.	Desert Creek, Sweetwater Range	38.60270	-119.34080	MVZ
NV-62	Churchill Co.	1 mi W Hazen	39.56320	-119.06350	MVZ
NV-63	Churchill Co.	3 mi SW Lahontan Dam	39.46320	-119.06770	MVZ
NV-63a	Churchill Co.	8 mi E & 3 mi S Fallon	39.42800	-118.60790	KU
NV-64	Churchill Co.	8 mi S and 1 mi E Eastgate	39.26720	-117.85320	MVZ
NV-65	Mineral Co.	Endowment Mine, Excelsior Mts.	38.28800	-118.36120	MVZ
NV-66	Elko Co.	3 mi S Jackpot	41.94693	-114.69157	MVZ
NV-67	Elko Co.	Goose Creek, 2 mi W Utah line	41.94800	-114.07610	MVZ
NV-68	Elko Co.	Tecoma	41.32020	-114.08140	MVZ
NV-69	Elko Co.	Cobre	41.11190	-114.40090	MVZ
NV-70	Elko Co.	0.5 mi W Debbs Creek, Pilot Peak	41.07280	-114.08090	MVZ
NV-71	Elko Co.	8 mi S Wendover [Utah]	40.65080	-114.07980	MVZ
NV-72	Lander Co.	Battle Mountain	40.64210	-116.93430	USNM

	County	Locality	Latitude	Longitude	Source
NV-73	Lander Co.	1.5 mi NW Cortez, Cortez Mts.	40.15190	-116.62100	MVZ
NV-74	Lander Co.	Silver Creek, 15 mi from Austin	39.73440	-117.18040	USNM
NV-75	Lander Co.	2.5 mi NE Smiths Creek Ranch	39.40390	-117.58200	MVZ
NV-76	Lander Co.	Campbell Creek	39.29210	-117.57590	MVZ
NV-77	Lander Co.	Peterson Creek, Shoshone Mts.	39.23630	-117.52650	MVZ
NV-78	Lander Co.	35 mi SW Austin	39.22870	-117.33820	MVZ
NV-79	Lander Co.	White Rock Valley, 30 mi SW Austin	39.29270	-117.26340	USNM
NV-80	Eureka Co.	8 mi W Eureka	39.51700	-116.02850	MVZ
NV-81	Eureka Co.	10 mi SE Eureka	39.40000	-115.92000	MVZ
NV-82	White Pine Co.	Cherry Creek	39.88740	-114.85450	MVZ
NV-83	White Pine Co.	Halstead Creek, 1 mi W Illipah	39.35100	-115.36590	MVZ
NV-84	White Pine Co.	2.5 mi SW Hamilton	39.24380	-115.51660	MVZ
NV-85	White Pine Co.	Water Canyon, 8 mi N Lund	38.98800	-114.96170	MVZ
NV-86	White Pine Co.	Deadman Creek, Mt. Moriah	39.33290	-114.12500	MVZ
NV-87	White Pine Co.	near Smith Creek Cave, Mt. Moriah	39.33570	-114.11470	MVZ
NV-88	White Pine Co.	Hendry Creek, 1.5 mi E Mt. Moriah	39.29180	-114.13210	MVZ
NV-89	White Pine Co.	2 mi E Smith Creek Cave, Mt. Moriah	39.26530	-114.02190	MVZ
NV-89	White Pine Co.	3 mi W Smith Creek Cave, Mt. Moriah	39.26530	-114.02190	MVZ
NV-90	White Pine Co.	Baker Creek	39.21910	-113.92500	MVZ
NV-91	White Pine Co.	Cleve Creek, Shell Creek Range	39.23330	-114.47310	MVZ
NV-92	White Pine Co.	Willard Creek, Spring Valley	39.01110	-114.40920	MVZ
NV-93	White Pine Co.	2 mi W Baker	39.01640	-114.15840	MVZ
NV-94	Nye Co.	Pancake Range	39.09039	-115.79312	MVZ
NV-95	Nye Co.	Ophir Creek	38.93600	-117.20450	MVZ
NV-96	Nye Co.	South Twin River	38.90880	-117.14680	MVZ
NV-97	Nye Co.	Wilson Canyon Toquima Range	38.92200	-116.79390	MVZ
NV-98	Nye Co.	Toquima Range, 1.5 mi E Jefferson	38.78900	-116.88870	MVZ
NV-98	Nye Co.	Toquima Range, 1 mi E Jefferson	38.78900	-116.88870	MVZ
NV-99	Nye Co.	Monitor Valley, 9 mi E Toquima Mt.	38.79560	-116.84880	MVZ
NV-100	Nye Co.	Peavine	38.57400	-117.27000	MVZ
NV-100	Nye Co.	Peavine Creek, 6000 ft	38.57280	-117.27130	USNM
NV-100	Nye Co.	S base Peavine Creek	38.57280	-117.27130	USNM

NV-101	Nye Co.	Monitor Valley, 5 mi N Belmont, 6900 ft	38.64680	-116.85840	USNM
NV-102	Nye Co.	Hot Creek Range, 4 mi N Hot Creek	38.54490	-116.33900	MVZ
NV-103	Nye Co.	Hot Creek Range, 7 mi W Tybo	38.38570	-116.44930	MVZ
NV-104	Nye Co.	Hot Creek Range, 8 mi W Tybo	38.37280	-116.42210	MVZ
NV-105	Nye Co.	San Antonio	38.16160	-117.23560	MVZ
NV-106	Nye Co.	McKinney Tank	38.10204	-116.91021	MVZ
NV-107	Nye Co.	Quinn Canyon Mts., Burned Corral Canyon	38.12050	-115.57680	MVZ
NV-108	Nye Co.	Garden Valley, 8.5 mi N Sharp	38.12430	-115.42060	MVZ
NV-109	Nye Co.	White River Valley, 15 mi SW Sunnyside	38.11060	-114.97420	MVZ
NV-110	Nye Co.	Breen Creek, Kawich Range	37.71740	-116.67280	MVZ
NV-111	Nye Co.	Phinney Canyon road	36.99910	-117.02510	MVZ
NV-112	Nye Co.	5 mi E, 1 mi N Grapevine Peak	36.97120	-117.05160	MVZ
NV-113	Nye Co.	8 mi E Grapevine Peak	36.98090	-117.02710	MVZ
NV-114	Nye Co.	2.5 mi E and 1 mi S Grapevine Peak	36.95980	-117.10530	MVZ
NV-115	Nye Co.	Currie Well	36.96910	-116.92670	MVZ
NV-116	Nye Co.	1.5 mi W Beatty	36.90888	-116.78607	MVZ
NV-117	Nye Co.	Amargosa River, 3.5 mi NE Beatty	36.93690	-116.73290	MVZ
NV-118	Nye Co.	15 mi N Ash Meadows	36.60770	-116.32660	USNM
NV-119	Nye Co.	Ash Meadows	36.42750	-116.35200	USNM
NV-120	Nye Co.	Pahrump Valley	36.00470	-115.86300	USNM
NV-121	Esmeralda Co.	Fish Lake	37.73800	-118.05450	MVZ
NV-122	Esmeralda Co.	Cave Spring, Silver Peak Range	37.79990	-117.84260	MVZ
NV-123	Esmeralda Co.	13.5 mi NW Goldfield	37.82920	-117.34160	MVZ
NV-124	Esmeralda Co.	Pigeon Spring	37.41720	-117.66590	MVZ
NV-125	Lincoln Co.	Coyote Spring, R63E, T2N	38.03190	-114.86220	MVZ
NV-126	Lincoln Co.	Eagle Valley, 3.5 mi N Ursine	38.02590	-114.20440	MVZ
NV-127	Lincoln Co.	Meadow Valley [Wash], 24 mi S Caliente	37.78490	-114.41400	MVZ
NV-127	Lincoln Co.	Meadow Valley Wash, 7 mi S Caliente	37.78490	-114.41400	MVZ
NV-128	Lincoln Co.	11 mi E Panaca	37.78600	-114.23160	MVZ
NV-129	Lincoln Co.	E slope Irish Mt.	37.61930	-115.35730	MVZ
NV-130	Lincoln Co.	9.2 mi NW Crystal Springs, Pahranagat Valley	37.60600	-115.34500	MVZ
NV-131	Lincoln Co.	16 mi E Groom Baldy	37.45000	-115.73480	MVZ

NV-132	Lincoln Co.	SW base of Groom Baldy	37.41160	-115.79280	MVZ
NV-133	Lincoln Co.	Crystal Springs, Pahranagat Valley	37.53160	-115.23390	MVZ
NV-134	Lincoln Co.	Pahranagat Valley, 4 mi NE Crystal Spring	37.54980	-115.19440	MVZ
NV-135	Lincoln Co.	Delamar Mountains, 10 mi W Caliente	37.59273	-114.75977	MVZ
NV-136	Lincoln Co.	1 mi S Crystal Spring, Pahranagat Valley	37.19210	-115.03540	MVZ
NV-137	Clark Co.	Bunkerville, Virgin River Valley	36.77300	-114.12800	USNM
NV-138	Clark Co.	west slope Virgin Mts.	36.58354	-114.20722	MVZ
NV-139	Clark Co.	south end Overton Ridge	36.47376	-114.45328	MVZ
NV-140	Clark Co.	Valley of Fire, 10 mi SE of I-15 [-By Hwy. 40]	36.43940	-114.65460	MVZ
NV-141	Clark Co.	Rocky Hill, 4.5 mi W Boulder City	35.98930	-114.89140	MVZ
NV-142a	Clark Co.	Searchlight	35.46530	-114.91970	CSULB
NV-142	Clark Co.	5.3 mi E Searchlight	35.47477	-114.85582	MVZ
NV-143	Clark Co.	7.8 mi W Searchlight, by Hwy. 68	35.48110	-114.78210	MVZ
NV-144	Clark Co.	Colorado River, 14 mi E Searchlight	35.48570	-114.70580	MVZ
NV-145	Clark Co.	Jap Ranch, Colorado River, 14 mi E Searchlight	35.49220	-114.68660	MVZ
NV-146	Clark Co.	12 mi S and 5 mi E Searchlight	35.26690	-114.76120	MVZ
NV-147	Clark Co.	6.5 mi W [by Hwy. 77] and 2.4 mi N Davis Dam	35.22610	-114.68270	MVZ
NV-148	Clark Co.	8 mi N Dead Mt.	35.16950	-114.73570	MVZ
NV-149	Clark Co.	3 mi SW Hiko Springs	35.15530	-114.72360	MVZ
NV-150	Clark Co.	Hiko Spring, 8 mi SSE Dead Mt.	35.09860	-114.71610	MVZ

United States: Utah

UT-1	Box Elder Co.	Kelton	41.74600	-113.10640	MVZ
UT-2	Box Elder Co.	E slope Promontory Mountains, 10 mi N	41.49130	-112.46910	MVZ
UT-3	Box Elder Co.	10 mi N S end Promontory Point	41.35670	-112.42970	USNM
UT-4	Box Elder Co.	Promontory	41.22220	-112.41140	USNM
UT-5	Tooele Co.	Carrington Island, Great Salt Lake	41.00630	-112.57050	USNM
UT-6	Tooele Co.	Stansbury Island, N end Great Salt Lake	40.83830	-112.50300	USNM
UT-7	Tooele Co.	Stansbury Island	40.79223	-112.52510	BYU
UT-8	Utah Co.	5.1 km S & 12.8 km W Lehi	40.32790	-111.95640	BYU
UT-9	Millard Co.	Lawson Cove Canyon, Wah Wah Mts.	38.62570	-113.55980	USNM
UT-10	Beaver Co.	Indian Peak	38.27500	-113.82033	BYU

UT-11a	Washington Co.	10.5 mi W Enterprise	37.62220	-113.99910	CSULB
UT-11	Washington Co.	Little Pine Valley, 6 mi SW Hebron	37.50950	-113.85050	USNM
UT-12	Washington Co.	Pine Valley, 6400 ft.	37.39110	-113.51410	USNM
UT-12a	Washington Co.	Diamond Valley	37.24360	-113.62080	CSULB
UT-13	Washington Co.	8 mi NNW St. George	37.18120	-113.62110	USNM
UT-14	Washington Co.	Santa Clara	37.13300	-113.65410	USNM
UT-15	Washington Co.	0.5 mi NW St. George	37.10810	-113.59420	USNM
UT-15a	Washington Co.	13 km N & 6 km W St. George	37.20000	-113.63333	UNT
UT-16	Washington Co.	St. George	37.10420	-113.58410	MVZ
UT-17	Washington Co.	S side Virgin River, St. George	37.08350	-113.54510	MVZ
UT-18	Washington Co.	10 mi SE St. George, Fort Pearce Wash	37.04500	-113.52720	MVZ
UT-19	Washington Co.	Fort Pierce	37.00000	-113.41666	LACM
UT-19	Washington Co.	Fort Pierce, Fort Pierce Wash	37.00000	-113.41666	BYU
UT-20	Washington Co.	8 mi. WSW St. George	37.06520	-113.67930	USNM
UT-21a	Washington Co.	Beaverdam Wash	37.00000	-113.99680	CSULB
UT-21	Washington Co.	Beaverdam Slope, 4 mi W Castle Cliff, 3850 ft	37.02560	-113.90800	USNM
UT-22	Washington Co.	Beaverdam Slope, 4 MI SE Castle Cliff	37.02370	-113.85540	USNM
UT-22a	Washington Co.	Ash Creek Bridge	37.41110	-113.23580	CSULB
UT-22b	Washington Co.	near Virgin	37.20830	-113.18880	CSULB
UT-22c	Washington Co.	8 km W & 13 km E Hurricane	37.10000	-113.15000	UNT
UT-23	Piute Co.	Marysvale	38.44940	-112.23020	USNM
UT-24	Kane Co.	Kanab	37.04750	-112.52630	MVZ, USNM
UT-25	San Juan Co.	mouth Cottonwood Canyon	37.11754	-111.85264	MVZ
UT-26	Kane Co.	59 km E & 25 km N Kanab	37.24840	-111.84410	BYU
UT-27	Kane Co.	42 mi S & 18 mi W Escalante	37.48980	-111.65220	USNM
UT-28	Garfield Co.	Henry Mountains	37.95972	-110.80972	BYU
UT-29	Garfield Co.	Mount Ellen, Henry Mts.	38.10970	-110.81380	USNM
UT-30	Wayne Co.	Loa	38.40280	-111.64300	USNM
UT-31	Wayne Co.	1 mi N Notom	38.24724	-111.11112	MVZ
UT-32	Emery Co.	Huntington Canyon, 13.2 km NW Huntington	39.43840	-111.01420	BYU
UT-33	Emery Co.	10 mi WSW Green River	38.95387	-110.37522	MVZ
UT-34	Grand Co.	Rock Canyon Corral	38.81880	-109.77280	MVZ

ID	Locality	Latitude	Longitude	Institution
UT-34	Grand Co.	0.5 mi E Rock Canyon Corral	38.83132	-109.76837	MVZ
UT-34a	Grant Co.	15 mi NW Moab	38.67060	-109.68520	CSULB

Mexico: Baja California

ID	Locality	Latitude	Longitude	Institution
BCN-1	0.5 mi S Monument 258 [US-Mexico boundary, Pacific Ocean]	32.48966	-117.13115	MVZ
BCN-2	Tijuana, 50 ft	32.53000	-117.02000	USNM
BCN-3	Nachoguero Valley	32.59866	-116.35027	MVZ
BCN-3	N end Nachoguero Valley	32.59866	-116.35027	MVZ
BCN-4	Rancho Viejo, 15 mi E Alamos, 3500 ft	32.40690	-116.50746	USNM
BCN-5	S end Valle de Las Palmas	32.36552	-116.52778	MVZ
BCN-5	Valle de Las Palmas	32.36552	-116.52778	MVZ
BCN-6	8 km SE Rosarito	32.31000	-116.94639	CIB
BCN-7	Las Palmas Canyon, W side Laguna Salada	32.36833	-115.80083	MVZ
BCN-8	Gaskill's Tank, near Laguna Salada	32.33333	-115.78333	SDNHM
BCN-8a	90 km S Mexicali	31.83333	-115.17500	UNT
BCN-9	Sierra Juarez, Laguna Hanson	32.05333	-115.90722	MVZ, SDNHM
BCN-10	Tesonaco, 15 mi N Laguna Hanson, Sierra Juarez	32.01862	-116.33184	MVZ
BCN-11	Sierra Juarez, Las Cruces	31.93860	-116.35938	MVZ
BCN-12	Ensenada, 50-100 ft	31.87000	-116.62000	USNM
BCN-13	Isla Todos Santos	31.79720	-116.89600	MVZ, SDNHM, USNM
BCN-14	north coast Punta Banda, 7.5 mi W Maneadero	31.73488	-116.76271	MVZ
BCN-15	Arroyo Seco, 20.8 mi. S San Vicente	31.32500	-116.25000	SDNHM
BCN-16	San Isidro Ranch	31.25175	-116.41600	USNM
BCN-17	4 mi. S Ejido Erendira	31.26667	-116.37500	SDNHM
BCN-18	9 km S & 7 km E San Vicente	31.23860	-116.15600	CIB
BCN-19	Vallecitos, San Pedro Martir Mts., 8000 ft	31.43357	-115.76000	USNM
BCN-20	Valle de la Trinidad	31.40000	-115.72306	MVZ, SDNHM
BCN-20	Valle de la Trinidad, Aguajito Spring	31.40000	-115.72306	SDNHM
BCN-20	Valle de la Trinidad; La Zapopita	31.40000	-115.72306	LACM

BCN-21	8 mi S & 10 mi E Valle de la Trinidad	31.33300	-115.75000	CIB
BCN-22	9.2 mi E & 6.6 mi S Colonia Cardenas	31.32867	-115.52000	MVZ
BCN-23	San Matias Pass (summit, near Diablito Spring)	31.31667	-115.53333	SDNHM
BCN-24	Rancho San Antonio, west base San Pedro Martir Mts.	31.20979	-115.57143	USNM
BCN-25	Cañon Esperanza, E base Sierra San Pedro Martir	31.16667	-115.39303	USNM
BCN-26	Johnson Ranch (=San Antonio del Mar)	31.10833	-116.27500	MVZ
BCN-27	Cabo Colonet	30.97333	-116.32750	MVZ
BCN-28	7 km S & 3 km W Colonet	30.96685	-116.27200	CIB
BCN-29	5 mi W, 1.25 mi S San Telmo de Abajo	30.95000	-116.16250	MVZ
BCN-30	San Telmo	30.96806	-116.09083	MVZ
BCN-30a	20 mi SE San Telmo	31.95833	-115.83333	USNM
BCN-31	San Francisquito	30.84483	-116.18725	USNM
BCN-32	3 mi. N Camalu	30.88333	-116.06667	SDNHM
BCN-33	La Huerta, west base Laguna Hanson Mts.	30.96550	-115.88000	USNM
BCN-34	San Jose	30.98450	-115.68200	MVZ
BCN-35	Sierra San Pedro Martir, La Grulla	31.02098	-115.48632	MVZ
BCN-36	5 mi. N San Felipe	31.11667	-114.90000	SDNHM
BCN-37	San Felipe	31.08444	-114.86750	MVZ, SDNHM
BCN-38	Bahia San Felipe	31.02000	-114.85000	USNM
BCN-38	Punta San Felipe	31.03389	-114.83361	MVZ
BCN-39	1 km W San Felipe	31.02500	-114.84161	CIB
BCN-40	Valladares	30.85951	-115.68393	MVZ
BCN-41	Santo Domingo, 25 ft	30.76817	-115.94821	USNM
BCN-42	San Antonio Ranch, Río Santo Domingo	30.77056	-115.93611	MVZ
BCN-43	ca. 20 mi SE San Telmo	30.72000	-115.68333	USNM
BCN-44	El Cajon Canyon, E base of San Pedro Martir Mts.	30.79417	-115.29167	MVZ
BCN-45	1 km E San Quintin	30.60500	-115.94800	CIB
BCN-46	4 km S & 10 km E San Quintín	30.56000	-115.94500	CIB
BCN-46	5 km S & 15 km E San Quintín	30.56000	-115.94500	CIB
BCN-46	Arroyo Nuevo York, 15 mi S of Santo Domingo	30.55472	-115.93611	MVZ
BCN-47	Rio San Simon, San Quintín	30.54167	-115.91270	USNM

BCN-48	San Quintín	30.48000	-115.95000	SDNHM, USNM
BCN-49	Isla San Martín	30.48333	-116.12361	MVZ, USNM
BCN-50	7 mi. N, 1.5 mi. W Cape San Quintín	30.37500	-115.97500	SDNHM
BCN-51	17 km S & 13 km E San Quintín	30.37500	-115.89030	CIB
BCN-52	Socorro, 20 mi S San Quintín	30.31278	-115.81444	MVZ
BCN-53	Santa Maria Valley, 10 mi. E San Quintín	30.55833	-115.50000	SDNHM
BCN-53	Santa Maria, near San Quintín	30.55833	-115.50000	SDNHM
BCN-54	El Palmarito, 40 mi. E San Quintín, 3000 ft	30.45000	-115.45833	SDNHM
BCN-55	8 mi N El Rosario	30.19097	-115.78571	MVZ
BCN-56	5 km N & 6 km E El Rosario	30.09028	-115.67460	CIB
BCN-57	1 km N & 16 km E El Rosario	30.06944	-115.57937	CIB
BCN-58	1 mi E El Rosario	30.05611	-115.70778	MVZ
BCN-59	10 km S El Rosario	29.96700	-115.78800	CIB
BCN-60	San Fernando, 1400 ft	29.98333	-115.28333	USNM
BCN-61	1 km W Mision San Fernando	29.99800	-115.25000	CIB
BCN-61	Mision San Fernando	29.99800	-115.24500	CIB, MVZ
BCN-62	Pozo San Augustin, 20 mi E San Fernando, 1800 ft	29.95139	-114.98333	USNM
BCN-62a	Las Arrastras, west of Bahia San Luis Gonzaga	29.54167	-114.33333	CSULB
BCN-63	23 km N & 21 km W Bahia San Luis Gonzaga	30.00300	-114.51500	CIB
BCN-64	11 km N & 8.5 km W Cataviña	29.84375	-114.49412	CIB
BCN-65	Papá Fernández, Bahia San Luis Gonzaga	29.83333	-114.39844	CIB
BCN-66	15 mi NW Mission Calamahue	29.76889	-114.37000	MVZ
BCN-67	Cataviña	29.73556	-114.70167	MVZ
BCN-68	24 km N & 20 km W Cataviña	29.67000	-114.81700	CIB
BCN-69	Calamahue, 950 ft	29.63333	-114.41667	USNM
BCN-70	Santa Catarina Landing	29.50000	-115.28333	SDNHM
BCN-71	Ubai, 30 mi SE Calamahue, 2000 ft	29.25965	-114.13953	MVZ
BCN-72	23 km S & 2 km E Punta Prieta	29.04386	-114.07108	CIB
BCN-73	26 km WSW (by rd) Bahia de los Angeles	29.02500	-113.77833	MVZ
BCN-74	Punta Prieta	28.92794	-114.16089	CIB
BCN-75	3 km N & 3 km W Bahia de los Angeles	28.97917	-113.58333	CIB

ID	Locality	Latitude	Longitude	Collection
BCN-76	7 km W (by road) Bahia de Los Angeles	28.96667	-113.61778	MVZ
BCN-77	Bahia de Los Angeles	28.94167	-113.55000	CIB, SDNHM
BCN-78	23 km S & 2 km W Punta Prieta	28.73200	-114.24000	CIB
BCN-78a	3.8 mi S El Cardón	28.91333	-114.43333	SDNHM
BCN-79	San Andres, 200 ft	28.71491	-114.29733	USNM
BCN-80	4 km S & 3 km W Nuevo Rosarito	28.66800	-114.11000	CIB
BCN-81	Santa Rosalitita	28.65278	-114.17054	USNM
BCN-82	17 mi N & 1 mi W Guerrero Negro Jct.	28.29861	-114.02326	USNM
BCN-83	Morro de Santo Domingo	28.27000	-114.12500	MVZ
BCN-84	Santo Domingo Landing	28.25000	-114.10000	SDNHM
BCN-85	Rancho Mesquital, 33 mi W Calmalli	28.26417	-113.80583	MVZ
BCN-86	4 mi N & 13 mi E Guerrero Negro Jct.	28.11806	-113.76336	USNM
BCN-87	10 mi SE Mesquital	28.10000	-113.82000	MVZ
BCN-88	Calmalli, 1200 ft	28.23333	-113.55000	USNM
BCN-89	Las Palomas	28.22500	-113.40000	SDNHM
BCN-90	24 km N & 24 km E El Arco	28.32000	-113.21800	CIB
BCN-91	El Barril	28.29167	-112.86667	SDNHM
BCN-91a	7 mi. W San Francisquito Bay	28.41667	-117.88333	SDNHM
BCN-92	40 km N & 40 km E El Arco	28.06167	-112.99306	CIB
BCN-93	12 mi. E El Arco, Rancho Mira Flores	28.04167	-113.25000	SDNHM
BCN-94	El Arco	28.01444	-113.36500	CIB
BCN-95	Isla Ángel de la Guarda, N end	29.51667	-113.53333	SDNHM
BCN-95	Isla Ángel de la Guarda, Puerto Refugio	29.51667	-113.53333	USNM
BCN-96	Isla Ángel de la Guarda	29.51667	-113.53333	USNM
BCN-96	Isla Ángel de la Guarda, S end	29.00000	-113.15000	UCLA, SDNHM
BCN-97	Isla Cedros, 2 mi S of N end of island	29.51667	-113.53333	MVZ
BCN-97	Isla Cedros, N end of island	28.32639	-115.21538	MVZ
BCN-97	Isla Cedros, Punta Norte	28.32639	-115.21538	USNM
BCN-98	Isla Cedros	28.16667	-115.23333	SDNHM, UCLA, USNM
BCN-99	Isla Cedros, San Isidro	28.05556	-115.22917	USNM

Code	Locality	Latitude	Longitude	Institution
BCN-99	Isla Cedros, south bay	28.05556	-115.22917	SDNHM, USNM
BCN-99	Isla Cedros, wash W of Ciudad Cedros	28.05556	-115.22917	USNM
BCN-100	Cerro Prieto, near Volcano Lake (head of Rio Hardy)	32.41667	-115.30208	USNM
BCN-101	Cerro Prieto, 20 mi SSE Mexicali	32.19417	-115.34333	MVZ
BCN-102	Sierra de Cocopah, 3000 ft.	32.25517	-115.37097	USNM
BCN-103a	10 mi. S Jesus Maria	not found		SDNHM
BCN-103b	3 mi. S El Marmol	29.82900	-114.80400	SDNHM
BCN-103c	3.5 mi. up Arroyo San Jose	27.57160	-114.36980	SDNHM
BCN-103d	Mesquital	not found		SDNHM
BCN-103e	San Andreas, 15 mi NE; Placeras de San Andreas	not found		LACM
BCN-103f	San Andreas, Viscaino Desert	not found		LACM
BCN-103g	San Bruno, 25 ft	not found		USNM

Mexico: Baja California Sur

Code	Locality	Latitude	Longitude	Institution
BCS-1	Viscaino Desert; San Rafael, 13 mi NW	27.91000	-113.71667	LACM
BCS-2	Turtle Bay	27.70000	-114.91667	MVZ
BCS-2	Vizcaino Desert; Tortugas Bay	27.70000	-114.91667	MVZ
BCS-2	Vizcaino Desert; Tortugas Bay, 6.5 mi NW	27.70000	-114.91667	LACM
BCS-3	1 mi SE Cabo Tortolo	27.66450	-114.85385	MVZ
BCS-3	1.5 mi E Cabo Tortolo	27.66450	-114.85385	MVZ
BCS-4	Vizcaino Desert; San Jose	27.53333	-114.45400	LACM
BCS-5	San Jose Arroyo (foot); Tortuga Bay, 12 mi E	27.55000	-114.30000	LACM
BCS-6	18 km S & 57 km W Vizcaino	27.49200	-114.03700	CIB
BCS-7	"El Madrazo" 2.7 km N & 14.2 km E Bahía Asunción	27.20750	-114.15222	CIB
BCS-7	6.2 km N & 18 km E Bahía Asunción	27.20750	-114.15222	CIB
BCS-8	16 mi W El Arco	27.55000	-113.65944	USNM
BCS-9	Valladares	27.55889	-113.52750	MVZ
BCS-10	6.7 mi SE & 1 mi E Ejido Vizcaino Jct.	27.58000	-113.32000	USNM
BCS-11	Rancho Santa Ana, 36 km E Vizcaino	27.68000	-113.09500	CIB

ID	Locality	Latitude	Longitude	Collection
BCS-12	San Francisco de la Sierra	27.60784	-113.04170	CIB
BCS-13	1 km S & 1 km W San Francisco de la Sierra	27.58824	-113.06065	CIB
BCS-14	33 km N & 5 km W San Ignacio	27.59500	-112.97500	CIB
BCS-15	23 km N & 36 km E San Ignacio	27.59184	-112.52308	CIB
BCS-16	10 km N & 14 km W Santa Rosalia	27.42857	-112.45385	CIB
BCS-17	20 mi W San Ignacio, 200 ft	27.36735	-113.12782	USNM
BCS-18	San Ignacio, 23 mi W; Rancho las Martiles, 2 mi SE	27.30000	-113.11667	LACM
BCS-19	1 km N & 43 km W San Ignacio	27.30300	-113.35000	CIB
BCS-20	San Ignacio, 800 ft	27.28194	-112.89583	MVZ, SDNHM
BCS-21	5 km S & 36 km W San Ignacio	27.26417	-113.26289	CIB
BCS-22	8 km S & 38 km W San Ignacio	27.22528	-113.26861	CIB
BCS-23	Santa Clara Mts., 1500 ft	27.16667	-113.58333	USNM
BCS-24	2 km S San Ignacio	27.25000	-112.48300	CIB
BCS-25	3 mi E & 1 mi S Santa Agueda	27.22000	-112.35000	USNM
BCS-26	19 km S & 1 km W San Ignacio	27.12056	-112.91472	CIB
BCS-27	26.5 km N & 19.5 km W Pta. Abreojos	26.95944	-113.75917	CIB
BCS-28	18.5 km N & 19.75 km W Pta. Abreojos	26.89694	-113.76500	CIB
BCS-29	14 km N & 16 km E Abreojos	26.70000	-113.58333	CIB
BCS-30	Isla San Marcos	27.21667	-112.10000	CIB, MVZ, SDNHM, UCLA
BCS-30	Isla San Marcos, canyon on W shore	27.21667	-112.10000	MVZ
BCS-31	23 km S & 15 km E Santa Rosalia	27.10781	-112.13397	CIB
BCS-31a	12 km S Santa Rosalia	27.22500	-112.20833	
BCS-32	El Potrero, near Mulege, 600 ft	26.95238	-112.09160	USNM
BCS-33	Bahia Concepción, 13 mi SE Mulege	26.73000	-111.88000	USNM
BCS-34	Bahia Concepción, El Coyote	26.71667	-111.91667	LACM
BCS-34	Playa El Coyote, Bahia Concepción	26.71667	-111.91667	UCLA
BCS-35	Bahia Concepción, 13 mi SE Mulege	26.73000	-111.88000	MVZ
BCS-35a	SE side Bahia Concepción	26.67000	-111.73000	USNM
BCS-36	Bahia Concepción, 20 mi S Mulege	26.63611	-111.82000	MVZ
BCS-37	Bahia Concepción, 24 mi S Mulege	26.61224	-111.89313	USNM